High Value Polymers

Special Publication No. 87

High Value Polymers

The Proceedings of a Symposium organized by the Macro Group in association with the Industrial Division, Ireland Region, of The Royal Society of Chemistry.

The Queen's University of Belfast, 9th–12th April 1990

Edited by

A. H. Fawcett
The Queen's University of Belfast

British Library Cataloguing in Publication Data

A symposium organized by the Macro Group in
association with the Industrial Division, Ireland
Region, of the Royal Society of Chemistry
(1990) Belfast
High-value polymers. (Special publication 87)
I. Fawcett, Dr A. H.   II. Series
547.8
ISBN 0-85186-867-3

©The Royal Society of Chemistry 1991

Published by The Royal Society of Chemistry,
Thomas Graham House, Science Park, Cambridge CB4 4WF

Printed by Redwood Press Ltd,
Melksham, Wiltshire

Contents

Introduction

Chemists are familiar with the concept of an enthalpy or an entropy change associated with a chemical reaction, and have recognised the importance of such thermodynamic quantities to the issue of getting the reaction to go; they do not in the same manner associate a difference in value with the chemical change, yet that is what determines whether the particular reaction will be employed in an industrial manner. The man who places a Δ £ figure along with Δ G to the right hand side of a chemical equation may receive a funny look from a university audience, as I well know, but potential value changes have often been the driving force of scientific enquiry. The activities of the early students of the chemical phenomena, the alchemists, are legendary. They lived in a mainly pastoral society, operating with the wildest of hopes within an almost metaphysical framework of knowledge to transmute the baser metals into gold, the basis then of the currency itself. This was at a time long before the concept of a macromolecule had been conceived, and materials such as commodity and speciality polymers had been made. What alchemist would have had the audacity to anticipate computers, the electronic coding of information and the latest form of transaction, plastic money? Polymers have not become the basis of value, but they do provide a modern means of exchange. This is one rather modest example of a high technology use for polymers.

Our industrial society is information-based, the ideas being examined for accuracy and reliability in a variety of ways, from passive peer review with tests on reproducibility of published material at one - the academic - extreme, to subjection to competition within

the market-place in the other. Wealth generation is
regulated generally by the profit motive both within a
particular business and across industries, success
being measured not merely by the ability to survive but
more by the ability to adapt and to grow within the
evolving economy. The concerns operate within a
framework of laws governing such things as customer
rights, environmental impact, accounting, taxation
and patents. These laws are primarily the creation of
the nation state, and the products and services a
company produces are priced in its currency, but as one
national economy after another emerges there is an
increasing tendency for the market to extend over the
world, and for standards of health and safety at work,
the treatment of proprietary information, the stock
market quotation of companies, and so on to become
uniform – the GATT-promoted concept of the level playing
field. A fiscal illustration of this on a local scale
is the existence throughout the European Community of
the Value Added Tax, which furthermore operates,
within the chemical industry, as a tax on the increase
in value of substances when they are subject to a
chemical change. If the price of a chemical product is
that at which a legal contract arranges exchange for
money, it depends upon a succession of influences: the
money spent on researching the product, the cost of the
raw materials, the efficiency of the chemical process
plant that produces it, the effectiveness of the
company that designed, owns and operates that plant,
the assistance of the services that the company calls
upon such as capital, the power supply, the transport
infrastructure and the skill and education of the work
force, and finally upon the price set upon similar
products by the company's competitors. In short, the
price of a product depends upon and provides for a wide
range of cost factors: value is not like a
thermodynamic function of state – an intrinsic property
of the product – but instead reflects the economy as it
operates in a complex manner in probably more than one
geographical location.

        In the published histories of the Imperial
Chemical Industries and Du Pont companies there is
described the systems of cartels and agreements that
prevailed in the period between the two world wars when
these companies were created by the amalgamation of
single business concerns[1,2]. Once such companies had
formed, they strove to extend their competence of understanding and of production over the whole field of

chemistry and of chemical engineering. They performed research and development that led, *inter alia*, to the discovery of what became the commodity polymer polyethylene, of the nylons, of the polyesters, of teflon and poly(chlorobutadiene). As a means of bringing about their new policy Du Pont recruited Carothers from academic life to direct the blue skies research that lead to the nylons, ICI adopted at an early stage the polyester work that had been initiated outside the company, and within ICI low density polyethylene was discovered by accident. This material was ignored for a time, and was then used as a speciality manner as a radar capacitor dielectric and cable insulator before being developed as a thermoplastic. From the discovery of polyethylene to its widespread use in that manner took two decades, for it took time for consumers to turn from 'metal bashing' to plastics extrusion. Engineers, designers and manufacturing workshops had to become familiar with the new materials, and as they did, so did the production capacity expand to meet the developing demand.

At the end of the second world war the United States emerged as the dominant feature in world economics. Under State Department influence the cartel system of regulated competition collapsed, and other companies joined in polyethylene production by licensing the ICI – Du Pont patents at the direction of Judge Ryan. Phillips announced their own catalytic process for high density polyethylene. There was a curious period in the Du Pont story, when new management from another background decided to provide only one grade of polyethylene, but it was soon realised that the market did require several grades for different end uses. Profits were low in the highly competitive field that resulted, for prices fell progressively.

Against the timescale of the fifty year period of a Kondratieff cycle for the rise and fall of a technical revolution, the bulk polymer industry is surviving very well. Certain materials such as viscose rayon and cellulose acetate are in decline and are no longer supported by research, but the consumption of the bulk polymers continues to grow, albeit at a modest rate. In periods of recession the production plant of certain materials such as PVC and phenol-formaldehyde has been rationalised. There is some movement of production towards countries with cheaper labour costs.

In the main market economies polypropylene consumption is expected to grow at 4.0% pa but production capacity at only 2.4% pa over the next decade to bring the two closer into line[3]. Plants are using their third generation of catalysts.

Not all polymers are now being produced on such a successful tonnage basis as polyethylene, or the other bulk polymers polystyrene, polyvinyl chloride and polypropylene. There is a class of polymers, the engineering plastics such as nylon 6, polyacetal, polycarbonates and the polyarylether sulphones that are less readily and less widely made but have properties that allow them to command a substantially higher price than polyethylene; the polycarbonates and the polyarylether sulphones, for example, have superior mechanical properties at high temperatures on account of their high glass transition temperatures and their relatively high thermal stabilities. When first introduced these polymers were specialities, but their success led to widespread use and to prices between the specialities and commodities[4]. There is a prevailing judgement that it is unlikely that many further similar polymers will come onto the market; Kevlar as a strong fibre (its tensile strength is nearly the theoretical value (page 392)), polyimides as thermosets (page 428) and the main chain liquid crystalline polymers as anisotropic moulding plastics (page 428) may be the last materials to emerge that are valued for their mechanical properties.

The industrial interest in macromolecules developed in parallel with their academic understanding. Each stimulated the other, once the macromolecular nature of useful natural substances such as rubber, cellulose and silk had been recognised. The chemistry of polymerization, polymerization kinetics and thermodynamics, polymer solution properties, chain microstructure (by spectroscopy), the reactions of polymers, the physics of their solid state, polymer engineering, both of production and use, have all developed as scientific subject areas. Lord Todd, when President of the Royal Society, stated in 1980: `the development of polymerization is perhaps the biggest thing chemistry has done, where it has had the biggest effect on everyday life'[5]. A vast amount of work

and intellectual achievement is recorded in the journals
that were established, and is available as a basis for
informing future work. It has been referred to as a
true `revolution enchains', in recognition of the
importance of arrangement and organization of linear and
network molecules.

          In the early years significant developments,
such as the discovery of polyethylene and
poly(tetrafluoroethylene)could be discovered by accident
- all that was required was a proper observation and a
curiosity to understand. A more recent example is the
alkylcyanoacrylates whose rapid polymerization and
adhesion properties led to the concept of 'Super Glue'
(page 352). Now much more effort is devoted to design.
One conceives of a substance with desirable effects or
properties, and attempts to produce it by chemistry
and processing using known science as far as is
possible. The chemical design content lies in the
choice of monomer and comonomer, and of the method of
polymerization so that the microstructure can be controlled; it lies in the control of the molecular
weight, the molecular weight distribution and the network structure, if any, and in the presence of groups
with a chemical or a physical function. A simple example of this idea is provided by the work of Datta who
here describes how a linear ethylene-propylene
copolymer chain may be formed that also contains a
third, functionalised monomer, whose heteroatom group
remains masked until the product is treated with dilute
mineral acid at the stage when the catalyst residues are
removed. From such olefin polymers functionalised with
acid groups that produce metal salts he may obtain an
ionomer (page 33). The presence of charged species on
the chain causes association within the solid state,
but since the sites are randomly arranged along the
chain the solid state is randomly organised by the
modification. These materials have to be assessed and
then may be tested on the market. Still at the research
stage are the electrode coatings of Grimshaw (p 256):
his aim is to produce electrochemically a conducting
film that carries groups capable of high selectivity in
an electrochemical reaction with a species in solution,
so that a sensor is obtained as an electrode. The
critical material, the coat, would have a weight of
merely a few milligrams. Of more sophistication is the
work of Laschewsky and Ringsdorf (p 247), who are
exploring the possibilities of utilising polymers
prepared from monomers self assembled in Langmuir-Blodgett

films; the idea is to produce at the molecular level a
structurally organised and functionally integrated
material,   the   illustrations   being   chosen   from
substances   that   are   capable   of   utilising   such
interaction   as   antibody-antigen   interactions   in
biological assay.

          Self assembly   giving rise to anisotropy,   is
a theme in the papers on liquid crystalline side chain
polymers (pages 131, 228), in the paper on liquid
crystal main chain thermoplastics (page 428) and in the
spinning of Kevlar fibres (page 392). The side chain
liquid crystalline polymers may be utilised for information storage if they can first be aligned by an
electric field,   and may be exploited for their non-
linear optical effects,   as well as for analysis of and
resolving racemic mixtures in a chromatographic column.
When liquid crystalline main chain polymers are
injection moulded they flow into shapes that have
anisotropic physical properties.   The problem is to
design the mould and the paths of flow within the mould
to control the anisotropy so that it creates the
stiffness and strength in the required orientation. Now
that the chemist has produced the material,   the next
task is to learn how best to use it. Anisotropy is also
a basic theme in polymers for aerospace, but here the
structural units are not molecules but fibres placed
within a matrix such as an epoxy or a polyimide resin
(pages 373, 420, 455).   At the moment preimpregnated
mats of aligned and woven carbon fibres with other
fibres are arranged by craftsmen to create the desired
orientation of reinforcement.   This method of laying up
the mats before the cure can be used for aerospace
components, where lightness is a premium and expense of
secondary importance (half the weight of a fighter
aircraft may come from such a material). At present 3/4
of present consumption of carbon fibres lies in the area
of aerospace.   If they are to spread to such applications as   the mass production of car body parts,
then automation must be found for the procedures of
cutting and arranging the woven fibre mats before or
after the matrix resin is applied.

          This message,   that the market for a product
must be developed by down stream research in co-
operation with the customer,   is promulgated by
Fitzgerald and Irwin (p 392). They describe how the use
of Kevlar in motor car tyres requires a new approach

because Kevlar, as a consequence of its molecular structure, suffers readily from compressive fatigue, which is a severe problem if it is to be employed to reinforce at the edge of the tread. Their `systems' approach required an adjustment of the cord twist level and a new design for the reinforcement belt. In a similar manner, to use Kevlar fibres in the riser-tensioners of deep water oil drilling platforms, an application that subjects the material to repetitive cycling through pulleys, the rope itself had to be redesigned. Internal lubrication, the use of three rather than just one size of strand, and a reduction in the twist angle of the rope were factors that were recognised, studied and optimised, to produce a rope that was five time as long lasting as steel under use. This amplified the advantage Kevlar has when wet of being 20 times as strong as steel, weight for weight.

Materials are not simply valued for their mechanical properties. The conducting behaviour of polymers is exploited in many ways, as Kathirgamanathan describes (page 174). Besides such simple uses as in a conducting adhesive, antistatic devices, and battery materials which utilise the bulk conducting properties, the discovery of conducting substances such as polypyr-role and polyaniline opens up the prospect of electronic devices designed at the molecular level. Returning to the larger scale, the prospect of switching the viscosity of a liquid by the application of an electric field to liquids containing polymers has been explored by Block (page 151). Should this prove possible, there are endless applications to motor car transmission and suspensions, and to hydraulic systems.

The location of research effort within a large organization may be within one central research station or it may be distributed to the divisions. Bell Telephones is one of the companies that have adopted the first strategy, that brings workers from many disciplines together to share costly facilities and allows mutual stimulation in collaborative efforts, with well known results for the development of good science and the production of innovative and money spinning ideas. The latter, divisional, strategy is perhaps more suitable for progressive improvements in the quality and performance of an established product, for it will operate in close contact with those familiar with the production process and will take into account the views of the sales force on the developing market. For small companies in well established areas, membership of a research association may be the way to keep up with technical developments, but it is unlikely to provide a significant advantage to one against competitors that are other members of the association. Such companies will devote a certain fraction of their income to research, a fraction that is probably higher the higher the scientific and technological design content of their products. The manager of research has to make choices between competing areas. He will have a number of investigative projects in train in new and familiar areas, and have watching briefs for other areas, so that they can take advantage of significant developments made elsewhere. In small high-technology companies, sometimes formed by management buy-outs from large firms, a large proportion of the staff are engaged in innovation. A suitable location might be a university science park.

Professor Neil Graham, who describes his hydrogel system here for sustained drug-release (page 79), is the only entrepreneur among the authors. Once the basic idea was produced and protected, he sought and eventually raised £250,000 in funds for development from the British Technology Group, and his university, as well as contributing himself. A company, Polysystems, was founded, a business plan was devised by Graham with a consultant, and was adopted, and pursued in accommodation within the Clydeside Business Park. After three years the company employed 17 people and was `in the black' as a result of the contacts and licence arrangements that it developed with a number of pharmaceutical companies. The business is founded upon approximately 20 patents. Related work expanding the science of the materials was performed within the Department of Applied Chemistry at Strathclyde University where Graham has a personal chair, thus fostering among young scientists an appreciation of the way in which science and business may be interdependent. At one time Hutchinson and Graham were colleagues within ICI. Graham's experience can be contrasted with that of Hutchinson (page 58) within the much less transparent management-controlled Pharmaceutical Division of ICI.

A guide to the price of a high value polymer drug-release system may be roughly computed as follows: suppose such a system might replace a month's course of pills `to be taken' three times a day; if each pill costs £1, one might charge £100. If the system were to weigh 20 mg, the cost of a gram of the new form of drug is thus £5,000. This method of reaching a price takes no account of such factors as the cost of research, trials or production of both drug and release system, but if these required a higher price the medical profession might not adopt the new treatment. This view implies, perhaps unfairly, that a Doctor's prescription for patient treatment does not reflect savings that would be achieved in a hospital by relieving staff of the task of frequent pill hand-outs nor of savings that might be made if such a treatment would allow a patient to be treated at home rather than in a hospital. A similar convenience would be available to a veterinary surgeon if he could treat a herd of cows, say for mastitis, by a single vaccination for each cow, rather than a full course of treatment for the whole herd that required several visits. The programmed drug delivery system may be of as much value as the drug itself for this reason. The factor most likely to lower the price of a course of treatment would be the availability of an alternative from a different manufacturer. In this area the patentability of the material is of prime importance, and profits are required from the drug and delivery systems systems that do sell to finance work on the much larger number systems that do not make the grade for various reasons. Currently the growth in this area is of the order of 100% pa, and the world market of bioerodible polymers may reach £1B by 1994. There are many other applications of polymers within medicine, from catheters, membranes for heart-lung machines, and implants, for which silicone polymers have often proved suitable (page 98), to bone cements, degradeable sutures, plaster of Paris substitutes, tissue adhesives (page 352), artificial heart valves and so on, and all of these are high cost articles.

        A further source of new ideas lies within the universities: these days staff are not expected to ignore a potential application of a discovery, and are encouraged to seek industrial involvement in their work. The work of Sherrington within Strathclyde University is a prime example of that type (page 1), though at a formative stage of the work he spent time on sabbatical in the industrial laboratory. The possible use of polyacetylene as a holographic device derives from Feast's Durham University sited demonstration that

linear one dimensionally conducting polyacetylene,
might be made by the reverse Diels-Alder decomposition
of a metathesis polymer (page 206). At the symposium the
audience was intrigued to understand how
photocrosslinking could control the decomposition to
create patterns. One of the most interesting aspects
about the discovery of polyethylene is that it was first
made in a university laboratory by Speed Marvel[6], but he
reported in 1930 that the white solid, a linear
polymer[6], `has not been studied further'[7]. However the
significance of the discovery of low density
polyethylene within ICI was not lost on the company's
executives: Eric Fawcett incautiously corrected the
expressed view of Staudinger that ethylene could not be
polymerised at a Faraday Discussion in 1935, and found
his career as a polymer scientist was consequently
terminated[8]. Fawcett was aware of the existence of
macromolecules, having worked within Du Pont with
Carothers[8], but so too was Marvel, of course, and
moreover he was a consultant to Du Pont at that time[9] and
must have been aware of their commercial interest in
making polymers. This can hardly be a case of a new
development being missed by the unprepared mind. At
scientific conferences on macromolecules academics rub
shoulders with industrial scientists; they read each
others papers and often their research proposals, and
share ideas on desirable developments. The fact remains
that significant discoveries are often not planned or
expected, they occur by serendipity and will be
recognised only by the lateral-thinker.

Besides the macromolecular drugs described in
Hutchinson's article (page 58) there are a number of
other polymers whose chemical properties are the significant factor. The polymeric materials considered by
Pethrick derive their value from their radiation
chemistry (page 267). Positive resists are required to
decompose readily: the known radiation chemistry of
poly(methylmethacrylate) and poly(olefin sulphone)s,
for example, led to their selection and to the
development of Electron Beam Systems to exploit them.
The cost of the polymers for such a purpose is of the
order of £600 for a litre of solution, which is substantially more than the cost of the monomers used.
Only a small amount is used in the production of each
micro chip but their role in the microlithography
process is one of the critical factors that governs the
smallness of each device, as Pethrick describes, and
that determines the speed, power consumption and cost

of the product. If an appreciation of the complexity of
the factors involved in a high value polymer area were
wanted,   this article provides it. It is no wonder that
Dick has been consulted by others producing resist
polymers (p 300). Though progress is foreseen for some
time yet,   at some stage progressive improvements in
miniaturisation will no longer be achievable,   and a
jump to another technology will be required. The search
is on already for materials and devices that will allow
switching of light and the optical storage of information.

        Sherrington describes the organic chemistry of
reactions on polymers (page 1). The Merrifield synthesis of polypeptides within beads is well known;   less
well known is the commercial production of the anti
knock additive methyl-*t*-butyl ether on a polystyrene
based sulphonic acid catalyst. GPC column packing
materials are a commonly used high value application
within polymer science itself (pages 1, 109).
Sherrington has produced a novel,   highly porous,
polystyrene material, Polyhype^R, and found how to make
it hydrophillic by grafting. He indicates that prices
of £20 to £3,000 per kilo can be commanded by polymer
supported species,   prices which will attract others
into this field.

        Money is not the only driving force of change:
considerations for the health of the workforce and the
effect of a particular practice or product on the
environment can be important.   The discovery that the
reaction solvent first chosen for Kevlar was a carcinogen caused a hectic search for a new system (page
392),   and similar though less urgent considerations
prompted the work on the new paint systems.   International Paints,   now called Courtaulds Coatings,   devised
an isocyanate-free paint system that has found a wide
outlet in Australia (page 297).   ICI Paints have
developed a new water-based system of paint suitable for
stoving applications such as car bodywork (page 326).
Encapsulation may be one way of handling a toxic
substance (page 1).   Other things being equal,   environmentally friendly materials will save consumers
costs by requiring less attention to cleaning up ef-
fluents,   and so will penetrate the market.   Legislation can stimulate such work by creating a time scale
within which undesirable technologies should be phased
out,   as with chlorofluorocarbons as foaming agents for

polystyrene and polyurethane. Perhaps governments would adopt this course more frequently if they had more expectation that the improvements could be made.

This book records the lectures that were given to a symposium sponsored by the Industrial Division of the Royal Society of Chemistry, at the Congress in Belfast in 1990. The topics covered are only a small fraction of the work that is being performed in this area. Speakers and authors were asked to produce as much commercial information as possible, so that the chemistry could be set in its economic framework, but the reader will find that hardly any has appeared. Perhaps academic authors are not familiar with published sources that provide an economic perspective to their activity, or are performing research in a field that cannot yet be linked to a clear economic gain. Authors from industry are reluctant to give much information on their views of the size of the markets and the cost structures for current products and anticipated developments, for fear of encouraging competition (if the project is successful). Only when the history of this period is written will the business plans and discounted cash flow projections be available for study. We were assured at the symposium that the reputedly high R & D costs of Kevlar have been recouped, and the positive cash flow is presumably now being used to develop new products and processes as well as to pay the dividends.

What is apparent, however, is that businesses can be founded upon high value polymers, that these have a low volume of production but a high research content, and that they require exceptional skills and education of their workforce. This book may encourage a wider appreciation of these ideas.

References

1)   Hounshell, D.A.; Smith, J.K. Jr, *Science and Corporate Strategy,   Du Pont R & D,   1902 - 1980,* Cambridge University Press, Cambridge, 1988.

2)   Reader, W.J., *Imperial Chemical Industries: a History, Volumes I and II,* Oxford University Press,   London,   1970.

3)   *Chemistry in Britain,* August, 1990, p 747.

4)   Lemstra, P.J.; Kirschbaum, R., *Polymer,* (1985) **26,** 1372.

5)   *Chemical and Engineering News,* (1989) **58,** 29

6)   Mark, H., in *History of Polyolefins,* Eds Seymour, R.B.;   Tai Cheng, Reidel, 1986, p vii.

7)   Frederics, M.E.P.;   Marvel C.S.,   *J. Amer. Chem. Soc.,* (1930) **52,** 376.

8)   Seymour, R.B., in *History of Polyolefins,* Eds Seymour, R.B.; Tai Cheng, Reidel, 1986, p 1.

9)   Reference 1; p 293.

Preparation, Characterization, and Application of Polymer-supported Species

D. C. Sherrington

DEPARTMENT OF PURE AND APPLIED CHEMISTRY, UNIVERSITY OF STRATHCLYDE, THOMAS GRAHAM BUILDING, 295 CATHEDRAL STREET, GLASGOW G1 1XL

1 INTRODUCTION

Over the last two decades great progress has been made in a new area of polymer chemistry which focuses essentially on the chemical attachment of reactive and interactive-functional groups to polymer backbones, and subsequent application of the supported or immobilised species in some chemical or biochemical process. The area has grown enormously during this period and this chapter will deal only with applications which are in the main chemical in nature. Other uses, for example in biochemistry, chromatography, diagnostics and sensors, will not be described.

The potential exploitation of polymer–supported species in chemical processes has attracted considerable industrial interest, and in recent years commercialisation of a number of systems has been achieved. These can be regarded as 'High Value Polymers' and the question of industrial application and value will be dealt with more fully later in this chapter.

A number of authoritative textbooks covering the detailed science of this area have already been published, and readers seeking more details are stongly urged to consult these[1-7]. In addition a series of international symposia have been held on this subject and the proceedings of each of these have also been published[8-12].

Although most applications of polymer–supported species have utilised spherical resin beads as the supporting polymer, in principle the polymer support might be a linear soluble species, a porous crosslinked resin, or indeed a macroscopic polymeric object. In the context of this chapter the chemically bound functional group might be a reagent, a protecting group, a ligand, a metal complex catalyst or some other catalytic species e.g. an acid or base, or a phase transfer or micellar catalyst.

Immobilisation of a reactive species on a support might provide a number of important advantages and some of these are listed in Table 1. Simply

TABLE 1   POTENTIAL ADVANTAGES IN USING

                 A POLYMER SUPPORT

1.   MACROSCOPIC "HANDLE" – IMPROVED SEPARATION,
     ISOLATION PURIFICATION

2.   CONVENIENT USE OF EXCESS REAGENT

3.   RETENTION OF PRECIOUS SPECIES

4.   RE–USE OR RE–CYCLING POSSIBILITIES

5.   "ENCAPSULATION" OF CORROSIVE, NOXIOUS OR
     TOXIC SPECIES

6.   BATCH OR COLUMN REACTORS

7.   GAS OR LIQUID PHASE REACTIONS

8.   SITE ISOLATION – INFINITE DILUTION

9.   REDUCED SIDE–REACTIONS

10. ENHANCED SELECTIVITY

11. ENHANCED REACTIVITY

12. STABILISATION OF REACTIVE SPECIES

TABLE 2   POSSIBLE DISADVANTAGES IN USING

                 A POLYMER SUPPORT

1.   INITIAL EXTRA COST

2.   NOT AVAILABLE "OFF–THE–SHELF"

3.   REDUCED REACTIVITY-SLOWER REACTIONS

4.   REDUCED SELECTIVITY

5.   INCREASED SIDE–REACTIONS

converting the reactive function to a macromolecular species immediately aids separation, isolation and purification procedures involving the bound species. Excess of a polymeric reagent can be readily employed without incurring a penalty in work–up, and precious species such as optically active catalysts or platinum group metal complexes might be efficiently retained for re–use or re–cycling. Corrosive noxious or toxic species might be effectively encapsulated, with obvious advantage in environmental terms. A number of processing gains may also be made since supported species lend themselves for use in gas and liquid phase reactions, and also in batch or continuous processes. Finally a number of significant chemical advantages might also accrue. Immobilised groups are essentially site–isolated, and so such groups can behave as if present in infinite dilution. Conversely two of more groups can be forced to remain in close proximity to each other. There are also good prospects for a reduction in side reactions, for enhanced reactivity and selectivity and indeed for producing and 'stabilising' reactive species which are unknown in homogeneous solutions, or at least whose existence is too transient to be usefully exploited.

It is also important of course to be realistic and appreciate that some potential disadvantages can arise and these are listed in Table 2. The first two are very obvious and very important. In any useful application, these two factors must be more than balanced by one or more of the advantages listed above. In addition the possible disadvantageous effects on the chemistry of the analogous homogeneous reaction must also been borne in mind at all times.

2 PREPARATION OF POLYMER SUPPORT

Although in principle a wide variety of physical forms of a polymer might be employed as the basis of supported species, in practice most useful applications have employed spherical crosslinked particulate resin beads. These are prepared in the laboratory and on an industrial scale by suspension polymerisation techniques[13]. Typically an organic phase, consisting of a vinyl monomer and divinyl crosslinking comonomer, is dispersed as droplets by stirring in an aqueous phase containing a water–soluble polymer as a suspension stabiliser (Figure 1). A free radical initiator is dissolved in the monomer droplets and so when the suspension is heated, polymerisation occurs, slowly converting the mobile liquid monomer droplets into rigid, solid polymer particles or 'beads'. These can be collected by filtration and washed free of the aqueous phase. They can also be readily freed of residual comonomers by extraction with a swelling solvent in a Soxhlet extractor. On drying free flowing polymer resin beads are obtained (Figure 2).

In practice the suspension polymerisation process can be quite tricky and the water–soluble suspension stabiliser can be very important for success with some comonomer species. The process is also aided by use of a flat–bottomed reactor with flanged sides. This arrangement maximises the level of turbulence produced for a given energy input from the stirrer system (Figure 3).

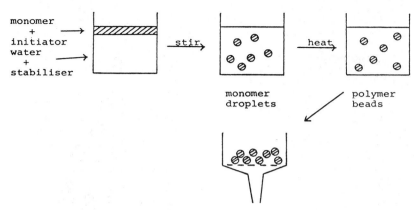

monomer
  +
initiator
water
  +
stabiliser

stir,   monomer droplets   heat,   polymer beads

Figure 1   Suspension Polymerization Scheme

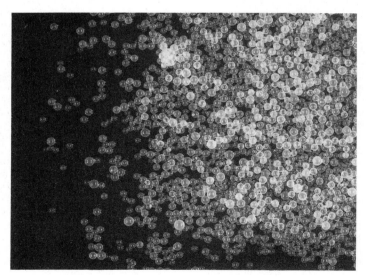

Figure 2   Optical Photograph of Suspension
            Polymer Beads ($\sim$ 200 μm diameter)

<u>Figure 3</u>   Flat-bottomed Flanged Suspension
Polymerization Reactor

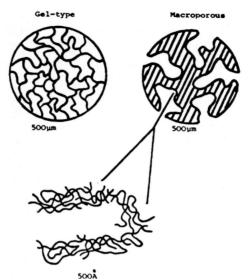

Figure 4   Gel-Type and Macroporous Resins

TABLE 3   **RESIN PARAMETERS**

Gel-type	Macroporous
Clear and glass-like	amorphous and opaque
Lightly crosslinked (often ~ 2%)	heavily crosslinked (often > 20%)
Microporous in dry state	permanent pores (up to ~ 10,000 Å diameter)
No internal surface	significant internal surface area
–	surface area $\propto 1/$pore size
Swellable in good solvents (solvent porosity)	pores accessible to all solvents
Non-swellable in poor solvents	–
Loading up to ~ 100% possible	loading often limited to ~ 40% maximum
Limited use in columns	suitable for column applications

The internal structure of the resin beads can also be controlled by other parameters in the polymerisation process, notably the percentage of crosslinking comonomer used, and the type and volume of any diluent or porogen (an organic solvent) included in the monomer phase. When no porogen is employed and the level of crosslinker is kept low, typically <5 volume % a so–called gel–type resin is formed. These are transparent and glassy in appearance. If a porogen is used and higher levels of crosslinker, typically >20 volume %, then permanently porous resins (macroporous or macroreticular) are formed. These are usually very opaque in appearance and often have a rough surface. Figure 4 shows these variants schematically. Gel–type species are relatively uniform in their structure whereas macroporous species have a permanent porous structure, with highly crosslinked and entangled regions as well. Table 3 summarises the main characteristics of each type, and particular species have been found to be more successful in some applications than in others.

The ability to control the internal structure of resins has become of increasing importance in recent years. High surface area materials with large number of small pores can be achieved using high levels of crosslinker (>50 volume %) and a porogen which is highly solvating for the polymer being formed. However low surface area species with a large number of large pores can be generated most easily using a precipitating porogen. The mechanism by which these morphologies are formed has been discussed in detail and a complex structural model has been proposed (Figure 5)[14].

One potentially very important parameter in polymer–supported species is the loading. This is the mole % of monomer segments which carry the required functional group for a particular application. With gel–type resins, particularly those which are very lightly crosslinked <2%, loadings approaching 100% are achievable. With macroporous species, however, inevitably many of the segments are located in the densely crosslinked environments of the matrix and many of these are inaccessible for chemical modification, or if made from a functional comonomer (see next section) have some active groups which are unavailable. Experience shows that the maximum usable loading of macroporous species is often limited to ~40%.

3. IMMOBILISATION OF REACTIVE SPECIES

In principle a reactive functional group can be introduced onto a polymer resin matrix using two different strategies. A non–functional resin can be synthesised then in second separate procedure the required chemical functionality can be introduced by chemical modification (one or more steps) of the polymer matrix. This is called the 'chemical modification route'. Alternatively the required functionality can be introduced into a polymerisable monomer and then this functional comonomer can be used as one component of the polymerising mixture during resin preparation. This is called the 'functional comonomer' route. Combination of these approachs is also possible.

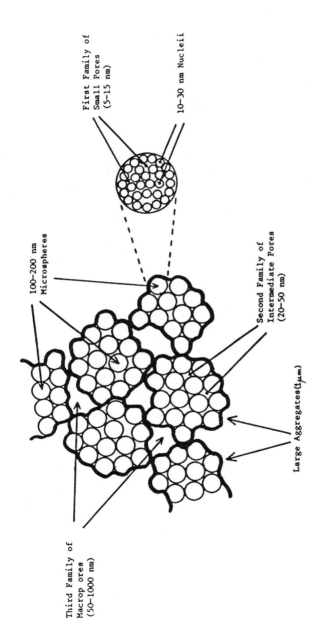

First Family of
Small Pores
(5-15 nm)

10-30 nm Nucleii

100-200 nm
Microspheres

Second Family of
Intermediate Pores
(20-50 nm)

Large Aggregates(1μm)

Third Family of
Macrop ores
(50-1000 nm)

Figure 5   More Detailed Model of Macroporous Resin

Each strategy has some advantages and some disadvantages and the various factors are listed in Table 4. Summarising these it is apparent that the comonomer route allows the structure of the functional group to be clearly defined and offers more control over the number and distribution of groups. Overall the chemistry is better defined. The chemical modification route offers an overall easier strategy, enabling off-the-shelf resins to be employed, and provides groups which are less likely to be located in inaccessible regions of the resin. Coupled with lower cost these factors have meant that the chemical modification route is the one which has been most widely employed. We shall see later, however, that this strategy has to be used carefully, if resin groups with even a reasonable level of purity are to be obtained.

Having made or purchased a non-functional precursor resin a wide variety of chemistry is available to generate specific functionality by chemical modification. In principle any chemical transformation which has been carried out on small molecules in solution can be achieved on anologous macromolecular structures. In practice extra care may be required in selecting suitable reaction conditions, and sometimes much experimentation is required before success is achieved. Typically for polystyrene-based resins, however, the various electrophilic substitutions shown in Figure 6 can be achieved quite readily. Some of these, for example, the chloromethylation and sulphonation reactions, are carried out on a large scale industrially. Sulphonation yields cation exchange resins, whereas amination of the chloromethylated species using trimethylamine yields anion exchange resins[16]. The latter is an example of an attack by a nucleophile, and this is the second important group of reactions used for introducing functionality into polystyrene-based resins. Some other examples are shown in Figure 7[15]. Though the chemistry of polystyrene is by the most well developed, great scope remains with other polymers. For example, rubber chemists have examined the derivatisation of polydienes in great detail, and a considerable literature exists on the derivatisation of mainchain alkene residues (Figure 8).

4. REACTION MODELS

Having introduced a desired functionality into a resin, the supported species is then used typically in some chemical reaction. More often than not, the ideal situation which is aimed for is called "pseudohomogeneous". In this the reactant penetrates the resin, undergoes a reaction induced by the immobilised species, and the product rapidly leaves the resin (Figure 9). The polymer matrix serves only to 'hold' the immobilised species and neither the reactant nor the product detects the presence of the resin. The thermodynamics and kinetics of the reaction are essentially unaffected by the resin, and for all practical purposes are identical to the analogous homogeneous reaction in isotropic solution.

Unfortunately, this ideal situation is seldom achieved in practice. More often, some additional complication arises as a result of the presence of the polymer support. For example, the reactant may be favourably sorbed into the resin and a rate acceleration observed as a result. Likewise, the reactant

TABLE 4   <u>DERIVATISATION OF RESINS VIA CHEMICAL</u>
                <u>MODIFICATION AND USE OF FUNCTIONAL</u>
                <u>COMONOMERS</u>

<u>Factor</u>	<u>Chemical Modification Route</u>	<u>Functional Comonomer Route</u>
availability	off-the-shelf product	comonomer and polymer to be prepared
structure of groups	maybe unclear	well defined
number of groups	poor control	better control
distribution of groups	poor control poorly defined	better control better defined
overall chemistry	maybe poorly defined	generally clear-cut
access to groups	generally good	possibly limited
cost/convenience	low/high	higher/lower

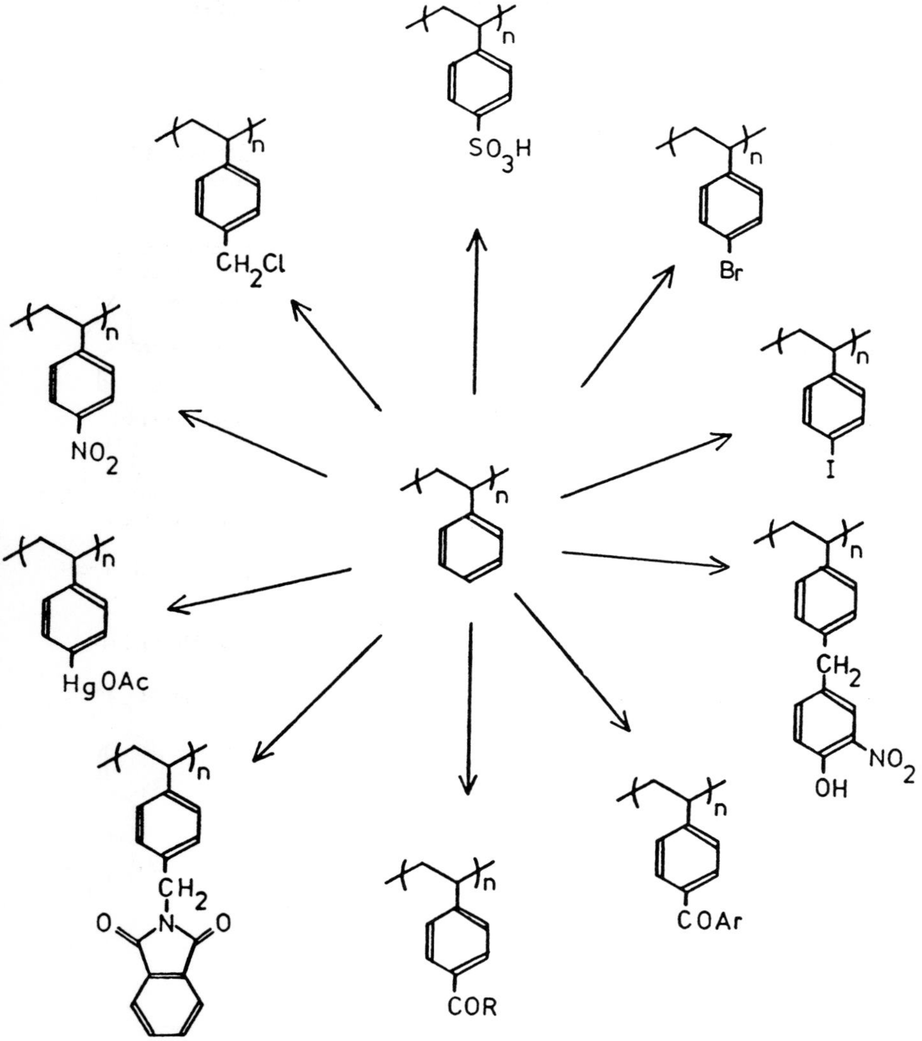

Figure 6   Electrophilic Substitution Reactions of Polystyrene

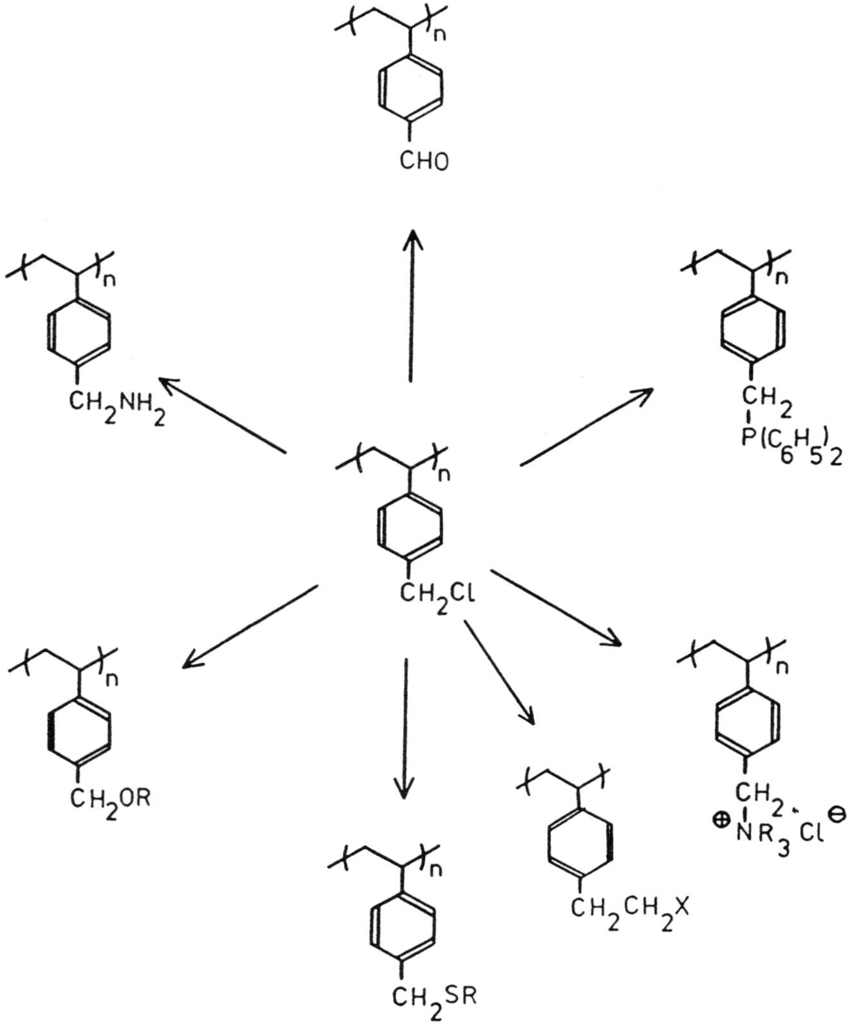

<u>Figure 7</u>   Nucleophilic Substitution Reaction of
            Poly(chloromethylstyrene)

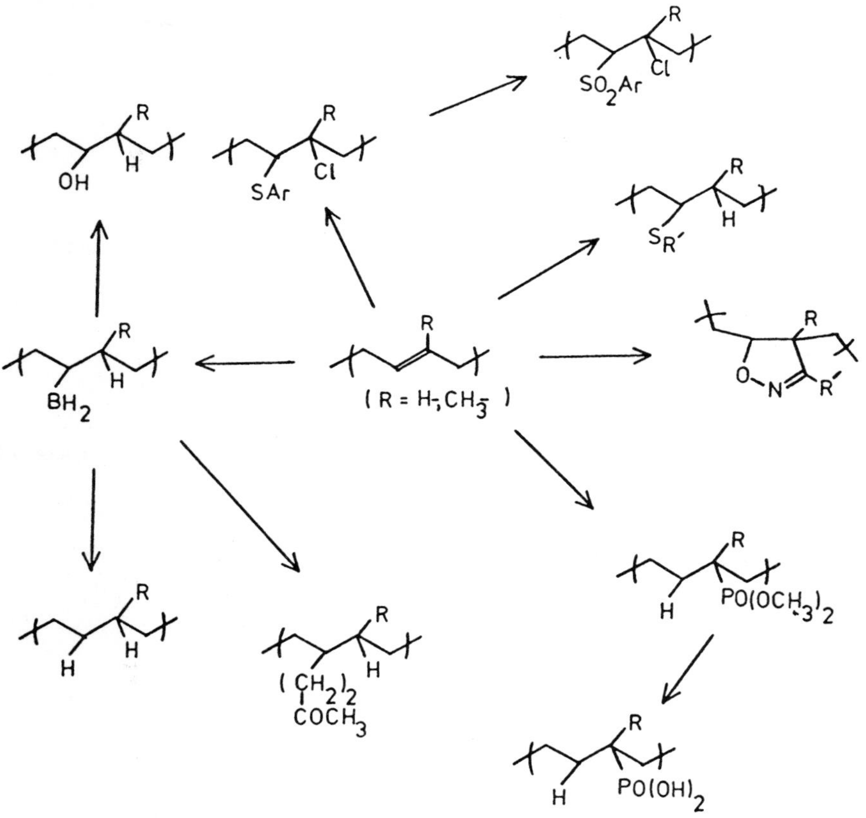

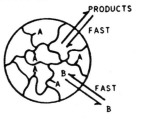

Figure 8   Chemical Modification of Polydienes

Figure 9   Pseudo-homogeneous Reaction of
             B with Immobilised A

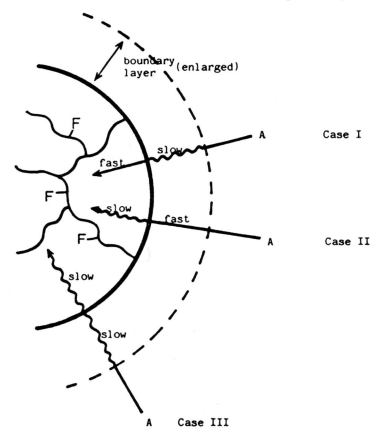

Figure 10   Kinetic Restrictions in Resin Reactions

Figure 11   Substrate Size Effects in Resin
             Reactions

may be partially excluded from the resin in which case the reaction rate is adversely affected.   Similar effects can arise with the product.   Furthermore, direct kinetic effects may arise.   The reactant may experience slow diffusion through the boundary layer of the resin, or through the resin matrix itself (or both) (Figure 10).   Reactants of different sizes may experience different degrees of kinetic restriction, and indeed large reactants may be totally excluded on simple steric grounds from the resin (Figure 11).

When the potential complexity of the resin morphology itself is considered, along with the above possibilities, it quickly becomes obvious how difficult it can be to understand the kinetics of supported reactions in great detail.   Many attempts have been made to do this, in order to optimise resin design, and these aspects have been reviewed recently[17].

5.   APPLICATIONS OF SUPPORTED SPECIES

A summary of the chemical applications of polymer–supported species is shown in Table 5.   Here also is indicated some of the optimum parameters which have been useful in each application.   The list is by no means exhaustive and new information is constantly being generated which requires the data to be up–dated.   For example, early work on the solid phase synthesis of peptides suggested that only very low loadings could be employed in order to minimise interaction between one growing peptide and another.   More recently it has been shown that with highly swollen polyamide supports some oligopeptides can be synthesised at very high loadings on supports quite effectively and in high purity[18].

Some of the optimised parameters are quite evident, others are not.   For example, when using a stoichiometric polymeric reagent in order to be as highly weight effective as possible, then the loading on the support should be as high as possible.   Similarly, when supporting an active transition metal complex catalyst which might deactivate by self–oligomerisation then clearly site–isolation via a very low loading on the support is necessary.

6.   POLYMERIC SULPHONIC ACID CATALYSTS

Typical cation and anion exchange resins are represented by structures I and II.

I   II

Polystyrene based species were developed quickly after World War II and the industrial scientists in the US, UK and Germany, who achieved their rapid

TABLE 5   AREAS OF APPLICATION OF POLYMER SUPPORTS AND CORRESPONDING PROPERTIES FOUND OPTIMUM FOR EACH AREA

Application	Hydrophobic — Hydrophilic	Gel–Type — Macroporous	Loading (Low — Medium — High)
Polymeric reagents	↕	↕	↕ (Low–High)
Polymeric acid catalysts	↕	↕	↕ (Low–Medium)
Supported metal complex catalysts	←--→	←--→	↕ (Low–Medium)
Supported phase transfer catalysts	↕	↕	↕ (Low–Medium)
Solid phase synthesis	←--→	↕	←--→ (Low–Medium)
Polymeric chelating ligands	←--→	↕	↕ (Medium–High)

Note   This table is limited to essentially chemical applications. Numerous biochemical and separation applications have also been described and optimised similarly under these headings.

commercialisation, equally quickly realised that these species might make useful solid acid and base catalysts. Indeed, between 1945–1965 several thousands of publications appeared describing such applications[19]. Potentially the most important were the sulphonic acid resins. These were shown to be at least as acidic as H_2SO_4 and could operate with good lifetime at temperatures ~100–120°C. The hydroxyl form of quaternary ammonium anion exchange resins proved much less thermally stable, and the disadvantage remains even today. Many of the advantages listed earlier for supported species were attractive in exploiting solid polymeric acids, but perhaps the most important are that the corrosive acid function is essentially encapsulated within the hydrocarbon polymer such that there is virtually no contact with the metal components of the plant; and secondly the solid form of the acid allows the use of both gas and liquid phase reactions flowing continuously through large columns.

Despite the early development work, successful application on a large scale has taken much longer to achieve. However, polymeric acid catalysts are now an important technology. Probably the most important use is in the synthesis of methyl t–butyl ether (MTBE) from methanol and isobutene.

$$CH_3OH \; + \; CH_2{=}C(CH_3)_2 \; \xrightarrow{\;\; \textcircled{P}-\!\!\!\langle\rangle\!\!-SO_3H \;\;} \; CH_3OC(CH_3)_3$$

MTBE is the major organic anti–knock replacing tetraethyl lead in petrol (gasoline). It has to be used in large quantities ~15 vol. % and so lead replacement has been possible only as quickly as MTBE plants could be built and brought on stream. In 1986 the production was ~ 6×10^6 tons from ~ 30 plants worldwide. Similar polymeric acid catalysts are now used in the production of higher ethers, in the hydration of propylene and but–1–ene to yield isopropanol and butanol, in the alkylation of phenol, in the dimerization of cyclohexanone, and in the syntheses of methyl isobutyl ketone, bis–phenol A and methacrylate esters.

The immobilisation of an acid on one resin and an alkali on another offers the unique prospects of mixing such an acid and an alkali in one 'pot' without direct neutralisation. This fact has been realised and exploited very cleverly in prostagladin synthesis[21] (Scheme 1). Acetal(III) is thermodynamically favoured over the aldehyde form (IV) and the equilibrium level of the latter in the presence of the acid resin is therefore small. This low concentration allows the base–resin–catalysed reaction to proceed almost exclusively as an intramolecular process to yield the product indicated (after dehydration). Furthermore this drain on (IV) graudally shifts the initial equilibrium and allows a steady conversion to the desired product.

In terms of further industrial application a major restriction remains the ultimate thermal stability of sulphonic acid resins. Aromatic sulphonation is a reversible reaction and in the presence of water at temperatues above ~120°C, gradual desulphonation of resins occurs. If resins with improved stability could be produced cost effectively, many other acid catalysed reactions would lend themselves to the use of polymeric acid technology. Many attempts have been

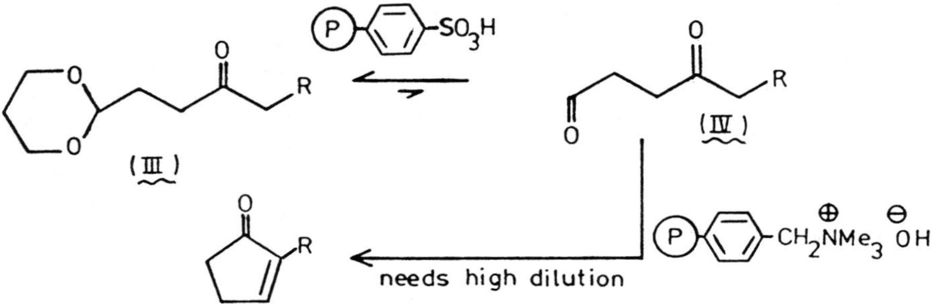

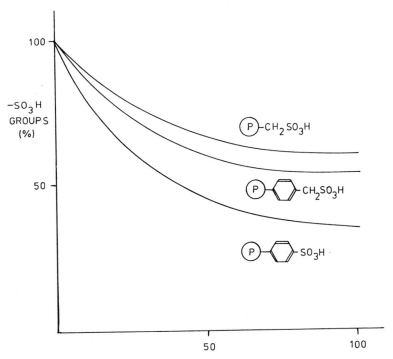

Scheme 1   Concurrent Acid and Base Catalysed Reactions

Figure 12   Sulphonic Acid Resin Stability
in Water at 200°C

made to improve stability and these have been reviewed recently[22]. It is known that aliphatic sulphonic acids desulphonate much less readily than aromatic ones, and so in principle an all–aliphatic hydrocarbon acid resin could provide the solution. In collaboration with Widdecke at the Technical University of Braunschweig we have synthesised such a resin by chemical modification of a methyl methacrylate resin crosslinked with divinylbenzene (Scheme 2)[23]. Treatment of the aliphatic polymeric acid with water at 200°C indeed confirmed that it was more stable than a conventional aromatic acid, and an aromatic species with a single methylene group spacer (Figure 12). However, the loading of the resin is poor and the synthetic route is too costly for exploitation.

7. POLYMER–SUPPORTED TRANSITION METAL COMPLEX CATALYSTS

It has been recognised for many years now that the immobilisation of highly active and highly selective homogeneous transition metal complex catalysts on polymers, such that they retain all of their catalytic properties in their "heterogeneous" state, has tremendous commercial potential. Most investigations have focused on alkene hydrogenation, hydroformylation, isomerisation and oligomerisation catalysts[4]. Much less work has been reported on oxidation catalysts[24]. This is because homogeneous metal complex oxidation catalysts are themselves less well developed, and also there has been a basic worry that organic polymers are unlikely to prove sufficiently chemically robust in strongly oxidising environments. Increasingly by the latter is proving not to be the case. In our own laboratory we have immobilised Mo(VI) and V(V) species on polystyrene and polymethacrylate based resins and used them successfully as catalysts in the epoxidation of cyclohexene, using t–butylhydroperoxide as the mono–oxygen source. For example a blue polymer complex was prepared by ligand exchange of MoO_2 $(acac)_2$ with the polymeric ligand (V). The blue complex

$\underline{V}$

proved to be inactive as a catalyst but treatment or activation with excess t–butylhydroperoxide produced a bright yellow polymer–supported complex which proved to be a potent catalyst (Figure 13). A major issue in this type of work concerns the stability of the supported system in terms of leaching of the metal complex. In this instance the loss of Mo(VI) was monitored carefully during successive re–cycling of the catalyst, and depending on the nature of the ligand, highly stable systems have been identified where leaching extrapolates virtually to zero after ten successive cycles (Figure 14)[25]. There remains, however, a need for more thermo–oxidatively stable polymer supports and this is an area which is now being actively researched.

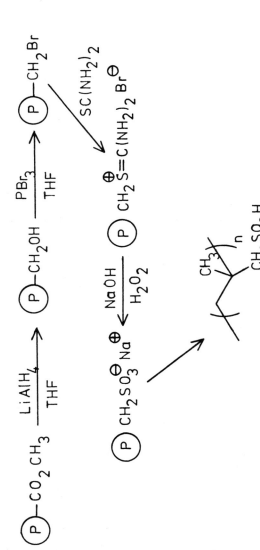

Scheme 2   Synthesis of Aliphatic Sulphonic Acid Resin

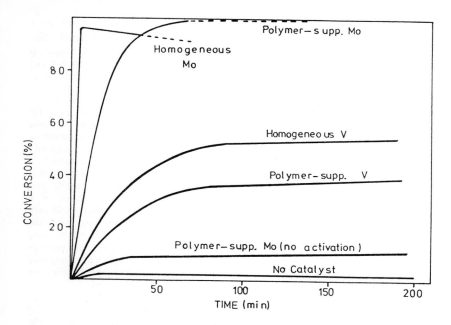

Figure 13   Epoxidation of Cyclohexene Using t-butyl
Hydroperoxide and Mo and V Catalysts

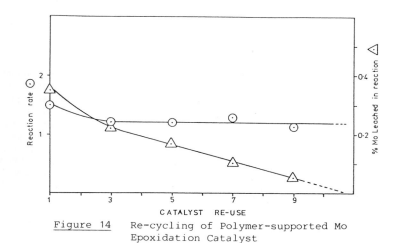

Figure 14   Re-cycling of Polymer-supported Mo
Epoxidation Catalyst

8.   SELECTIVE CHELATING ION EXCHANGE RESINS

Another area of application in which it has long been appreciated that significant industrial potential exists is that of selective chelating ion exchange resins.   In principle a selective homogeneous ligand might be attached to a polymer support and used to capture a particular metal ion very selectively. In practice this approach has proved singularly unsuccessful and there is growing belief that two factors are responsible for this.   Firstly most of the resins which have been used are hydrophobic, and almost certainly intimate molecular contact between the immobilised ligand and the mobile aqueous phase does not occur.   The complexation conditions which prevail in isotropic aqueous solution with free ligands have not therfore been achieved.   Secondly most of the chelating resins reported in the literature[26] have involved a number of chemical steps performed on the resin in order to construct the specific ligand.   These multi–step chemical modifications carry inherent and rather general problems with them.   Supposing for example the synthesis of a ligand or other bound species requires four chemical steps, and that each step is achieved with a very reasonable yield of ~70%.   This means that the proportion of the required structure on the final resin is $(0.7)^4$ i.e. ~25%.   As a result not only is the loading of the required group rather low, but perhaps more importantly, the resin is also contaminated with ~50% of "impurity groups".   Not surprisingly therefore resins designed to chelate specifically one metal ion more often than not perform very badly.

To try to minimise this problem we have synthesised resins based on glycidyl methacrylate monomer, which yields reactive epoxide groups in the resin.   These allow the introduction, in a single chemical reaction, of a wide variety of ligand species[27].   For example reaction with 2–amino methylpyridine yields the chelating resin (VI).   This ligand

is known to be selective for the binding of Cu(II)[28], but when immobilised on a polystyrene resin its selectivity is reduced.   However, methacrylate resin (VI) retains the remarkable selectivity, and also shows very rapid exchange kinetics. For example the resin removes virtually all the Cu(II) from a solution containing 1 g of Cu(II) and 150 g of Zn(II) per litre, while removing very little of the dominant Zn(II) (Table 6)[29].   The favourable kinetic performance almost certainly arises from the relative hydrophilicity of the resin, and the batch extraction behaviour is reproduced exactly in column extraction[30]. Indeed the removal of Cu(II) from such solutions is necessary as a pre–treatment of the feed solutions used in the electro–winning of Zn(II), and B.P. researchers have taken this resin to a small pilot scale, and shown its performance to be totally suitable for commercial use[31].

TABLE 6   SEPARATION OF Cu(II) FROM A SOLUTION CONTAINING 1g l^{-1} Cu(II) and 150g l^{-1} Zn(II) SULPHATES at pH5, 25°C

| Resin | Resin Capacity (g metal g^{-1} resin) | | |
	Cu(II) On Loading	On Elution	Zn(II) On Elution
Ps/AMPy	15	11	1
Ps/PyIm	12	5	3
GMA/AMPy	16	14	5
GMA/PyIm	14	16	7

Ps = polystyrene resin;   GMA = polyglycidyl methacrylate resin;
PyIm = 2-pyridylimidazole ligand;   AMPy = 2-aminomethylpyridine ligand.

TABLE 7   DEMANDS MADE ON A SUPPORT

Microenvironmental

appropriate chemical functionality
loading level
pore size
surface area
local polarity
response to solvent
local flexibility
local "space"
site isolation/interaction

Macroscopic

particle size
particle shape
shear strength
compressive strength
solvent resistance
chemical stability
thermal stability

9. DESIGN OF POLYMER SUPPORT — A CASE STUDY

For every successful application of a polymer–supported system there are many which have fallen for short of expectation, and over the last few years we have been trying to design supports <u>ab initio</u> for particular applications, in attempt to improve our own success rate. In doing this we have come to realise that the demands we make of a support can really be grouped into two categories. The first are microenvironmental or chemical demands, and the second are macroscopic or mechanical demands. Table 7 lists these. Obviously in some instances the distinction between these is not clear cut, but the division is a useful conceptual one. To date virtually all applications of polymer–supported species have utilised one polymer to fulfil all these demands, and in retrospect it is not surprising that one polymer has often been found lacking.

For some time now we have been pursuing the concept of de–coupling the macroscopic and microenvironmental requirements in a support by designing composite systems in which one macromolecule acts as a porous rigid 'scaffold' to provide the overall mechanical properties, and a second macromolecular species functions as an internal matrix to provide the optimum reaction microenvironment (Figure 15). A recent success with a support for solid phase synthesis provides a useful case study[32].

Sheppard and his coworkers were probably the first to recognise and point out the dichotomy of using a highly non–polar polystyrene based support on which to construct highly polar oligopeptides. Rather simplistically we can imagine that solvents which expand the polystyrene backbone (e.g. toluene) will tend to collapse peptide chains (Figure 16a) and similarly solvents which expand the peptide (e.g. dimethylformamide, DMF) will tend to collapse the support backbone (Figure 16b). An ideal situation might be one in which both support and peptide are expanded (Figure 16c) and this led Sheppard to develop a new resin based on N,N–dimethylacrylamide (NNDMA)[33]. Such supports are generally used at relatively low capacity, and are soft and highly swollen in solvents appropriate for peptide synthesis. As such they are useful only for batch operation work, and are best applied by skilled scientists. Epton and his coworkers[18] also developed a polyamide–based support, but demonstrated that it could also be used very successfully at relatively high capacities. Its soft swollen form, however, again restricted its use to small scale batch reactions, at least at that time.

In late 1984, we became aware of an entirely novel form of polystyrene discovered by Unilever scientists at their Port Sunlight Laboratory[34]. By polymerizing the continuous phase of a high internal phase emulsion and washing out the aqueous internal phase (~90% v/v) using ethanol followed by vacuum drying, a Polyhipe ^{R} polymer is formed. Such materials are characterised by an extremely low bulk density and very high pore volume (74–99% v/v), generally higher than conventional gas blown polymer foams. Furthermore, by appropriate choice of surfactant a matrix can be formed consisting of spherical cells 5–30 μm in diameter, which are totally interconnected by a plurality of holes. Figure 17 shows a scanning electron

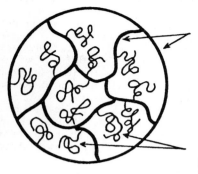

Primary Scaffold
Macroscopic Properties

Secondary Matrix
Microenvironmental
Properties

Figure 15   Schematic Representation of a
Composite Polymer Resin

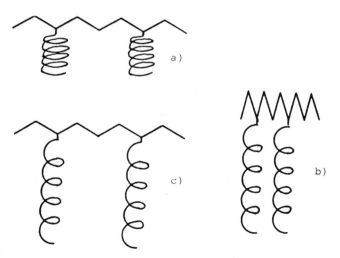

Figure 16   Response of Solid Phase Peptide Resin
to Solvents a) Toluene b) DMF
c) ideal situation

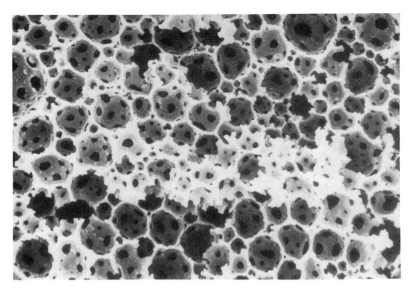

<u>Figure 17</u>   Scanning Electron Micrograph of ·
                   Polyhipe R Polymer, ⊢————⊣ =10 μm

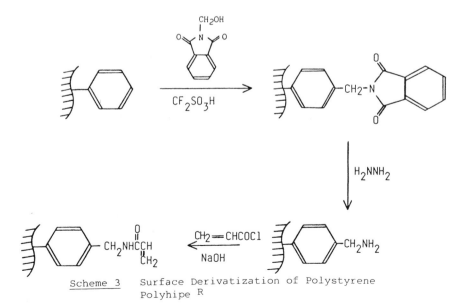

<u>Scheme 3</u>   Surface Derivatization of Polystyrene
               Polyhipe R

micrograph of a typical poly(styrene–co–divinylbenzene) Polyhipe R. Despite the very low bulk density of such materials the spherical nature of the cells, and their relatively small size distribution, gives rise overall to remarkable mechanical rigidity. Indeed, this macroscopic property together with the very high pore volume, made us realise that these materials were excellent candidates to function as "scaffold" structures for composite supports. In particular, the very high pore volume seems to offer the prospect of producing composite supports with a volume fraction of the internal matrix so high as to approach very closely to the state of the isolated matrix itself.

The strategy we involved therefore was to choose an appropriate Polyhipe R polymer in granular form, then to bind within the entirety of its pore structure a highly solvent swollen secondary network conforming very closely to the structure of the soft polyamide gels already well developed as solid phase peptide synthesis supports. To ensure that the secondary network was chemically bound within the Polyhipe R scaffold the surface of the latter was derivatised with acrylamido groups using conventional chemistry, as shown in Scheme 3.

Derivatised Polyhipe R polymer then formed the starting scaffold for the introduction of a Sheppard[33] type polyamide gel, shown schematically in Figure 18. Grafting of this matrix was achieved simply by imbibing the scaffold with a DMF solution of monomers and a free radical initiator, and heating at 60°C. Low molecular weight residues were removed by exhaustive extraction. In this example the sarcosine residues are the sites at which peptide assembly is carried out. The composite was slurry packed in DMF into a small column for use in a commercial continuous flow peptide synthesiser. No column back pressure was observed under standard flow conditions. Using normal Fmoc protection strategy a test synthesis of a segment of the acyl carrier protein, ACP(65–74) was carried out. For comparison the same segment was also synthesised on a commercially available polyamide impregnated kieselguhr support and in batch on a conventional Merrifield polystyrene support. The results are shown in Table 8.

These indicate clearly that the yield of peptide on the new composite is about an order of magnitude higher than that on kieselguhr without any loss of purity. Additionally the advantage of the polyamide microenvironment is confirmed with the result from the conventional batch polystyrene methodology where the peptide purity was only ∼ 35%.

The new composite has been shown to have a number of distinct advantages over existing supports when applied in continuous flow synthesis. The structure provides for excellent flow properties and reagent/solvent accessibility under low pressure. The chemical binding of the secondary matrix ensures no loss of gel even under extreme flow conditions and the high porosity of the primary scaffold allows high synthetic capacities to be attained with appropriate secondary solid phase gels.

We believe the results demonstrate that appropriate design of support architecture with, in particular, due regard being taken of the macroscopic and

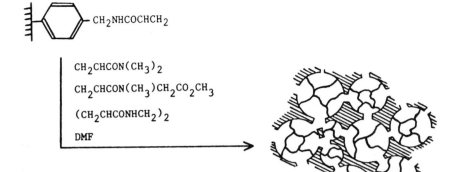

Figure 18   Grafting of Polyamide Gel onto
             Polystyrene Polyhipe [R]

TABLE 8   **SYNTHESIS OF ACYL CARRIER PROTEIN AMINO ACID**
           SEQUENCE 65-74 ON VARIOUS SUPPORTS [a]

Support	Methodology	Yield (a)	Purity (%)[b]
Polyhipe[R]/ polyamide composite [c]	Fmoc/Column	0.91	97
kieselguhr/ polyamide composite [a]	Fmoc/Column	0.07	96
polystyrene [e]	tBoc/Batch	0.35	35

a) 1g of support;   b) from HPLC analysis;
c) capacity, 1 mmole g^{-1};   d) capacity, 0.1 mmole g^{-1}
e) capacity, 0.7 mmole g^{-1}.

microenvironmental requirements of the support, can lead to considerably enhanced performance in use. These composite species provide for the first time a high capacity rigid support with excellent physical stability, offering a real prospect for a large–scale continuous–flow low pressure column technology.

10. FUTURE PROSPECTS

By the early 1980's it began to appear that the area of polymer–supported reactive species had perhaps run its course. However, this quickly proved not to be the case as the decade moved on. All the early hard work, failure and frustration gradually started to bear fruit, and with growing environmental pressure on industrial chemists the prospect of commercialisation of polymer–supported systems became a reality. The most important examples are the polymeric sulphonic acid catalysts (see Section 6) but there is evidence of increasing use of supported species, particularly in downstream processing. There is now little doubt that the technology is accepted, and when it is appropriate, and is cost effective in the broadest sense against other approaches, it will be used with confidence and enthusiasm.

In the laboratory there is a good prospect of producing longer oligopeptides by the solid phase method, possibly up to 100 amino acid residues. The laboratory methodology will undoubtedly be scaled–up for commercial use, if this is not already the case. Recent advances in the solution coupling of saccharides, suggests that it is timely to re–examine the solid phase synthesis of oligosaccharides. A few groups attempted this in the early 70's with very little success, but with new chemistry the situation looks optimistic again.

Research on polymer–supported metal complex catalysts continues, with emphasis on a search for non–leaching and more thermo–oxidatively stable systems. Polymers with appropriate properties do exist already and a few groups including our own are looking closely at these. Likewise sulphonic acid resins with better thermal stability remain an important target, and with ingenuity new materials will be developed here as well.

Exploitation of chiral catalysts and mediators is a vital area for provision of pharmaceutical precursors, and polymer–supported analogues could play an important role. Recent advances with a simple continuous flow technology here in the UK[35] suggests that, with some imagination and conviction, industrial application could be achieved quite easily.

Somewhat more speculatively there has always been an excellent possibility of generating reactive species on polymer–supports which either have no homogeneous analogue, or which possess analogues too short–lived to be exploitable. Some examples have been reported in the literature, but so far these have proved to be largely curiosities. Again more imagination and conviction is needed to produce systems with a real prospect for application. The possible marriage of polymer–supported reactive species and ordered molecular films might also provide exciting technology in the future, such films might be located on electrodes, or could essentially represent two–dimensional crystals. Likewise the application of the principles of biochemical interactions to polymer–supported species offers considerable scope for scientific development. This will require the crossing of discipline barriers but the rewards could prove very handsome.

Finally of course there is little doubt that the need to harmonise not only the activities of the chemical industry, but the activities of all humanity, with nature and the environment will stimulate further exploitation of supported systems. As already indicated this will become increasingly evident in effluent treatment and in product isolation and purification, but will also move upstream steadily to eliminate or minimise environmental problems of source in the chemical reactor.

11. COST AND VALUE

It goes almost without saying that the value of supported species to society could be immense, but in the end clean processes must be cost effective to succeed. Realistic costing, including long–term disposal, will obviously help the case of supported species. Many of the polymers described here fall into the category of 'high value' materials when measured against commodity plastics. Ironically, however, in many cases further expansion may hinge on reducing costs, and hence monetary value. Clearly there is a balance to be struck by industry as potential producers and potential users of polymer–supported systems.

Of the materials currently used commercially a wide range of costs are involved. Generally large volume use e.g. polymeric sulphonic acids, requires correspondingly low costs since the petrochemical processes involved are themselves low cost high volume activities. At the other extreme in small scale applications yielding very high value products e.g. solid phase synthesis of oligonucleotides and oligopeptides, the cost of the polymer–supported species can be negligible compared with other raw materials costs, and hence a large premium can be asked of these materials. Typically therefore a polymer–supported species commands laboratory prices ranging from £18 to £3000 per kilogramme. Interestingly therefore there is obviously considerable scope in terms of the value of such polymers. To some extent a continuity of price might be demanded related only to the application, and more precisely to the value of the product from a particular process. This situation is not new of course and is certainly not restricted to polymer–supported species. It is important, however, because it suggests that considerable scope remains for the development of novel polymers, but that this will only be possible in the light of the knowledge of the costing of existing processes. In general such knowledge is not readily available to academic scientists, and suggests that closer interaction between industrial and academic researchers will be necessary, if sensible areas of exploitation are to be identified.

12. ACKNOWLEDGEMENTS

The work described here has been carried out by a succession of young coworkers over the years, and I am grateful to them not only for all their efforts, but also for their camaraderie during the difficult times as well as the productive ones.

REFERENCES

1. P. Hodge and D.C. Sherrington, Eds., 'Polymer-supported Reactions in Organic Synthesis', Wiley, Chichester, 1980.
2. N.K. Mathur, C.K. Narang and R.E. Williams, 'Polymers as Aids in Organic Chemistry', Academic Press, New York, 1980.
3. W.T. Ford, Ed., 'Polymeric Reagents and Catalysts', Amer. Chem. Soc., Washington, D.C., 1986.
4. F.R. Hartley, 'Supported Metal Complexes', Reidel, Dordrecht, 1985.
5. P. Laszlo, Ed., 'Preparative Chemistry Using Supported Reagents' Academic Press, London, 1987.
6. D.C. Sherrington and P. Hodge, Eds., 'Syntheses and Separations Using Functional Polymers', J. Wiley, Chichester, 1988.
7. A.D. Pomogailo, 'Polymeric Immobilised Metal Complexes Catalysts' Naukar, Moscow, 1988 (Russ).
8. 1st International Symposium on Polymer-supported Reactions, Lyon, France, 1982; Proceedings Nouv. J. Chemie., 1982, 6, 12.
9. 2nd International Symposium on Polymer-supported Reactions, Lancaster, UK, 1984; Proceedings, Brit. Pol. J., 1984, 16, 4.
10. 3rd International Symposium on Polymer-supported Reactions, Jerusalem, Israel, 1986; Proceedings, Reactive Pol., 1987, 6, 2-3.
11. IUPAC Microsymposium on Reactive Polymers, Prague, Czechoslovakia, 1987; Plenary Lectures, Pure and Appl. Chem., 1988, 60, 3; Lectures and Posters, Reative Pol., 1988, 9, 3.
12. 4th International Symposium on Polymer-supported Reactions, Barcelona, Spain, 1988; Proceedings, Reactive Pol., 1989, 10, 103.
13. See Appendix in reference 2. p469.
14. A. Guyot, 'Synthesis and Structure of Polymer Supports' in reference 6, Chap. 1, p1.
15. P. Hodge, 'Organic Reactions Using Polymer-supported Catalysts, Reagents or Substrates' in reference 6, Chap. 2, p43.
16. R.M. Wheaton and M.J. Hatch, in 'Ion Exchange', Ed. J.A. Marinsky, Marcel Dekker, New York, 1969, Vol. 2, Chap. 6.
17. D.C. Sherrington, 'Reactions of Polymers' in 'Encyclopedia of Polymer Science and Engineering', Eds. M.F. Mark, N. Bikales, C.G. overbenger, G. Menges and J.I. Kroschwitz, J. Wiley and Sons, New York, 1988, Vol. 14, p101.
18. R. Epton, G. Marr, B.J. McGinn, P.W. Small, D.A. Wellings and A. Williams, Int. J. Biol. Macromol., 1985, 7, 289.
19. N.G. Polyanskii, Russ. Chem. Rev., 1962, 31, 496; 1970, 39, 244.
20. D.C. Sherrington, 'Catalysis by Ion-Exchange Resins and Related Materials' in Reference 1, Chapter 3, p.157.

21. J.C. Stowell and H.F. Mauck, Jr., J. Org. Chem., 1981, 46, 2429.
22. H. Widdecke, 'Design and Industrial Application of Polymeric Acid Catalysts', in Reference 6, Chapter 4, p.149.
23. Z. Yang, K. Teng, H. Widdecke and D.C. Sherrington, Polymer, submitted for publication.
24. D.C. Sherrington, Pure and Appl. Chem., 1988, 60, 401.
25. S.J. Simpson and D.C. Sherrington, Polymer, in preparation.
26. A. Warskawsky, 'Polymeric Ligands in hydrometallurgy', in Reference 6, Chapter 10, p.325.
27. D. Lindsay and D.C. Sherrington, Reactive Polym., 1985, 3, 327.
28. R.R. Grinstead, 'New Developments in the Chemistry of XFS4195 and XF343084 chelating ion exchange resins' in 'Ion Exchange Technology', D. Naden and M. Streat, Eds., Ellis Harwood, Chichester, UK, 1984, p. 509.
29. D. Lindsay, D.C. Sherrington, J.A. Greig and R.D. Hancock, Reactive Polym., 1990, 12, 59.
30. D. Lindsay, D.C. Sherrington, J.A. Greig and R.D. Hancock, Reactive Polym., 1990, 12, 75.
31. J.A. Greig and D. Lindsay, 'Copper Recovery from Zinc-bearing Leach Liquors' in 'Ion Exchange for Industry', M. Steat (Ed), Ellis Harwood, Chichester, UK, 1988, p.337.
32. P.W. Small and D.C. Sherrington, Chem. Comm., 1989, 1589.
33. E. Atherton, P.L.J. Clive and R.C. Sheppard, J. Amer. Chem. Soc., 1975, 97, 6584.
34. D. Barley and Z. Haq, Europ-Patent 0, 060, 138; 1982 (to Unilever).
35. P. Hodge, Dept. of Chemistry, University of Manchester, private communication.

Directly Functionalized Ethene–Propene Copolymers: Synthesis and Applications

S. Datta

POLYMERS DIVISION, EXXON CHEMICAL CO., P.O. BOX 45, LINDEN, NJ 07036, USA

1 INTRODUCTION

This chapter[1] describes a synthetic procedure we have developed for the synthesis of ethene-propene (E-P) copolymers containing reactive organic functionality. The synthetic procedure is direct polymerization of ethene, propene and the functional monomer in hexane solution. The functionality needs to be deactivated ('masked') prior to introduction into the polymerization reactor: the functionality is regenerated during the deashing operation which is an integral part of the E-P polymerization process. The functionalities that can be introduced using this synthetic procedure include carboxylic acids, alcohols and amines. These polymers have been characterized: we find that the functionality is uniformly distributed across the molecular weight distribution and we believe, but have no definite proof, that the distribution of the functional groups along a chain is random.

2 FUNCTIONAL E-P POLYMERS: SYNTHESIS

The direct synthesis of functionalised polyolefins containing reactive organic functionality on a polyolefin backbone has been the goal of considerable synthetic effort[2] in the last three decades. The incentives for this effort are obvious: these materials would combine the stability, durability and economy of polyolefins with new properties such as chemical reactivity or polar interaction. This is expected to increase the range of potential applications for these polymers. Directly functionalised polymers made with Ziegler-Natta catalysts

are a particularly challenging area since most organic
functionalities react with the polymerization catalyst and
terminate polymerization. Seemingly harmless functional
monomers, such as acrylonitrile, are surprisingly
efficient at terminating[3] Ziegler polymerization. A
solution was suggested by Langer[4a] who found that
sterically hindered amines, in particular -N(isopropyl)$_2$,
could participate in Ziegler-Natta polymerization without
interacting with the catalyst center. Similar experiments,
using t-butyl ethers, have been suggested by Pruett[4b].
This shows that isolation of the functionality from the
catalyst center is important and can determine the success
or failure of these polymerizations. Recently, a number of
independent efforts[5] have described combinations of steric
and electronic isolation techniques (masking) to
synthesize functionalised Ziegler polymers. These
procedures suffer from one or more of the following
problems: (i) low level of functionality incorporation,
(ii) concentration of the functionality in the low
molecular weight fraction of the polymer and (iii) the use
of non reactive organic functionalities such as esters,
ethers and amides. Therefore, these synthetic procedures
are of limited practical use and need improving.

We recognized that improvement in this synthesis is
possible only if the functionality contains **at least one
active hydrogen.** This structural requirement is the key to
our synthetic success where others have failed. The
presence of active hydrogen on the functionality benefits
both the polymer synthesis - points (i) and (ii) above -
and the chemical reactivity in potential applications of
this polymer - point (iii) above. Easily polymerizable
monomers which contain only ester, amido, ether or
tertiary amine functionality are not acceptable for this
synthetic procedure. Prior efforts[5] have almost always
concentrated on attempts to mask functionalities such as
the ones listed above. These are less potent poisons for
Ziegler catalysts compared to the proton donors which we
prefer; however we will show that, compared to proton
donors, the amides, esters or ethers are poorly masked.
Effective masking leads to a more stable polymerization.
We admit that this appears, at first sight, to be contrary
to intuition.

The polyolefin we use as the backbone for the organic
functionality is a ethene-propene copolymer (E-P). We
choose this polymer for the following reasons:[6]

(i) E-P is usually polymerised in a hydrocarbon

solution process with soluble vanadium based polymerization catalysts. The accessibility of the bulky masked, functional monomers to the polymerization center will not be restricted by diffusion through bulk polymer in this process.

(ii) E-P solution polymerisation process is typically run at high monomer conversion. This is due to the inherent expense of recycling unreacted monomer. The functional ethene-propene polymerization procedure uses this to introduce a high concentration of functionality in the polymer.

(iii) E-P polymerization process requires the use of a deashing stage (usually hydrolysis) for the removal of catalyst residues which are principally vanadium and aluminium compounds. Our functional ethene-propene polymerization procedure uses this stage also to remove the mask on the functionality.

(iv) E-P polymers cover a wide range of monomer composition: this leads to difference in the polymer properties such as crystallinity. We believe that this versatility can be exploited in applications of functional polymers.

The functionalised EP polymer contains the functionality pendant from the backbone. The reason for this architecture, in contrast to the case where the functionality is a part of the polymer backbone such as the ethylene - carbon monoxide copolymers[2b,c] is simple: we do not want reactions of the functional group to lead to polymer chain scission. This architecture needs the functionality to be polymerised into the chain through a olefin to which it is chemically linked. This olefin has to be easily polymerisable. We will show later that this requirement dictates some of the structural features of the functional olefins.

The functional E-P polymers which we can synthesize are not significantly limited in either composition or molecular weight by the presence of the functionality. Practical limitations of the EP polymerization catalyst and the process exist in both of these areas and require that the polymer have a ethylene content of between 30 and 80 percent by weight and a molecular weight (Mn) of less than 400 000. These are approximately the range of limitations of these parameters for the functionalised EP polymers. We believe, based on our experience[6] of the kinetics of olefin incorporation on soluble vanadium

catalysts, that the distribution of functionalities along
each chain is random. The number of functional groups per
chain is controlled by the number of hetero atoms in the
functional group. Under the best of synthetic conditions,
about 30 hetero atoms per polymer of molecular weight 100
000 can be polymerized: this is equivalent to 15
carboxylic acid groups or 30 alcohol or amine
functionalities. Polymer chains of higher molecular weight
contain a proportionately larger number of functional
groups. These ranges and limitations are only approximate:
they represent accessible goals without compromising the
versatility of the EP polymerization process in the areas
of molecular weight and composition. Experiments to
introduce significantly higher levels of functionality
will produce functionalised polymers but the control and
reproducibility of such polymerizations are poor.

3 POLYMERIZATION

The polymerization scheme used for the functional EP
polymers is shown in Figure 1. We have used a hypothetical
monomer containing a single carboxylic acid functionality
'olefin-CO_2H' (1) to show the most important operations in
the synthetic scheme. The scheme could be demonstrated for
the alcohol or the amine functionality with the monomers
'olefin-CH_2OH' or 'olefin-CH_2NH_2': the principle and the
sequence of operation would be the same. The changes in
the nature of the functionality do not affect the level or
the distribution of hetero atoms in the polymer.

In the first step of the synthesis - Masking - which
is done outside the polymerization reactor 1, is reacted
with one equivalent of a metal alkyl (M-R, 2) in
hydrocarbon solution. The metal alkyl needs to react
rapidly with active hydrogen of the carboxylic acid group
to form the masked monomer (3) and evolve a mole of the
hydrocarbon, R-H. This limits the choice of the metal (M
in 2) to Groups 1,2,3 of the Periodic Table since the
alkyls of the later elements are not as active. The masked
monomer also needs to be soluble in the hydrocarbon
polymerization solvent - usually hexane or the
incorporation of the monomer does not occur. Solubility of
Group 1 metal carboxylates in hexane is poor and we are
limited to alkyls of Groups 2 and 3 of which the Ethyl
Aluminium (Et_3Al, 2a), Ethyl Zinc (Et_2Zn, 2b) and Hexyl
Magnesium ((C_6H_{13})$_2$Mg, 2c) are representative and the most
common examples. 3 is therefore, most commonly, 'olefin-
CO_2AlEt_2' (3a), 'olefin-CO_2ZnEt' (3b) or 'olefin-$CO_2MgC_6H_{13}$'
(3c). In this masked form the active hydrogen of 1 has

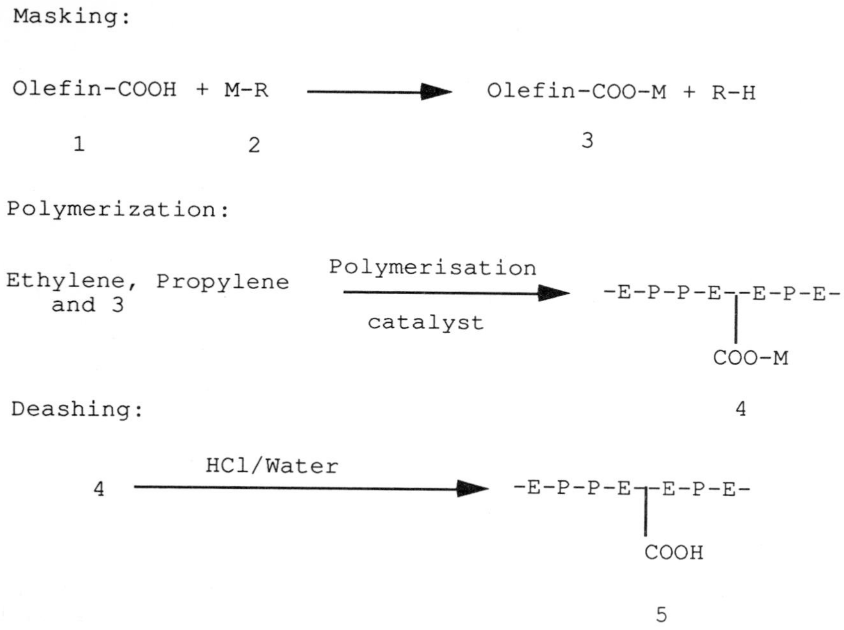

Masking:

Olefin-COOH + M-R   ⟶   Olefin-COO-M + R-H

    1   2   3

Polymerization:

Ethylene, Propylene   Polymerisation
        and 3   ⟶   -E-P-P-E-⌐-E-P-E-
                   catalyst |
                                                COO-M

Deashing: 4

       HCl/Water
    4   ⟶   -E-P-P-E-⌐-E-P-E-
                                        |

                                        COOH

                                       5

Figure 1: Synthetic Scheme for Polymerization

been lost and the carboxylate group is bound to the metal
atom (M) by at least one covalent bond. This reaction is
monitored by the evolution of alkane (if R-H is volatile),
by the heat of reaction or by observing the appearance of
3 by 13C NMR. We show NMR data, which is diagnostic, in a
later section. 3 is shown as a unimolecular species for
convenience of representation, although in reality this
masked monomer exists as a equilibrium[7] of the
unimolecular species and its oligomeric aggregate as shown
in Figure 2. The degree of aggregation as well as the
kinetics of the monomer-oligomer equlibrium must depend on
the temperature, solvent and concentration: we have little
information in this area which pertains to the dilute
alkane solution of the masked monomer. 3 is shown without
the ancillary ligands on M for the purpose of visual
clarity, however their presence is chemically necessary
and their identity plays an important role in the
formation and stability of 3.

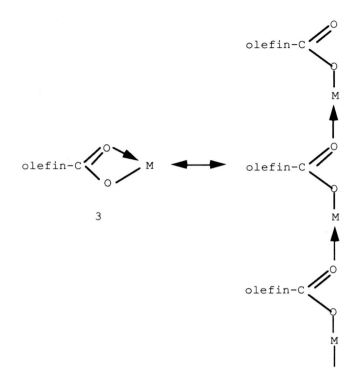

<u>Figure 2</u>: Masked monomer: Monomer-Oligomer equilibria
other ligands on M have been omitted for clarity

    The hexane solution of 3 is introduced into a
continuous polymerization reactor where in the presence of
ethene, propene and an <u>in situ</u> generated, soluble Ziegler-
Natta polymerization catalyst (from the reaction of
vanadium halides and ethyl aluminium chlorides)
polymerization of the available olefins rapidly occurs at
20 - 55 °C. Incorporation of the olefin in 3 into the EP
backbone occurs during the polymerization reaction. The
functionalised polymer is hexane soluble. In this process
there are two independent sources of metal alkyls - first,
the metal alkyl (2) used in the mask and second the
aluminium alkyl halides used as the cocatalyst for the
vanadium halides. These two metals probably exchange alkyl
and halide groups once they are inside the polymerization
reactor. Although we have not directly observed this
phenomenon in our polymerization systems, this reaction is

anticipated [8,9] for 3a. This feature causes some problems in the polymerization since ligand exchange changes the effective composition of the cocatalyst as shown in Equation 1 for the masked monomer 3a.

$$\text{olefin-CO}_2\text{AlEt}_2 + \text{AlEt}_x\text{Cl}_{3-x} =$$
$$\text{olefin-CO}_2\text{AlEtCl} + \text{AlEt}_{x+1}\text{Cl}_{2-x} \ldots (1)$$

Note that while the cocatalyst introduced into the polymerization is $\text{AlEt}_x\text{Cl}_{3-x}$ the effective composition of the cocatalyst tends to be $\text{AlEt}_{x+1}\text{Cl}_{2-x}$. This effects the molecular weight and the molecular weight distribution of the polymer. The extent of this cocatalyst alkylation reaction and the position of the equilibrium depends on the concentration of the masked monomer in the reaction zone as well as the nature of the functionality being masked. We compensate for this reaction by starting out with a cocatalyst which has a lower degree of alkylation (Et/Al ratio) than is normally required for the vanadium catalyst and allowing the equilibrium above to generate the appropriate cocatalyst in the polymerization zone. Thus for a vanadium polymerization catalyst which requires the use of Ethyl Aluminium Sesquichloride $\text{AlEt}_{1.5}\text{Cl}_{1.5}$ we would use Ethyl Aluminium Dichloride AlEtCl_2 as the cocatalyst introduced into the reactor.

Although exchange of alkyl and halide groups does occur between the mask and the cocatalyst during the polymerization they have to be very distinctive compounds prior to the polymerization reaction. This is crucial to the success of this synthetic scheme. We will show data that the choice of ligands around the masking metal atom leads to changes in the stability of the masked monomer and its utility in the polymerization reaction.

The product of the polymerization is the ethene-propene copolymer (4) containing the incorporated masked functionality. The hexane solution of 4 is deashed repeatedly by treating with dilute mineral acid. This deashing stage is a normal part of the polymer recovery[6] to remove the catalyst metal residues. We use this existing step to demask the functionality and is a different use of a existing procedure in the EP polymerization process. The thermodynamic driving force for this masking-demasking cycle for the functional group is the hydrolysis of masking metal alkyl and the use of 2a, 2b or 2c as the masking agent, in contrast to the corresponding compounds of the late transition metals, ensures that the demasking is thermodynamically possible.

Since practical polymerization reactions cannot be
conducted to complete conversion of any monomer the
demasking process occurs both for the polymer incorporated
functionality as well for the residual unincorporated
monomer. Unincorporated monomer needs to be removed from
the polymer since it has the same chemical reactivity as
the functionalised polymer and will compete with it in the
critical application and end use reactions. Our choice of
protic functionalities is fortunate since the low
molecular weight (less than 200) functional monomers,
which we prefer, are sufficiently soluble in water that
they can be removed by this procedure. A important
consideration is that the functionality should not
hydrolyse during the deashing/demasking process. Thus
primary and secondary amides, which contain active
hydrogens suitable for masking, is a poor choice of
functionality as it will hydrolyse to the carboxylic acid
and the corresponding amine during deashing. The last
traces of the unreacted monomer are removed by the final
recovery process for the EP polymer which is the removal
of the solvent by steam distillation. The resulting
polymer is a clear, colorless elastomer (5) containing
free functionality.

4 CHOICE OF OLEFIN AND MASK

We have not described either the possible variations for
the olefin in 1 or the masking metal alkyl 2. These
choices affect the extent of incorporation of the
functionality as well as its distribution in the polymer.

The olefin should be (i) easily polymerizable, (ii)
synthetically accessible and (iii) less than 200 in
molecular weight. For E-P polymers made with a soluble
vanadium polymerization catalyst we observe that the
relative ease of polymerization decreases in the order[10]
ethene $\geq$ norbornene $\gg$ propene > alpha-olefins $\gg$ internal
olefins. In a normal well controlled polymerization the
conversion of the masked carboxylic acid norbornene
monomer 3 to an enchained masked functionality 4 is about
70 to 90%: the remaining 10 to 30% of the masked monomer
stays unincorporated. The corresponding incorporation for
the alpha-olefin carboxylic acids is usually only about 10
to 30%. Norbornene olefins are therefore preferred, though
alpha olefins can be substituted. The most easily
available functionalised norbornene olefins are 5-
Norbornene-2-carboxylic acid (1a), 5-Norbornene-2,3-
dicarboxylic acid (1b) and the analogous alcohol and amine

compounds. They are easily made by the Diels-Alder reactions of cyclopentadiene and corresponding functionalised olefin. The conditions for such additions are usually mild and the yields are correspondingly high. All of these norbornene monomers fulfil the molecular weight restrictions which are needed to remove the unincorporated monomer from the polymer. A suitable alpha-olefin functional monomer would be 3-butenoic acid (1c).

The metal alkyl masking agent 2 should (i) form a stable mask on the functionality, (ii) not react with the norbornene olefin and (iii) completely suppress the reactivity of the functional group towards the polymerization catalyst center. The masking of a protic functionality such as carboxylic acid in 1a with a reactive metal alkyl such as 2a leads to the formation of a sigma bond between the oxygen and aluminium by the elimination of a mole of alkane. This bond is further augmented and the nucleophilicity of the oxygen diminished by the O -> M bond. Estimates of the bond strength for this combined interaction are about 100 kcal/mole depending on the nature of M. The importance of having an active hydrogen on the functionality is apparent: in the absence of the stabilizing sigma bond the interaction between the metal and the functionality is only through the O -> M dative bond: this is thermodynamically weak[8] with typical values ΔH = 20-30 kcal/mole. Further, kinetic studies[9] on the dissociation of ligand -> metal complexes for Group 2 and 3 metals where the only interaction is through a ligand to metal bond, show that the rates of dissociation and recombination are usually very fast. Conversely, the rates are slower in the cases where more than one pair of electrons, such as in 3, participates in the bond formation. Thus single dative bonds are both thermodynamically weak and kinetically labile; this combination of adverse properties makes a single donor O -> M bond, as would be formed by attempts at masking ethers and esters, essentially useless. At the typical conditions used for solution polymerization of E-P polymers (2-10 mmolar in masked functionality) there exists a adequate concentration of unprotected oxygen functionality which can interact with the Ziegler catalyst center. This is contrary to the desired role of the mask and provides an explanation for the poor results in polymerization from attempts at masking ethers and esters.[5]

The most common reactions of metal alkyls with norbornene olefins are cationic/anionic polymerization or rearrangement of the strained ring structure. Both of

Table 1: Masking reaction for 1c

13C-NMR in hexane: resonance in ppm from TMS

$CH_2= CH - CH_2 - COOH + AlEt_3$ ⟶

  C-4   C-3 C-2   C-1 $CH_2= CH - CH_2 - COOAlEt_2$

         1c

                                             C-4   C-3 C-2   C-1

                                                  3c

		C-4	C-3	C-2	C-1
1c	13C NMR	117.8(t)	129.7(d)	37.8(t)	177.1(s)
	J(C-H)	165 Hz	172 Hz	119 Hz	
3c	13C NMR	118.0(t)	127.7(d)	41.1(t)	178.8(s)
	J(C-H)	168 Hz	170 Hz	122 Hz	
6	13C NMR		122 (broad)		179(m)

these reactions are catastrophic since they affect the
availability of easily polymerizable norbornene double
bondson which the synthetic scheme depends. The masking
reaction should only react with the functionality. 13C-NMR
data assigns the structure of the unsaturated organic
fragment both before and after masking. Table 1 shows the
data from the C-13 NMR spectrum of the olefinic carbon (C-
4 and C-3), the methylene carbon C-2 as well as the

Table 2: Masking reaction for 1a:
13C nmr in hexane: resonances are ppm from TMS

		C-5	C-6	C-1
1a	13C NMR	136.3(d)	131.6(d)	180
	J(C–H)	168 Hz	170 Hz	
3a	13C NMR	137.0(d)	131.0(d)	181
	J(C–H)	175 Hz	169 Hz	
7	13C NMR	138 (broad)		186

carboxylic carbon for both the monomer 1c as well as the
masked monomer 3c formed by masking with triethyl
aluminium. Table 2 shows similar data for the masking of
1a with triethyl aluminium to form the masked monomer 3a.
In this case we show data for the olefinic carbons (C-5
and C-6) and the carboxylic carbon. Both of these spectra
were recorded in hexane solution at room temperature. The
olefinic carbons of 1a and 1c are essentially unaffected
by the masking procedure   both in the position of the
resonance and in the multiplicity of lines while the
carboxylic carbon shows a slight shift in the resonance on
complexation to the aluminium. Further evidence that the
double bond is not destroyed by skeletal rearrangement or

polymerization is obtained from the coupling constants (J_{C-H}) for the olefinic carbons. It is reasonable to suppose that any prepolymerization of the norbornene double bond during the masking process will be through these carbon atoms which will change to a sp3 hybridization (J_{C-H} of 125 Hz) from a existing sp2 state (J_{C-H} of 170Hz).[11] NMR data for the either 1a or 1c does not show any evidence for this polymerization on masking with 2a.

We have described the masking procedure only for homoleptic metal alkyls 2a,2b and 2c. Commonly available alkyl halides of these metals are not suitable masking agents for 1a or 1c since they invariably lead to the polymerization of the masked monomer. This is a important distinction from the historical literature[5] which invariably recommends these materials as the masking agents of choice. In one such reaction 1a and 1c are reacted with one equivalent of diethyl aluminium chloride (DEAC) in hexane solution. The initial reaction is the loss of a mole of ethane as in the masking reaction, however this material cannot be stabilized. Concentrated solutions of this reaction mixture (above 0.1 molar) slowly precipitate insoluble polymeric material. 13C-NMR spectral data for the more dilute solutions are shown as 6 and 7 in Table 1 and 2 respectively. We have not identified all the reaction products in this case but the olefinic carbons are reduced in total intensity (to approximately 30 - 10% of the original) and also separated into several pairs of independent signals. All of the resonances lie under a single broad envelope in the olefin region. Coalescence of these signals does not occur up to 60 °C and they appear to arise from several different species. The signal of the carboxylic carbon while undiminished in intensity is also split into several distinct resonances. The insolubility of the adduct (which could account for some of the 13C-NMR data) does not arise from the change of the ligands on the masking aluminium alkyl. Thus the reaction of diethyl aluminium chloride with hexanoic acid, which is a reasonable surrogate for either 1a or 1c does not lead to the formation of solids. These results are best explained by a irreversible polymerization or rearrangement of the organic fragment in the intermediate masked monomers. This is a process distinct from the reversible oligomerization of the masked monomer through the mask which is shown in Figure 2: the reversible process maintains the concentration of the easily polymerizable norbornene or alpha-olefin double bonds while the irreversible polymerization or rearrangement of the olefin does not. The formation of polymers from the interaction of the mask and the olefin is supported by other workers[12] who have shown similar reactions on attempting to react unsaturated carboxylic acids with aluminium halides. The exact mechanism for the rearrangement or polymerization of the norbornene nucleus is dependent on the solvent, temperature and the identity of the aluminium alkyl halide. However, the effect of either of these two reactions on our synthetic scheme is similar: the resulting material does not readily incorporate into the growing E-P chain because it lacks the reactive norbornene double bond.

The primary function of the mask is to react with a active hydrogen compound to form a stable sigma bond. The ability (Lewis acidity) of the mask to withdraw electron density from the hetero atoms of the functional group protects the catalyst center from these hetero atoms. This occurs through a $p\pi$-$p\pi$ back bonding of the carboxylic acid to the aluminium atom in masked monomer 3a. This relationship between the Lewis acidity of the mask and the effectiveness of the mask is quite general and can be monitored by the yield of functional polymer with different masks on the functionality under otherwise identical conditions. Masking with the strongly Lewis acidic 2a (aluminium) or 2b (zinc) leads to effective protection of the catalyst center from the functionality while 2c (magnesium) is much less efficient in masking the functionality. This data is shown in Table 3. Further, attempts at masking with DEAC lead to strong polymerization though little or no incorporation of the masked monomer takes place. This is an important principle in this procedure: polymerization of the backbone and incorporation of functionality are not necessarily correlated. Further, the balance of Lewis acidity of the masking agent is crucial in determining the success of the synthetic scheme: too strong a Lewis acid while providing a excellent mask may lead to rearrangement/polymerization of the reactive olefin monomer, too weak a Lewis acid may not be able to protect the catalyst center from the hetero atoms.

5 EFFECT ON POLYMERIZATION

The introduction of a functional monomer masked with either 2a or 2b has two important effects on the polymerization - (i) masks lead to some lowering of the molecular weight of the polymer and (ii) the yield of the polymer slowly decreases as more masked monomer is introduced into the polymerization reactor.

Table 3: Polymerization of 1a after masking with
2a, 2b, 2c and DEAC.
Polymerizations are conducted under the same conditions
with changes in the masking agent only

Masking agent	Polymerization rate gm/hr	Polymer Molecular weight (Mn)
control: no 1a	258	72 x 10E3
2a	222	62 x 10E3
2b	206	12 x 10E3
2c	158	32 x 10E3
DEAC	239	

The lowering of molecular weight appears to be due to
an increase in the rate of chain transfer because of the
higher concentration of metal alkyl bonds. Data reported
in Table 3 supports this: masking with zinc alkyls leads
to polymers having low molecular weights (compared to
polymers made with aluminium alkyls 2a) since zinc alkyls
are known to act as very effective chain transfer agents
in olefin polymerization.

At high concentration of the masked monomer in the
polymerization reactor the yield of the polymer decreases.
The data is in Table 4 and shows that the mask is not
perfect: interaction of the oxygen atoms of the
functionality with the catalyst center although
significantly reduced has not been completely eliminated.
The use of 2a or 2b as masks is not the perfect solution:
it is only the current best.

6 REARRANGEMENT / REDUCTION OF MASKED MONOMER ?

We consider the possibility that the presence of the
vanadium halide - aluminium alkyl chloride catalysts in
the polymerization reactor will lead to rearrangement or
oligomerisation of the masked functional olefin 3 before
it is incorporated into the EP backbone. This is possible

Table 4: Effect of increasing concentration of masked
        monomer: Masking with 2a on monomer 1a.

Concentration (by weight) of masked monomer 3a in total monomer	Polymerisation rate
none	239
0.3%	222
0.6%	211
1.0%	200
1.5%	167

since these catalysts are active for hydride transfer and
aluminium alkyl halides lead to rearrangement of 3. We
find no data to support our speculation. This is probably
because the rate of Ziegler polymerization of norbornene
double bonds is much faster than the times required for a
rearrangement or oligomerisation. Further evidence that
these mechanisms are slow is obtained by identifying the
unreacted monomer in rapidly quenched samples of the
polymerization mixture containing 3a as the masked
monomer: GC analysis of these samples after suitable
derivitization (silanation) for ease of GC-MS analysis
indicates that less than 10 % of the unincorporated
monomer appears as a variety of products with the same
molecular weight as 1a. Higher molecular weight oligomers
of 1a were not observed. We believe that these materials
arise from rearrangement of the functionalised olefin
monomer 3a. The proportion of the rearranged material
rises steeply as a function of time: reaction product
samples examined after several tens of minutes delay
typically show greater than 50% of rearranged products in
the unincorporated monomer.

We consider the possibility that a significant
fraction of the functionality may be reduced during the
masking or the subsequent storage since a excess of
aluminium trialkyl can be used for reduction of
carboxylate groups to the corresponding ketones and
alcohols. Significant amounts of reduction are unlikely,
however, since we use only equimolar amounts. The
experiment to quantitate the possible reduction of the

carboxylate group is the following. Octanoic acid is
masked with 2a in the manner we would use for 1a and is
introduced into the polymerization reactor: incorporation
of the octanoic acid into the E-P polymer does not occur.
The polymer solution is deashed and freed of the high
molecular weight polymer by precipitation with acetone.
The resulting aggregate of the low molecular weight
compounds is analysed by gas chromatography after suitable
derivatization (silanation) of the carboxylic acids. The
results show that between 95 - 97% of the octanoic acid
was recovered unchanged; about 3 - 5% of the reduction
products (principally the corresponding ketone, 3-
decanone) are observed. This indicates that under the
conditions of masking and polymerization the extent of
reduction of the carboxylic acid functionality is minimal.

7 POLYMER CHARACTERISATION

The recovered polymers from this synthetic procedure are
clean, colorless materials. Composition of the backbone is
determined by infra red[13] and the molecular weight is
determined by GPC. Additional characterisation is needed
for the level of functionality and the distribution of the
functionality across the molecular weight distribution.

The level of acid functionality in the polymer was
determined by several methods: these are summarized in
Table 5 for a single sample of ethene-propene polymer made
with the functional monomer 1a. The presence of carboxylic
acid groups in this polymer is easily determined from the
strong absorbance at $1710cm^{-1}$ in the infra red
spectra, which cannot be removed from the polymer by
solvent extraction or precipitation. This is the same as
that of the starting carboxylic acid monomer. This
absorbance is a quantitative measure of the concentration
of acid functionality. The carboxylic acid content of the
polymer can also be estimated from a non aqueous acid-base
titration of the the polymer in THF solution using
potassium t-butoxide as the titrant. This result
corresponds closely (88%) to the results obtained from the
infra red spectra. Further evidence for the presence of
acid groups in the polymer is obtained by transforming the
carboxylic acids to esters or amides. The polymer is
reacted in separate experiments with large excess (4000
fold based on the estimated amount of carboxylic acid) of
1-hexanol or 1-hexylamine in refluxing xylene solution for
24 hours. At the end of this period the recovered polymer
shows a quantitative transformation of the acid
functionality into the ester (1745 cm^{-1}) or amide

Table 5: Carboxylic acid content for polymer
made with masked monomer 3a.

Analytical Method	Result: (mmoles/100 gms)
Infra red analysis of acid	6.1
" " " hexyl ester	6.5
Titration of Carboxylic acid in polymer	5.4
1H NMR of phenyl amide	5.1 - 5.8
Refractive index	5.8

functionality ($1675 cm^{-1}$), respectively, in the infra red spectra. The concentration of the ester functionality is estimated to be the same (107%) as the level of carboxylic acid functionality in the starting polymer. Direct observation of the carboxylic acid functionality was not possible in the NMR spectra of the polymer solution because resonances due to the single proton are too weak and too broad. However, the amidation of this polymer with 2-phenyl-1-aminoethane shows distinct phenyl resonances in the 1H-NMR. This indicates a concentration of carboxylic acid groups close (75 - 87%), depending on the features of the backbone spectra that are used as the reference to the values obtained by other methods. Direct observation of the norbornene ring is not possible in the 1H-NMR spectra because the resonances due to the norbornene are obscured by the signals from the E-P polymer backbone. Previous authors[14] have shown that norbornene residues incorporated in the E-P backbone lead to a increase in the refractive index. These increases are proportional to the mole percent of the norbornene residues present in the polymer. We find that the carboxylic acid functionalised polymers made with monomer 1a have a higher refractive index than the normal nonfunctionalised E-P: the amount of norbornenes is consistent (84%) with the amount of carboxylic acid present in the polymer. This result indicates that the carboxylic acid groups are actually polymerised through the norbornene residues without any rearrangement.

The distribution of the functionality in the polymer
is essentially uniform across the molecular weight
distribution of the polymer. This is determined by
reacting the carboxylic acid functionality polymer in
toluene solution with 2-phenyl-1-aminoethane under typical
amidation conditions. It is expected that under these
conditions a random fraction of the carboxylic acid groups
is converted to the corresponding amide. This polymer is
analysed in THF solution by size exclusion chromatography
(SEC) with sequential differential refractive index
detector (which is sensitive to the concentration of the
polymer in solution) and a UV detector (which is sensitive
to the concentration of aromatic groups) operating at 260
nm; the absorbance of a nonfunctionalised E-P polymer at
this wavelength is nearly zero across the molecular weight
distribution. Elution time data was correlated to
molecular weights by comparison to standard polystyrene
samples obtaining results consistent[6] with previous
publications from our laboratories. The response of the
two detectors for this sample of the phenyl tagged
functionalised E-P polymer in the range of 10 E+6.5 to
10E+3.5 amu is nearly linearly proportional: this
indicates that for the polymeric portion of the sample the
functionality is associated with the polymer. Quantitative
measure of the distribution of the functionality is
obtained by measuring the UV absorbance of the polymer per
unit weight of polymer (a measure of the extinction
coefficient) across the molecular weight distribution.
This is shown in Figure 3 for a EP polymer with a
carboxylic acid content of 5 mmoles per 100 gms of polymer
and a number average molecular weight of 62000 and a Mw/Mn
of 2.1. In the range of the polymeric molecular weight of
10E+6 to 10E+4 amu where approximately 95% of the
molecular mass is concentrated in the distribution of the
concentration of functionality across the molecular weight
distribution is essentially uniform with only a small
(less than 30%) variation in this range. There appears to
be strong absorbance in the molecular weight range below
10E+3.5 amu: this could arise from a combination of some
small quantity of the oligomerised monomer as well as
phenolic antioxidants and impurities in the solvent. A
feature of the above technique is that the data is least
accurate at the extremes of the molecular weight
distribution where the amount of the polymer is small and
interdetector spreading effects are large. Further, the
data can be affected by the presence of other chromophores
which can absorb at the wavelength of analysis. We also
show results of analysis for the functionality
distribution of the polymer by a solvent-nonsolvent
fractionation technique which does not suffer from

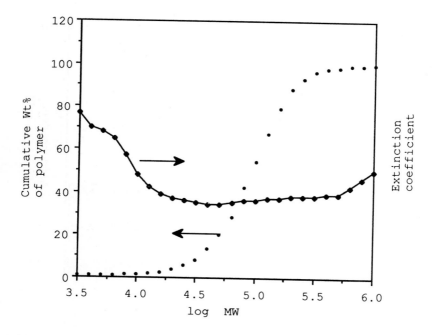

Figure 3: Extinction coefficient vs Molecular wt
for Ethylene-Propylene-Carboxylic acid polymer
containing incorporated monomer 1a.

interdetector effects.

In this procedure a sample of the same polymer is
dissolved in hexane at about 3 wt % concentration. To this
homogeneous solution is added aliquots of isopropanol
sufficient to precipitate about 10% of the total weight of
the polymer: the precipitate is allowed to equilibrate
with the remainder of the polymer solution for about two
hours before being removed. The process is continued with
the filtrate to isolate 8 polymer samples which differ
in their solubility and molecular weight: we anticipate
that the lower molecular weight fractions are most
soluble. The last fraction stays in solution and is
isolated by evaporation of the solvent. All of the
fractions are of similar weight (+/- 30% of the mean)

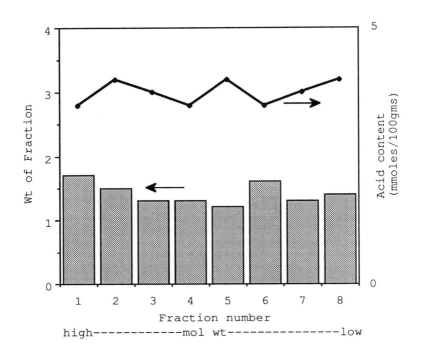

Figure 4: Distribution of functionality by Solvent-Nonsolvent Fractionation Technique: masked monomer is 3a.

and are analysed by infra red for the carboxylic acid content and the molar ratio of ethene/propene residues in the backbone. The results of the analysis are shown in the Figure 4. The composition of all the fractions is essentially constant at about 49 mole% ethene residues in the backbone. We believe that in a case of such a uniform distribution of composition the solvent - nonsolvent fractionation procedure we have employed is a reliable separation by molecular weight with the high molecular weight material being the least soluble. The data shows that the distribution of functionality across these fractions is uniform - however individual samples can vary by as much as 20% of the mean. The low molecular weight material does have a higher acid content than the bulk of the material but the actual amount of carboxylic acid present in this fraction is small - about 8% of the total.

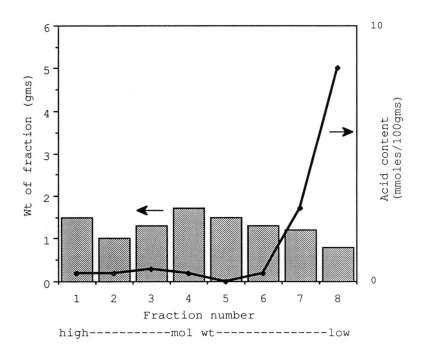

Figure 5: Fractionation of polymer by solvent-nonsolvent technique: masking of monomer 1a with DEAC.(masked monomer 7)

In conclusion it appears that the overwhelming amount – greater than 90% – of the carboxylic acid is actually polymerised into the polymer with a essentially uniform distribution of the functionality across the molecular weight distribution.

This result is compared to the distribution of functionality observed in the case of a polymerization conducted with a masked monomer which has undergone prepolymerization or rearrangement during the masking procedure or subsequent storage. The polymer was made by masking monomer 1a with DEAC and then attempting to polymerize this material into a ethene-propene backbone. The resulting polymer was analysed for the

distribution of functionality on the polymer; this is
extremely non uniform. Both size exclusion chromatography
with sequential differential refractive index detectors
and UV detectors operating at 260nm as well as the
fractionation of the polymer into several portions by the
nonsolvent-solvent technique (Figure 5) support this
distribution. The SEC data indicate that about 79% of the
functionality is concentrated in a low molecular weight
fraction of only few thousand number average molecular
weight. Some functionality does appear on the E-P polymer
but we estimate that this is about 20% of the total
functionality present in the polymer. These quantitative
conclusions are supported by the fractionation data which
show almost all of the functionality to be in the last,
low molecular weight fraction. We should point out that
this observation is not unprecedented: a similar study by
Lindsey,[12] who analysed the polymerization of unsaturated
carboxylic acid masked with aluminium trichloride in
Ziegler systems, came to similar conclusions about the
distribution of carboxylic acid functionality in his
polymers.

8 POLYMER PROPERTIES

It is expected that these polymers which contain enchained
carboxylic acid groups should form stable ionomers with
distinctive rheological properties on neutralization of
the acid with bases. We stress this form of
characterisation because it is   definite proof for the
incorporation of _some_ functionality in the polymer.
Ionomers are made with sodium and zinc ions by reacting a
hexane solution of the carboxylic acid functionalised
polymer made with the masked monomer 3a with aqueous
sodium hydroxide or zinc acetate followed by removal of
the solvent and volatiles under vacuum at 100 °C. The bulk
rheological properties of these polymers were evaluated
using gravity driven creep[15] at 100 °C. Creep data is
translated into zero shear viscosity which is shown in
Table 6. Compared to the parent polymer containing the
free acid functionality both the zinc and sodium ionomers
have a much higher zero shear viscosity: this is expected.
Note that the addition of a alcohol (1,2-hexanediol) leads
to a significant drop in the measured zero shear
viscosity. This result is parallel to that observed for E-
P polymers containing acid functionality[16] which has been
made by post polymerization grafting reactions and
provides strong support for our characterisation.

Table 6: Zero shear viscosities of E-P carboxylic
acid polymer (polymer 5a) and some ionomers

Cation	Zero Shear Viscosity
H+	1.21 x 10E+6
Na+	greater than E+8
Na+ with 1,2 hexanediol	2.5 x 10E+6
Zn++	greater than E+8

9 CONCLUSION

We have shown the synthesis and convincing
characterisation of a new class of functionalised ethene-
propene polymers. The distribution of functionality in
these polymers is random along a chain and uniform across
the molecular weight distribution. The synthesis is
practical and leads to the direct polymerization of a
range of chemically reactive polymers. This represents the
first such synthetic procedure for the practical synthesis
of amine, carboxy and hydroxyl containing Ziegler-Natta
polyolefins.

These polymers are different than the regular ethene-
propene copolymers in that they have sites suitable for a
variety of chemical reactions and/or polar interactions.
This is in addition to the expected properties of E-P
copolymers which are elastomers with a low Tg and
resistance to atmospheric degradation. Chemical reactions
through the functional groups can lead to the formation of
new polymeric structures. This has been explored by other
authors.[2,5] These polymeric structures may have
differences in the surface properties and viscoelasticity,
particularly as a function of temperature and shear which
cannot be attained with any one single polymer. Further
these reactive sites can be used to generate a non-bonded
polar interaction between these polymers and substrates
even in the absence of a discrete chemical bond. Polarity
of the reactive sites also leads to fundamental changes in
the viscoelastic properties of the polymer. These are new
properties of ethene-propene polymers and for polyolefins

in general and point to new areas of application.

This project could not have been possible without the thoughtful advice of the following: Dr. C. Cozewith, Mr R. Hazelton, Dr. E. N. Kresge, Dr. G. Ver Strate and Dr. R. West. NMR data was provided by Dr D. M. Cheng. I wish to thank Dr. J. J. O'Malley, Dr. B. M. Rosenbaum and Dr. F. Pasterczyk for their support and encouragement. This work was done entirely at the Linden Technology Center and I thank the management of the Polymers Group, Exxon Chemical Co. for permission to contribute this chapter.

REFERENCES

1. This is a preliminary report of this work. A more detailed account will be published in Macromolecules.
2. Notable advances in this otherwise futile effort have been shown by (a) U. Klabunde, J.C. Calabrese, W.C. Fultz, T. Herskovitz, A.H. Janowicz, R. Mulhaupt, D.C. Roe, T.H. Tulip and S.D. Ittel, Int. Symp. Trans. Met. Catal. Polymn., Akron, Ohio, 1986, (Ethylene-polar alpha olefin copolymerisation using a nickel catalyst), (b) A. Sen, Adv. Poly. Sci. 1986, 73/74, 126 (Ethylene-Carbon Monoxide polymers) and (c) Shell's extensive series of patents - US patents 3 694 412, 3 689 460, 3 835 123, 3 984 388, 4 740 625 and approximately 50 in this series-in the area of copolymerising carbon monoxide with olefins.
3. (a) A.M. Korshun and V.V. Mazurek, Vysokomol. Soedin., Ser. A, 1975, 17, 2657, (b) Kashiwa N. German Offen. DE 1963799, July 9, 1970, assigned to Mitsui Petrochemical.
4. (a) A.W. Langer Jr. U.S. Patent 4 094 818, June 13 1978, assigned to Exxon Research and Engineering. (b) R. Pruett, personal communication, June 1984.
5. The most detailed descriptions of these procedures which have appeared in the patent literature are the following: (a) U. S. Patent 3 492 277, (b) U. S. Patent 4 423 196, (c) U. S. Patents 3 796 687, 3 884 888, 3 901 860 and 4 017 669, (d) U. S. Patent 3 761 458, (e) U. S. Patent 4 139 417, (f) Japan Patent JA 7337756-R, (g) Japan Kokai JO 1259-012-A and (h) Japan Kokai 188996.
6. G. Ver Strate, 'Encyclopedia of Polymer Science and Engineering', Wiley & Sons, New York. 1986, Vol. 6, 522
7. (a) J. Weidlin, J. Organometal. Chem., 1969, 16, P33,

(b) J. Weidlin, <u>Angew. Chem. Intern. Ed. Engl.</u>, 1969, <u>8</u>, 927.

8. (a) T. Mole and E.A. Jeffrey, <u>'Organoaluminium Compounds'</u>, Elsevier Publishing Company, New York, 1972.

9. J.J. Eisch '<u>Comprehensive Organometallic Chemistry</u>' ed G. Wilkinson, F.G.A. Stone and E.W. Abel, Pergamon Press, New York, 1982, <u>1</u>, 555.

10. There exist some discussion about the relative reactivity of norbornene and ethene towards soluble polymerization catalysts. Our ordering does not intend to distinguish between these two but rather show them to be much more reactive than any other olefins.

11. H.G. Preston, Jr., and J.C. Davis, Jr., <u>J. Am. Chem Soc.</u>, 1966, <u>88</u>, 1585.

12. G.A. Lindsey, <u>Block Polymer: Science and Technology</u>, Harwood Publishers, 1979, <u>3</u>, 53.

13. Ethene content of the polymer is determined according to the ASTM method D-3900.

14. I.J. Gardiner and G. Ver Strate, <u>Rubber. Chem. Tech.</u>, 1973, <u>46</u>, 1019.

15. W. Graessley and G. Ver Strate, <u>Rubber Chem. Tech.</u>, 1982, <u>53</u>, 842.

16. (a) T.R. Earnest and W.J. MacKnight, <u>Polym. Prepr.</u>, 1978, <u>19</u>, 383; (b) R.A. Weiss and P.K. Agarwal, <u>J. Appl. Polym. Sci.</u>, 1981, <u>26</u>, 449; (c) D.G. Pfeiffer, J. Kaladas, I.Duvdevani, J.S. Higgins, <u>Macromolecules</u>, 1987, <u>20</u>, 1397.

Biodegradable Polymer Systems for the Sustained Release of Polypeptides

F. G. Hutchinson[1] and B. J. A. Furr[2]

[1]PHARMACEUTICAL AND [2]BIOSCIENCE DEPARTMENTS, ICI
PHARMACEUTICALS, MERESIDE, ALDERLEY PARK, MACCLESFIELD,
CHESHIRE, SK10 4TG

1 INTRODUCTION

In recent years there have been major advances in genetic
engineering and consequently the production of many interesting
and pharmacologically active polypeptides. There have also been
concurrent improvements in procedures for total chemical synthesis
of lower molecular weight peptides such as 'Zoladex'* (ICI 118,630:
D—Ser (Bu^t)^6—Azgly^{10}—LHRH: Figure 1) which is a highly potent
synthetic analogue of luteinizing hormone—releasing hormone (LHRH).
However, the therapeutic and commercial potential of this and other
polypeptide drugs will only be fully realised if these advances are
accompanied by improvements in the design of dosage forms, leading
to practical and effective formulations.

The use of polypeptides in human and animal diseases is fraught
with problems. These macromolecular drugs are usually ineffective
by the oral route as they are rapidly degraded and deactivated by

$^\llcorner$Glu-His-Trp-Ser-Tyr-Gly-Leu-Arg-Pro-Gly-NH$_2$

LH-RH

$^\llcorner$Glu-His-Trp-Ser-Tyr-D-Ser(Bu^t)-Leu-Arg-Pro-Azgly-NH$_2$

Zoladex ICI 118,630

Figure 1. Structures of LHRH and 'Zoladex'

---

* 'Zoladex' is a trademark, the property of Imperial Chemical
Industries PLC

[1] to whom correspondence should be addressed

proteolytic enzymes in the alimentary tract. Even if stable to enzymatic digestion, their molecular weights are too high for absorption through the intestinal wall to occur. Other routes of administration including intranasal[1,2] buccal[3], intravaginal[4-7] and rectal have been used, but these are all associated with a low and variable bioavailability and none of these offers a general solution applicable to all polypeptides. Consequently, polypeptides and proteins are normally administered parenterally (subcutaneous, intramuscular and intravenous injection) but since these drugs have very short elimination half-lives frequent injections are required to produce an effective therapy. For polypeptide hormones, where the pharmacology of the agent is compatible with sustained release, the most appropriate dosage form is one that is capable of releasing drug continuously at a controlled rate over a period of weeks or even months. If the carrier providing for such release is polymeric then it is preferred that it should be biodegradable and so would ultimately disappear from the site of administration. Currently, a number of biodegradable polymers are being evaluated as carriers for the sustained release of low molecular weight drugs (Figure 2).

⁎ { Polylactic acid (Polylactide)
Polyglycolic acid (Polyglycolide)
Poly (lactic acid—co—glycolic acid)
Poly (lactide—co—glycolide)

Poly (ε—caprolactone)
Poly (hydroxybutyric acid)

Poly ortho—esters
Poly acetals

Poly dihydropyrans

Synthetic polypeptides
Cross—linked proteins

Poly cyanoacrylates

Hydrogels (i) cross—linked
⁎   (ii) amphipathic block copolymers

Figure 2. Biodegradable polymers used in drug delivery.

Long experience with homo— and co—polymers of lactic and glycolic acids has shown that these materials are inert and biocompatible in the physiological environment and degrade to toxicologically acceptable products[8]. Consequently, these polymers are invariably the materials of choice in the initial design of parenteral sustained—delivery systems using a biodegradable carrier, particularly when release over many weeks is required. We have adopted this approach and were the first group to identify and characterise the mechanisms of transport which allow movement of these polypeptide drugs from biodegradable formulations based on these polyesters[9],[10].

The successful development of sustained—release biodegradable delivery systems for peptide drugs such as 'Zoladex' requires recognition and resolution of a number of major problems posed by these macromolecular agents. Firstly, the mechanism most commonly used to achieve sustained namely controlled diffusion through a polymeric matrix or membrane, may not be appropriate for a high molecular weight polypeptide. Design of a sustained—release dosage form must take into account both the properties of the rate—controlling polymer and the drug. For diffusion of the drug to occur it must have some limited solubility in the polymer; this is often the case with low molecular weight drugs. In contrast, it is well established that, in the absence of specific chemical interactions, polypeptides will either be insoluble in, or incompatible with, any polymer such as polyester, which has a totally dissimilar structure, because of entropic and enthalpic factors[11].

Consequently, low or negligible solubility of the macromolecular drug in a polymer, such as a polyester, will prevent diffusional transport of the agent through the polymer phase. With regard to the properties of the drug the most important of these are its size, shape and solubility[12]. There is an approximate log—log correlation between molecular weight (M) and diffusion coefficient (D) where: $\log D = a - b \log M$ (where a and b are arbitrary constants) such that D decreases as molecular weight increases. For polypeptides M is large and the diffusion coefficient becomes vanishingly small because the diffusant cannot be accommodated by the free volume of polymer arising from rotational and translational segmental mobility. Consequently, polymers such as polyesters are not likely to allow partition—dependent diffusion of polypeptides through the polyester phase to occur.

Secondly, polypeptides are biologically labile and can be readily degraded by tissue enzymes. They must, therefore, be effectively protected at the depot site if active drug is to be released continuously. The difficulty of achieving this is emphasised by the fact that synthetic polypeptides have actually been used as biodegradable carriers for drugs such as steroids and narcotic antagonists[13],[14].

Thirdly, excipients used to achieve sustained release of macromolecular drugs might provoke an adjuvant—induced immunological response, which may be related to the nature of the excipient, the delivery rate, or profile of release. There is some evidence that sustained release of large proteins may be an effective means of raising antibodies to them[15]. Finally, long-lasting depots might become encapsulated by fibrous tissue, thus inhibiting further release of drug. This is certainly the case for non-degradable silicone elastomer implants[16].

These imposing problems opposite sustained polypeptide delivery have been resolved by the design of biodegradable delivery systems based on polyesters such as poly (d,l-lactide) and poly (d,l-lactide-co-glycolide) to give formulations which allow release of polypeptides over an extended period of time. This work has been extended to include amphipathic block co-polymers consisting in part of biodegradable polyesters and which, in a physiological environment, behave as hydrogels. It has been shown that these technologies can be applied to polypeptides and proteins having a range of molecular weights[9,10].

2 MATERIALS AND METHODS

Lactic/glycolic acid (co-)polymers and depot preparation:

These simple biodegradable homo- and co-polymers were prepared at elevated temperature by the ring-opening polymerisation of dry, freshly prepared acid dimers, d,l-lactide and glycolide, using organo-tin compounds as catalysts. Control of molecular weight was achieved by using a chain transfer agent such as d,l-lactic acid. In this way polymers (Figure 3) of variable composition, having intrinsic viscosities from 0.1 to >1, can be prepared. The polymers can be further characterised by size exclusion chromatography relative to polystyrene standards to define number average molecular weight (M_n), weight average molecular weight (M_w) and polydispersity

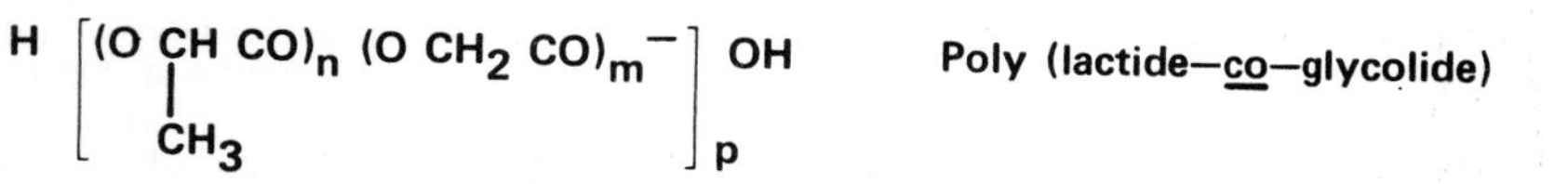

$$H \ (O \ \underset{\underset{CH_3}{|}}{CH} \ CO)_N \ OH \qquad \text{Polylactic acid/Polylactide}$$

$$H \ (O \ CH_2 \ CO)_M \ OH \qquad \text{Polyglycolic acid/Polyglycolide}$$

$$H \ \left[(O \ \underset{\underset{CH_3}{|}}{CH} \ CO)_n \ (O \ CH_2 \ CO)_m^- \right]_p \ OH \qquad \text{Poly (lactide-\underline{co}-glycolide)}$$

Figure 3. Polymers and co-polymers of lactic and glycolic acids.

$(p = M_w/M_n)$. Additionally, the polymers can be characterised by
^{13}C-NMR to define the distribution of co-monomers and polymer
structure, that is, the average values for n and m of copolymers shown in Figure 3. Viscosities (units in decilitres/gram)
were usually measured in chloroform at a temperature of 25°C.

The methods of preparing these polyesters using polycondensation
techniques or by ring-opening polymerisation have been described
previously[9]. Depots were prepared generally using conventional
polymer melt fabrication techniques such as extrusion or compression
moulding[9].

   Degradation studies. Degradation of the polymers in vitro in
buffer at pH 7.4 at 37°C was carried out using polymer films or
slabs of known thickness and weight. Prior to incubation in
buffer the polymer samples were dried thoroughly in vacuo at 60°C
for 2 days and weighed (W_1). The polymer samples were then
incubated in buffer at pH 7.4 at 37°C for various periods of time,
removed from the incubation medium and surface dried using blotting
paper. The hydrated and degraded polymer sample was weighed when
wet (W_2) and again when dry (W_3) after all the water had been
removed from the degraded product by drying in vacuo at room
temperature initially for 1 day and then at 70°C for 1 day.

The % weight of degraded polymer remaining is given by W_3/W_1 x 100
and the % water uptake is given by ($W_2 - W_3$)/W_3 x 100

   The internal morphology of the dried (at room temperature)
degraded polymers was investigated by scanning electron microscopy
following freeze-fracture of the specimen after immersion of the
sample in liquid nitrogen and shadowing of the fracture surface with
gold. The surface structure may be examined similarly but without
the need for freeze-fracturing.

   The molecular weight of the degraded polymers was measured by
size exclusion chromatography and by viscosity measurements[17].

   Release and biological studies. Continuous release of 'Zoladex'
in vivo can be measured qualitatively by the biological effect
elicited in adult female rats showing regular oestrous cycles.
Normally these rats have an oestrous cycle of 4 days and the
occurence of oestrus is indicated by the presence of cornified
cells in vaginal smears. In rats given subdermal depots of
'Zoladex', release of drug at an effective rate will cause a fall
in circulating oestrogens, which in turn leads to a suppression of
oestrus and absence of cornified smears. Female rats therefore show
an extended period of dioestrus.

Release, in vivo and in vitro, can be measured quantitatively. Implantation of depots followed by their excision at various time points and measurement of residual drug in the depot using high performance liquid chromatographic techniques gives the profile of release. The in vitro release of 'Zoladex' into an external aqueous medium at a pH of 7.4 can be measured by high performance liquid chromatographic analysis of the aqueous phase to give a quantitative measure of the amount released. Although not reported here, release in vivo and in vitro can be controlled to give a pseudo—zero order profile over most of the target delivery period.

The methods used, in performing these release studies and in the evaluation of 'Zoladex' sustained release depots in animal models for mammary cancer and prostatic carcinoma, have been described previously[18].

Results

The findings of an in vitro degradation study using 50/50 molar poly (d,l—lactide—co—glycolide) are shown in Figure 4. Polymers ranging from very low molecular weight (low intrinsic viscosity) to very high molecular weight (high viscosity) have been studied. These polymers have a normal polydispersity (p approximately 2) and it can be seen that for the high molecular weight types there is an induction period prior to weight loss. However, molecular weight decreases immediately following incuba—tion in the aqueous medium. In a related experiment the degradation of a high molecular weight 50/50 molar poly(d,l—lactide—co—glycolide) again having an essentially normal polydispersity has been studied and the results are shown in Figure 5. The results show an apparent discontinuity in the plot of the logarithm of number average molecular weight versus time late time points.

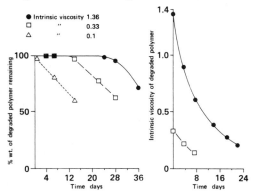

Figure 4.   In vitro degradation of 50/50 molar poly (d,l—lactide—co—glycolide) at 37°C in buffer at pH 7.4.

Polymer film 0.02cm thick.

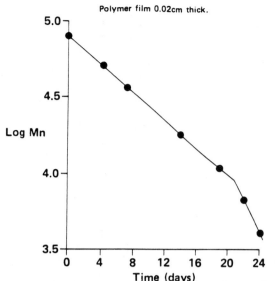

Figure 5. In vitro degradation of 50/50 molar poly (d,l—lactide—co—glycolide) at 37°C in buffer at pH 7.4 and dependence of number average molecular weight on time of degradagion.

Polymer film 0.02cm thick.

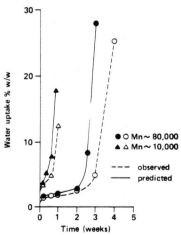

Figure 6. In vitro degradation of poly (d,l—lactide—co—glycolide) at 37°C in buffer at pH 7.4. Effect of degradation on water—uptake by polymer.

The water uptake by 50/50 molar poly (d,l—lactide—co—glycolide) polymers of different molecular weights is shown in Figure 6. These polymers were again of normal polydispersity and similar results and profiles of water uptake are seen for polymers containing higher lactide contents. However, the initial phase of water uptake for the higher molecular weight variants of these high lactide content polymers occurs to a lower degree (depending on composition) and the time scales for the phases of water uptake are more extended. Electronmicrographical studies (Figure 7) show unequivocal evidence of generation of microporosity. Initiation of pore formation, size of micropores generated, and contiguity or continuity of pores are a function of polymer molecular weight and polydispersity as well as polymer composition and structure.

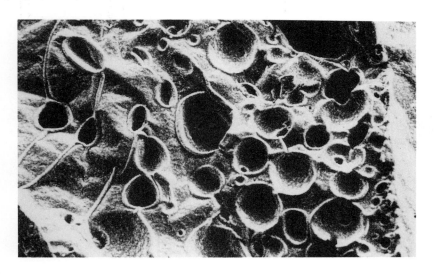

Figure 7. Electron photomicrograph of interior of degraded polymer. Polymer film (0.02cm thick) incubated in pH 7.4 buffer at 37°C for 14 days. Polymer is 50/50 poly(d,l—l actide—co—glycolide) having a molecular weight of 20000.

Release studies (Figure 8—10) show that release of drug (which is effective when all the female rats are in dioestrus) is a function of polymer composition, polymer molecular weight and drug loading. Other studies (results not presented here) have shown firm evidence that release is also controlled by the polydispersity (and perhaps even by the type of molecular weight distribution) of the polymer used to control release as well as the structure of (co—) polymers (degree of heterogeneity and segmental length of co—monomer units).

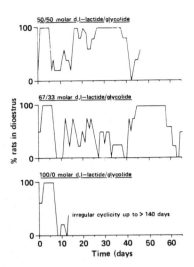

Figure 8. Effects of treatment with 100µg 'Zoladex' using
subcutaneous depots containing 3% w/w drug on oestrous cycles in
female rats. Molecular weight of polymers >100000.

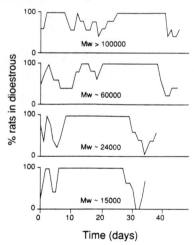

Figure 9. Effects of treatment with 300 µg 'Zoladex' using
subcutaneous depots containing 10% drug in 50/50 molar
poly(d,l-lactide-co-glycolide) polymers of different molecular
weight on oestrous cycles in female rats.

By adjustment of all these parameters continuous release of 'Zoladex' can be achieved over 28 days (Figure 12). Indeed, it is possible to design systems giving release over very much longer time periods.

The efficacy of this depot formulation of 'Zoladex' was tested in two sex hormone—responsive tumour models. The first was the dimethyl—benzanthracene (DMBA)—induced rat mammary carcinoma 21, which is known to be dependent on both oestrogen and prolactin 22. A single subcutaneous depot containing 500 µg 'Zoladex' caused an inhibition of oestrogen secretion, the disappearance of cornified cells from vaginal smears and regression of DMBA—induced mammary tumours (Figure 13). Half of the tumours present at the start of the experiment were not palpable at 28 days but all save one of them reappeared between 40 and 60 days as the single 'Zoladex' depot became exhausted. In contrast, the DMBA—induced mammary tumours increased in size by more than 50% in control animals given a placebo depot.

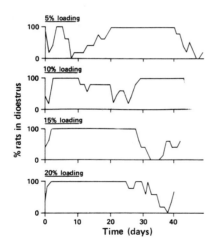

Figure 10. Effects of treatment with 300 µg 'Zoladex' using subcutaneous depots containing different drug loadings in a high molecular weight 50/50 molar poly (d,1—lactide—co—glycolide) on oestrous cycles in female rats.

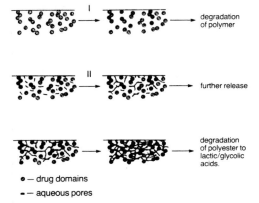

Figure 11. Mechanisms of release of polypeptides from formulations based on poly (d,l—lactide) and poly (d,l—lactide—co—glycolide).

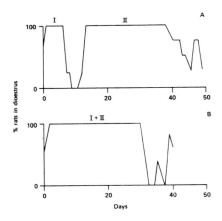

Figure 12. Effect of subdermal depots containing 300 µg 'Zoladex' on oestrous cycles in female rats. A—depots containing 3% w/w drug in high molecular weight polymer. B—depots containing 20% w/w drug in low molecular weight polymer. I—initial release due to leaching from surface, II—degradation—induced release.

If single depots of the drug were given subcutaneously at weeks 0, 4 and 8 of the study, the regression was more impressive (Figure 14). No tumour was palapable by week 11. Again, by week 16 regrowth of the tumours occurred because treatment stopped at week 8. By week 20, the tumours has re-attained pretreatment size.

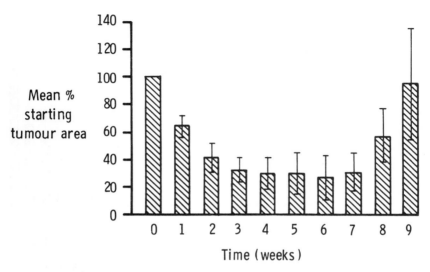

Figure 13. Effect of a single subcutaneous depot containing 500 µg 'Zoladex' given at time 0 on growth of DMBA-induced rat mammary tumours. The values shown are means ± SEM for ten rats.

The second tumour we have used is the Dunning R3327H rat transplantable prostate adenocarcinoma, which is androgen responsive and has been used extensively as a model for the human disease 23. Single subcutaneous depots containing 1 mg 'Zoladex' given every 28 days to rats bearing Dunning R3327H prostate tumours implanted on each flank caused a marked inhibition of tumour growth indistinguishable from that in surgically castrated rats (Figure 15). Twenty-one days after the eighth depot was given, the rats were killed and the weights of the sex organs recorded and serum hormone concentrations measured by radioimmunoassay (Table 1). Testes weights were about 10% those of control rats of a similar age and weight and showed atrophic histological changes. Ventral prostate gland and seminal vesicle weights were identical to those in the surgically castrated group and, histologically, were also completely atrophic.

Serum LH and testosterone (Table 1) were undetectable in the
group given 'Zoladex' depot and serum FSH was decreased by 60–70%.
Serum prolactin doubled in rats given 'Zoladex', as it did in
surgically castrated animals, probably as a consequence of androgen
withdrawal. This contrasts with the effect of 'Zoladex' in female
rats, where there is a significant reduction in serum prolactin
following oestrogen withdrawal[24].

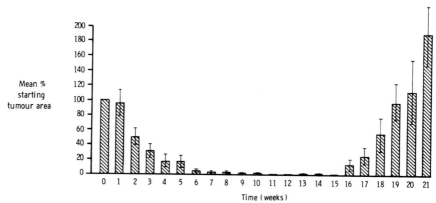

Figure 14. Effect of single subcutaneous depots containing 500 µg
'Zoladex' given at 0, 4 and 8 weeks on growth of DMBA–induced rat
mammary tumours. The values shown are the means ± SEM for ten
rats.

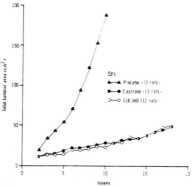

Figure 15. Effect of single subcutaneous depots containing 1 mg
'Zoladex' on growth of Dunning R3327H transplantable rat prostate
tumours. The depots given every 4 weeks. Treatment groups are
placebo controls (▲: 13 rats), 'Zoladex' (O:12 rats) and surgically
castrated animals (●:13 rats).

	Zoladex	Castrate	Control
Testes wt (mg)	366.5 ± 10.2	–	~ 3,500
Ventral prostate wt (mg)	21.3 ± 1.1	19.9 ± 0.7	~ 250
Seminal vesicle wt (mg)	54.3 ± 1.0	53.8 ± 0.8	~ 350
Serum LH (ng/ml)	< 0.2	12.9 ± 1.0	~ 1.5
Serum FSH (ng/ml)	174 ± 6	1,413 ± 24	~ 400
Serum prolactin (ng/ml)	63.2 ± 4.5	60.9 ± 7.5	~ 30
Serum testosterone (ng/ml)	< 0.25	< 0.26	~ 3

Table 1. Sex organ weights and serum hormone concentrations in 'Zoladex' — treated, and surgically castrated rats bearing Dunning R3327H prostate tumours. Control values for rats of a similar age and weight are shown for comparison.

3 DISCUSSION

Because polypeptides have high molecular weight and are generally water soluble, their release from these polyesters by classical partition—dependent diffusion is unlikely to occur. Consequently, degradation of the poly (d,l—lactide) or poly (d,l—lactide—co—glycolide) will be a critical factor in determining transport of the high molecular weight polypeptide from the dosage form. The degradation of these polymers in the absence of drug has been characterised in terms of molecular weight and its distribution, weight loss, water uptake and morphology of the hydrated and degraded polymer.

Degradation of the polymer in vitro in buffer at pH 7.4 results in progressive changes in molecular weight (and sometimes molecular weight distribution depending on polymer structure). Under these conditions degradation is not enzyme mediated and must occur by simple hydrolytic cleavage of ester groups; the profile of weight loss and change in molecular weight are consistent with this. High molecular weight polymers degrade to lower molecular weights, as measured by viscosity, yet retain their water insolubility. Only after an extended time of degradation does any weight loss occur. In contrast, very low molecular weight polymers can degrade with weight loss immediately. Similar results are obtained with high lactide–containing polymers except that the time scale of events is more extended for these more hydrolytically stable polymers. These results are consistent with bulk hydrolysis under in vitro conditions and this correlates broadly with degradation of these polymers in vivo suggesting that even in subcutaneous tissue, enzyme–mediated degradation is significantly less important than simple hydrolysis. In this event, polylactides could effectively protect polypeptides at the depot site from the influence of degradative enzymes.

For these degradation experiments, if the logarithm of the number average molecular weight is plotted as a function of time, then for high molecular weight polymers an essentially linear relationship is seen to hold except at extended times of degradation where a discontinuity arises. Pitt and Schindler[17], studying poly(d,l–lactide), have seen a similar behaviour but have ignored the nature of, and reasons for, the discontinuity. In fact, this arises because of water uptake by the degrading polymer. For an amorphous polyester, water uptake will be governed, empirically, by the intrinsic hydrophilicity of the repeat units and by end–group effects. For these polyesters, the end groups are alkoxylic and carboxylic and these increase as molecular weight falls. That is, as degradation proceeds the essentially hydrophobic polymer becomes more hydrophilic. The profile of water uptake at 37°C in buffer at pH 7.4, for polymers which have been dried rigorously is determined by two events. The first is simple diffusional ingress of water into the dried material and in the absence of degradation this would occur to a level that would be characteristic of the equilibrium swelling of this kind of material. However, these polymers are hydrolytically unstable and following, or even during this initial diffusional phase, the polymer can degrade and so take up more water. For high molecular weight polymers, having a normal distribution (p approximately 2), these two phases of water uptake are separated by an interval during which water uptake increases hardly at all. In contrast, low molecular weight polymers have an essentially continuous water uptake.

It can be shown empirically that the water uptake for a thin polymer film having a molecular weight M_n and a polydispersity p, in the absence of significant hydrolytic degradation, is described approximately by the hyperbolic function:

$$[H_2O] = a + b/p \, M_n$$

where a and b are constants related to polymer composition.

If the initial diffusional ingress into thin films is assumed to be instantaneous then approximate expressions can be derived for degradation—induced change of molecular weight and water uptake with time using a similar but modified model to that proposed by Pitt and Schindler[17].

Degradation of poly (lactide—co—glycolide) and polylactide proceeds by hydrolytic scission of ester groups generating polymers containing one terminal carboxyl group/chain. Defining degradation as appearance of $- CO_2H$ and applying the normal kinetic equation governing ester hydrolysis:

$$d[CO_2H]/dt = K[H_2O][ester][CO_2H] \qquad (1)$$

and $[CO_2H] \propto 1/M_{n,t}$
where $M_{n,t}$ is the number average molecular weight at time t. For all practical purposes [ester] can be considered a constant and as

$$[H_2O) = a + b/p.M_n$$
equation (1) reduces to

$$d[1/M_{n,t}]/dt = k.a + b/p.M_{n,t}.1/M_{n,t} \qquad (2)$$

where $k = K[ester]$.

At t = 0, $M_{n,t} = M_{n,o}$ where $M_{n,o}$ is initial number average molecular weight and the solution to equation (2) is:

$$M_{n,t} = M_{n,o} \, e^{-akt} - b (1 - e^{-akt})/a.p \qquad (3)$$

and $[H_2O]_t = a [1 + \dfrac{b}{a.p.M_{n,o} \, e^{-akt} - b(1 - e^{-akt})}] \qquad (4)$

It should be noted that these derived expressions relate to the condition where the polymers have initially a normal or most probable distribution (i.e. p approximately 2) and hydrolysis of the polymer chains is essentially a random process. However, it does confirm the importance of polydispersity in determining the course of degradation. This kinetic model is confirmed, at least in part, as the derived equation for water uptake correlates broadly with experimentally determined events. Thus, hydrolytic degradation is characterised by reduction in molecular weight, enhanced water uptake and ultimately weight loss of polymer. All these events occur at a temperature which is below or near the glass transition temperature of the polyester. This in turn implies that morphological changes are likely to occur within the polymer whilst hydrolysis is occurring. This is confirmed by scanning electron—microscopy of the degraded products which shows the development of porosity within the degrading polyester.

These studies have shown that degradation of poly (d,l-lactide) and poly (d,l-lactide-co-glycolide) is dependent on molecular weight, polydispersity, geometry, polymer composition and polymer structure. Degradation ultimately leads to enhanced water uptake and the generation of porosity. Thus, water-soluble polypeptides may be released from these biodegradable polyesters since enhanced water uptake and the generation of porosity should facilitate transport of polypeptide from the dosage form. This is likely to involve diffusion through aqueous pores generated in the drug polymer matrix. In this event, the release of polypeptide will differ mechanistically from the processes thought to occur during release of steroids, narcotic antagonists and antimalarials from poly (d,l-lactide) and poly (d,l-lactide-co-glycolide)[8]. Whereas these low molecular weight drugs will diffuse, by a simple partition-dependent process, through intact polymer membranes in diffusion cell experiments, these same polymer membranes are totally impermeable to polypeptides and proteins.

Research was focused on the release of 'Zoladex' from solid depots because this was thought more likely to afford a clearer understanding of the physicochemical parameters which allow transport of drug from the dosage form. Parameters which govern release of drug can be controlled more easily with solid depots.

On the basis of degradation studies, transport of drug from these depots is likely to be governed by various properties of the rate controlling polyester. These properties include polymer composition, molecular weight and distribution. Additionally, level of drug incorporation and morphology of the drug/polymer mixture exert controls on release. It has been shown that release of 'Zoladex' (and other peptides) from these bio-degradable polyesters occurs by diffusion through aqueous pores generated in the dosage form. These aqueous channels which facilitate drug release are generated by two distinct and separate mechanisms (Figure 11). The first involves leaching of drug from polypeptide domains at or near the surface and essentially is a diffusion/dissolution controlled event through aqueous pores. However, drug within the body of the depot, existing in isolated domains not continuous or contiguous with the surface, cannot be released until the second mechanism becomes operative. This second mechanism involves degradation of the polyester and is associated with the generation of microporosity in, and enhanced water uptake by, the degrading polymer and ultimately erosion of the polymer carrier.

Typical parameters controlling the initial phase of release are, for example, drug loading, morphology and geometry, whereas the second phase is intrinsically related to the degradation properties of the polyester. When these two phases of release do not overlap, discontinuous release is observed. However, by controlling the properties of the polymer, the initial phase of release can be made to overlap with the second phase and depots can be defined which give continuous release over not less than 28 days.

For amorphous homo— and co—polymers of approximately the same high molecular weight, increasing lactide concentration results in slower degradation. This is reflected in the biological effect elicited in rats using sub—dermal depots containing small amounts of drug in high molecular weight carriers.

When administered to female rats, some drug is released initially as judged by biological response. There then follows a period during which either drug is not released at all or is released at an ineffective rate and so the rats return to oestrus. At some later time point release recommences, the biological response returns and continues until the depot is fully depleted of drug. For polymers of similar molecular weight and distribution it can be seen that the interval between the two phases of release is shortest for the most rapidly degradable polymer.

However, for polymers of given composition and having normal polydispersity but of varying molecular weight, the interval between the two phases can be shortened by reducing the molecular weight. Indeed, the beginning of the second interval of release correlates with degradation events such as enhanced water uptake and appearance of microporosity within the controlling polymer. Some element of control over the initial phase of release can be achieved by varying drug concentration. Thus, for a polymer of given molecular weight (M_w approximately 100000) and defined polydispersity, increasing drug loading can extend the initial phase of release such that it overlaps with degradation—induced release.

It is clear from the studies in a model of advanced breast cancer that 'Zoladex' depots are likely to have utility in the treatment of advanced mammary carcinoma in premenopausal women. Similarly, in an animal model for prostatic carcinoma, chemical castration using 'Zoladex' depots causes inhibition of tumour growth to values indistinguishable from those in surgically castrated animals which suggests that the drug should prove effective in the clinical treatment of prostate cancer.

Indeed, depot formulations which are administered by subcutaneous injection and which release drug over at least 28 days have been developed for clinical use. The depots are based on a poly (d,l—lactide—co—glycolide) matrix in which 'Zoladex' (3.6 mg) is uniformly dispersed. The depot is in the form of a cylindrical rod 1.1 cm long and 1.1 mm in diameter and is injected using a pre—loaded applicator. The promising results achieved in animal studies have now been fully substantiated in clinical trials in men

suffering from prostatic carcinoma[25-31], and in premenopausal women
with advanced breast cancer[32]. The 'Zoladex' 1 month depot
formulation has been approved for the treatment of prostate cancer
by a number of regulatory authorities.

4 CONCLUSIONS

Although this paper has focused on the sustained delivery of a
synthetic analogue of luteinizing hormone releasing hormone it has
been demonstrated that the technology can be applied equally well to
other polypeptides and proteins[9]. Thus using acceptable biodegradable polymers based on lactic and glycolic acids it is
possible to design a diversity of polymer types which allow
continuous release of many different polypeptides. However as this
paper shows a much more profound appreciation of the nuances of
polymer chemistry and physics is demanded if these polymers, or
indeed any other class of synthetic macromolecule, are to be used as
drug carriers.

ACKNOWLEDGEMENTS

NIAMDD, Professor L.E. Reichert and G.D. Niswender for reagents for
FSH, LH and prolactin radioimmunoassays. The US National Prostate
Cancer Agency for supplies of rats bearing Dunning R3327H tumours.
The many colleagues in ICI Pharmaceuticals Research and
Pharmaceutical Departments who have made this research and
development possible, particularly J.R. Churchill, P Wreglesworth,
A.S. Dutta, J.R. Woodburn, B. Valcaccia, and B.Curry.

REFERENCES

(1) S.T. Anik, L.M. Sanders, M.D. Chaplin, S. Kushinsky and C.
    Nerenberg, LHRH and its Analogs: Contraceptive and Therapeutic
    Applications, (B.H. Vickery, J.J. Nestor Jr. and E.S.E Hafez,
    editors). MTP Press Limited, Boston, USA,1984, p. 421.

(2) W. Petri, R. Seidel and J. Sandow, Int. Cong. Series — Excerpta
    Medica, 1984, 656, 634.

(3) R. Anders, H.P. Merkel, W. Schurr and R. Ziegler, J. Pharm.
    Sci., 1983, 72, 1481.

(4) H. Okada, I. Yamazaki, Y. Ogawa, S. Hirai, T. Yashiki and H.
    Mima, J. Pharm. Sci., 1982, 71, 1367.

(5) H. Okada, I. Yamazaki, T. Yashiki and H. Mima, J. Pharm. Sci.,
    1983, 72, 75.

(6) H. Okada, T. Yashiki and H. Mima, J. Pharm. Sci., 1983, 72, 173.

(7) H. Okada, I. Yamazaki, T. Yashiki, T. Shimamoto and H. Mima, J. Pharm. Sci., 1982, 73, 298.

(8) D.L. Wise, T.D. Fellman, J.E. Sanderson and R.L. Wentworth, Drug Carriers in Biology and Medicine, (G. Gregoriadis, editor) Academic Press, London, England, 1979, p. 237.

(9) F.G. Hutchinson, European Patent Application 58481, August 25, 1982.

(10) J.R. Churchill and F.G. Hutchinson, United States Patent 4526938, July 2, 1985.

(11) L. Bohn, Polymer Handbook, 2nd edn, (J. Brandrup and E.H. Immrqut, editors) John Wiley and Sons, New York, III., 1975, p.211.

(12) R.W. Baker and H.R. Lonsdale, Controlled Release of Biologically Active Agents, vol. 47, Advances in Experimental Medicine and Biology, (A.C Tanquary and R.E Lacey, editors) Plenum Press, New York, 1974, p. 15.

(13) S. Mitra, M. Van Dress, J.M. Anderson, R.V. Peterson, D. Gregonis and J. Feijen, Polymer Preparation, American Chemical Society, Division Polymer Chemistry 1974, 20(2), 32.

(14) K.R. Sidman, A.C. Schwope, W.D. Steber and S.E. Rudolph, NIDA Research Monographs, 1981, 15, 889.

(15) R. Langer, Methods in Enzymol., 1981, 73, 57.

(16) J.M. Anderson, H. Niven, J. Pelagalli, L.S. Olanoff and R.D. Jones, J. Biomed. Mat. Res., 1981, 15, 889.

(17) C.G. Pitt and A. Schindler, Progress in Contraceptive Delivery Systems, (E.S.E Hafez, and W.A.A. van Os, editors), 1, MTP Press, Lancaster, England, 1980, p. 17.

(18) B.J.A. Furr and F.G. Hutchinson, F.H. Schroeder and B. Richards (Eds), EORTC Genitourinary Group Monograph 2, Part A: Therapeutic Principles in Metastatic Prostatic Cancer, (F.H Schroeder and B. Richards, editor) Alan R. Liss, New York, 1985, p. 143.

(19) G. Williams, D.J. Kerle, S.M. Roe, T. Yeo and S.R. Bloom, EORTC Genitourinary Group Monograph 2. Part A: Therapeutic Principles in Metastatic Prostatic Cancer, (F. H Schroeder and B Richards, editors) 1985, p. 287.

(20) J.B.F. Grant, S.R. Ahmed, S.M. Shalet, C.B. Costello, A. Howell and N.J. Blacklock, Br. J. Urol., 1986, 58, 539.

(21) C. Huggins, G. Briziarelli and H. Sutton, J. Exp. Med., 1959, 25

(22) V.C. Jordan, Clin. Oncol. 1982, 1, 21

(23) J.K. Smolev, W.D.W. Heston, W.W. Scott and D.S. Coffey, Cancer Treat. Rev., 1977, 61, 273.

(24) B.J.A. Furr and R.I. Nicholson, J. Reprod. Fertil., 1982, 64, 529

(25) S.R. Ahmed, P.J.C. Brooman, S.M. Shalet, A. Howell, N.J. Blacklock and P. Rickards, Lancet, 1983, 2, 415.

(26) J.M. Allen, J.P. O'Shea, K. Mashiter, G. Williams and S.R. Bloom, Br. Med. J., 1983, 286, 1607.

(27) K.J. Walker, R.I.Nicholson, A.O. Turkes, K. Griffiths, M. Robinson, Z. Crispin and S. Dris, Lancet, 1983, 2, 413.

(28) K.J. Walker, A.O. Turkes, R. Zwink, C. Beacock, A.C. Buck, W.B. Peeling and K Griffiths, J. Endocrinol., 1984, 103, RI—R4.

(29) M.R.G. Robinson, L. Denis, C. Mahler, K. Walker, R. Stich and G. Lunglmayr, Eur. J. Surg. Oncol., 1985, 11, 159—165.

(30) R.J. Donnelly, J. Steroid Biochem., 1984, 20, A21, 1375.

(31) B.J.A Furr and R.A.V. Milsted. Endocrine Management of Cancer 2. Contemporary Therapy, (B.A Stoll, editor) S. Karger, Basle 1988, p. 16.

(32) M.R. Williams, K.J. Walker, A. Turkes, R.W. Blamey, and R.I. Nicholson, Br. J. Cancer, 1986, 53, 629—639.

Hydrogels for Useful Therapy

N. B. Graham

DEPARTMENT OF CHEMISTRY, UNIVERSITY OF STRATHCLYDE, GLASGOW
G1 1XL

1 INTRODUCTION

Poly(ethylene oxide) is a most useful and biocompatible polymer. It has found application in industry for many years and is widely used in surface active agents which are approved for incorporation into foodstuffs and in pharmaceutical products for topical, oral and injectable uses. It was patented by Union Carbide[1] and eventually was marketed by Bard as a wound healing agent in a water swollen sheet form. W. R. Grace[2] market isocyanate terminated poly(ethylene glycols) as foamable hydrogels and these form the basis of a contraceptive sponge releasing nonoxynol phenol (itself a poly(ethylene oxide) surface active agent) as a spermicide. The use of poly(ethylene glycol) as a protecting agent for enzymes[3] has been used and commercialised as a technique for evading the natural body defences which would otherwise destroy the enzyme and allowing the treatment of enzyme deficiency conditions. ICI have a patent on block copolymers of poly(ethylene glycol)[4] as soluble polymers of use in controlled drug delivery and the author's team has for many years researched the use of crosslinked poly(ethylene oxides) as drug delivery vehicles with unusual properties. This paper discusses some aspects of the research and development towards a commercial product based on these systems.

2 THE CHEMISTRY

The Preparation of Crosslinked Poly(Ethylene Oxides)

In this paper the terms poly(ethylene oxide) and poly(ethylene glycol) are frequently used in a synonymous manner. The term poly(ethylene glycol) is used when the material under discussion contains terminal hydroxyl groups on a poly(ethylene oxide) chain. Terminally crosslinked poly(ethylene glycols) may be prepared by reacting them with a diisocyanate and a triol as illustrated in Figure 1 where the isocyanate of choice is methylene bis-(4-phenyl isocyanate).

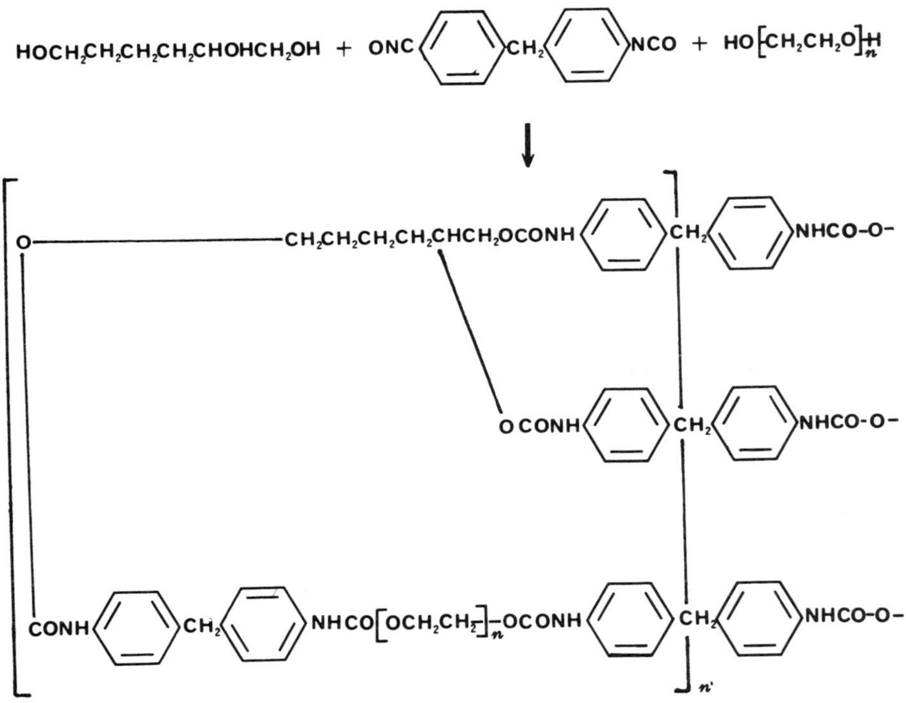

Figure 1   Schematic representation of the formation of a urethane crosslinked poly(ethylene oxide) hydrogel using an aromatic isocyanate.

Though this isocyanate is a component of polyurethanes for use as part of prosthetic devices for implantation it was felt, and later also indicated by mouse fibroblast studies, that materials made from the saturated aliphatic analogue dicyclohexylmethane-4,4'-diisocyanate would be more biocompatible and it was subsequently utilised for synthesis of the polymers as shown in Figure 2 .

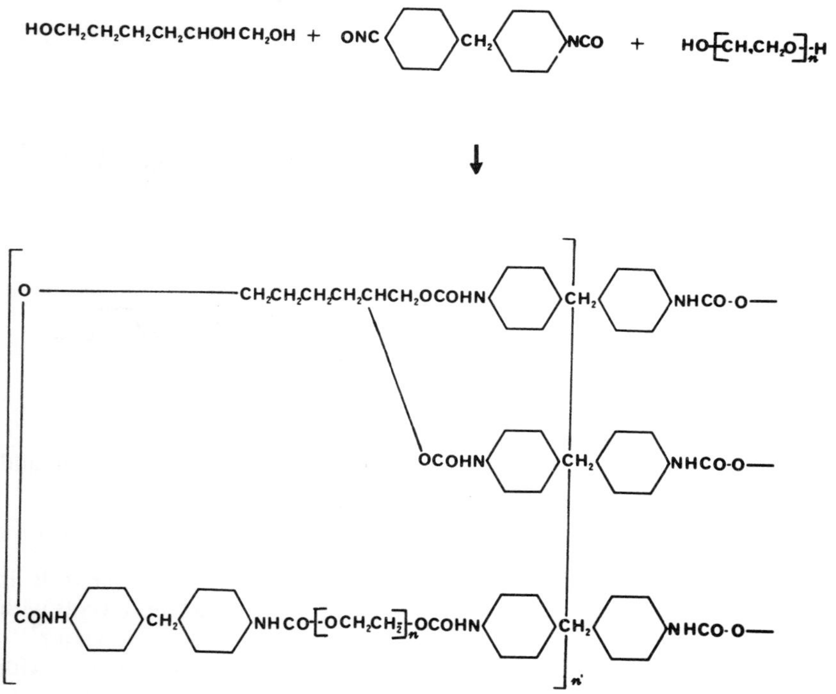

Figure 2   Schematic representation of the formation of a urethane crosslinked poly(ethylene oxide) hydrogel using an aliphatic isocyanate.

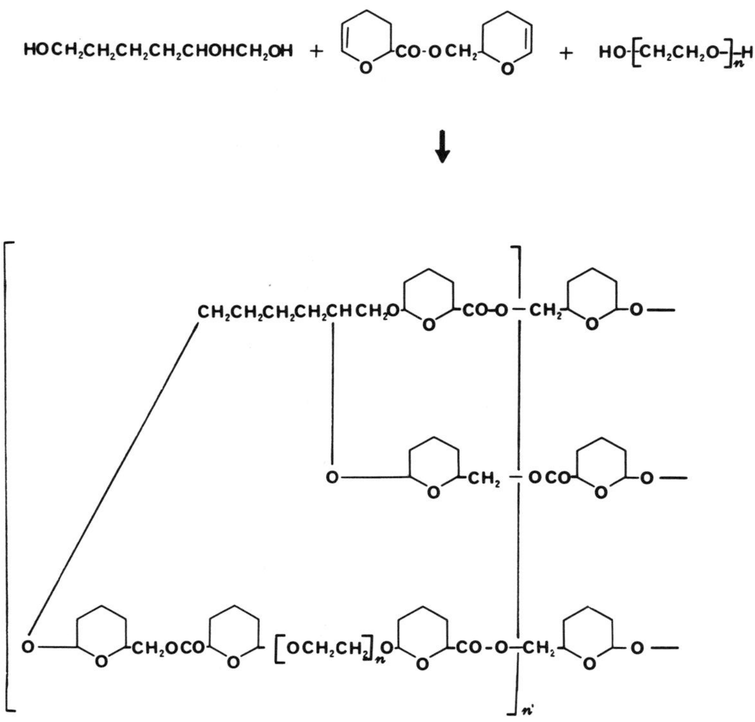

HOCH₂CH₂CH₂CH₂CHOHCH₂OH + [dihydropyran-CO-O-CH₂-dihydropyran] + HO⁻[CH₂CH₂O]ₙ⁻H

↓

[structure of crosslinked polymer network]

Figure 3 Schematic representation of the formation of an acetal and ester crosslinked poly(ethylene oxide) hydrogel.

Where biodegradable poly(ethylene oxide) hydrogels are desired it is possible to synthesise them by crosslinking the poly(ethylene glycol) with 3,4-dihydro-2H-pyran-2-methyl-(3,4-dihydro-2H-pyran-2-carboxylate)[5] as shown in Figure 3 . As the poly(ethylene glycols) are available with a range of molecular weights varying from a few hundred daltons to around 10,000 daltons, and these products have a wide variation in their crystallinities and crystalline melting ranges, they provide the basis of a large and varied family of polymers. The properties of these materials can also be controlled by their degree of crosslinking or by the copolymerisation with other glycols. Though they can be readily prepared to an approximate standard their manufacture to the standards

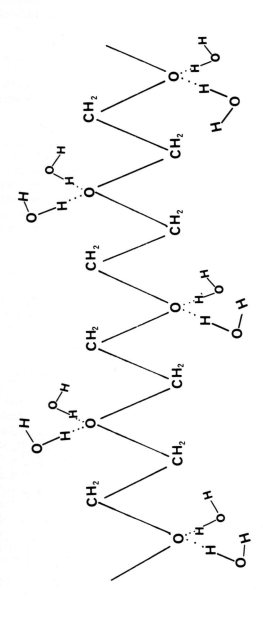

Figure 4 Representation of the hydrogen bonding interaction of two molecules of water per ether group in poly(ethylene oxide).

of consistency demanded by the pharmaceutical industry is much more difficult though attainable. These products in the dry state are termed xerogels while in the water swollen state they are called hydrogels. It seems obvious that these polymer structures would associate with water by hydrogel bonding two water molecules to each ether group as shown in Figure 4 . This however was apparently not the case and fundamental studies[6] pointed to the existence of a trihydrate which was tentatively allocated a helical structure. This helical array has a most interesting proposed configuration with the hydrophobic methylene groups turned outwards and the ether groups turned inwards where there is a hole along the centre of the cylinder. When the xerogels are made with poly(ethylene glycols) that are crystalline the products are themselves crystalline as seen under the crossed polarisers of an optical microscope. The crystallinity of the xerogel is reduced and at a lower melting temperature when compared with the values for the unreacted starting poly(ethylene glycol) as shown in Figure 5 . The ability of the compositions to be designed with various crystallinities and melting points is shown by Figure 6 which presents the plots of three different poly(ethylene glycols) converted into xerogels using different degrees of crosslinking. These xerogels are very unusual amongst water swellable materials comprising a low T_g rubber toughened by crystallites which are destroyed by the imbibition of water. The water acts initially as a strong plasticiser and reduces the T_g to lower values at levels up to one mole per ether group. At higher levels however it acts as an antiplasticiser and causes an increase in the glass transition temperature[7]. This behaviour is shown in Figure 7 for the linear poly(ethylene glycol). The dip in the curve at 3 moles of water per ether group is probably related to the formation of the trihydrate. The trihydrate is more clearly evidenced by the differential scanning calorimetric spectrum on the linear poly(ethylene glycol) containing various levels of water. At molecular weights greater than 950 daltons the trihydrate melting peak at between -10 and -20°C is clearly and almost always seen next to the crystalline melting peak of pure water. A slice of the xerogel will swell in water as shown in Figure 8 . Unusually the equilibrium swelling does not appear to be governed by the crosslinking density but rather by the percentage composition by weight of the polymer[8]. A large family of differently crosslinked polymers made with a variety of different molecular weight poly(ethylene glycols) gave a smooth curve when the degree of swelling was plotted against the weight fraction of poly(ethylene glycol) in the polymer. Water is in fact not an optimum swelling solvent for these gels. The solubility parameter is a valuable guide to the swelling and the plot of swelling against solubility parameter for a particular gel is presented in Figure 9 .

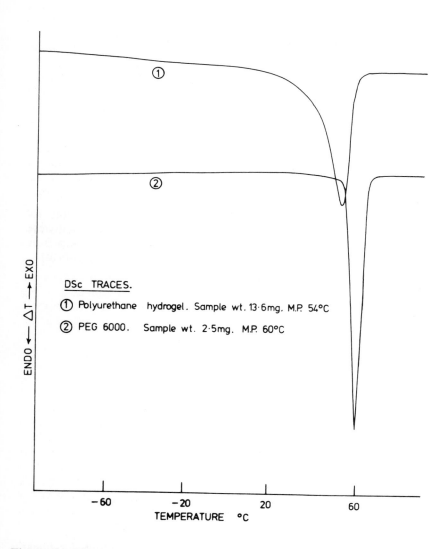

DSc TRACES.

① Polyurethane hydrogel. Sample wt. 13·6mg. M.P. 54°C

② PEG 6000. Sample wt. 2·5mg. M.P. 60°C

Figure 5 The crystallinity of linear and crosslinked poly(ethylene glycol) PEG 6000 as shown by differential scanning calorimetry.

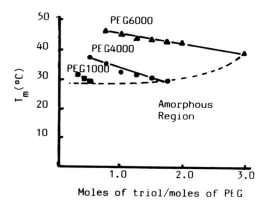

<u>Figure 6</u>   Melting point temperature versus triol concentration for hydrogels made from three poly(ethylene glycols), methylene bis(4-phenyl isocyanate) and 1,2,6-hexane triol. (Taken from <u>Polymer</u>, 1989, <u>30</u>, 2132 with permision from Butterworth).

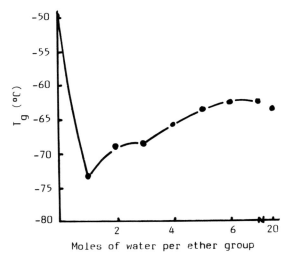

<u>Figure 7</u> T_g variation with moles of water per ether group for linear poly(ethylene glycol) of measured $\overline{M}_n$ = 5700. (Taken from <u>Polymer</u>, 1989, <u>30</u>, 530 with permission from Butterworth).

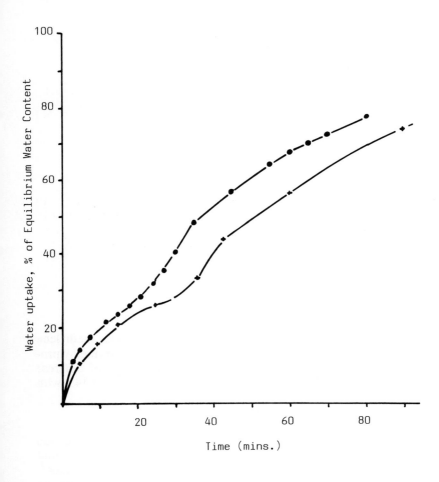

<u>Figure 8</u> The swelling of a 1.45mm thickness hydrogel slice made from
         PEG 8490/1HT in water at ₒ 37°C, + 25°C. (Taken from
         M. E. McNeill, PhD Thesis, University of Strathclyde, 1986).

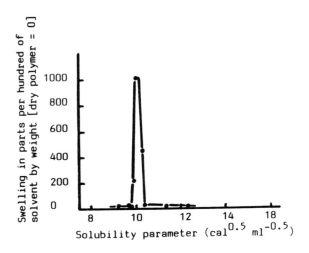

Figure 9 The swelling of a hydrogel based on PEG 6000, methylene bis-(4-phenyl isocyanate) and 2-ethyl-2-hydroxymethylpropane-1,3-diol, in a variety of hydroxylic solvents of different solubility parameters. (From *Polymer*, 1982, *23*, 1345 with permission from Butterworth).

In the case of hydrogen bonding solvents a maximum of swelling was obtained at a solubility parameter value close to that of benzyl alcohol which is a very good swelling solvent and much better than water[9]. When following the swelling of a crystalline xerogel with time an unusual undulation of the swelling curve is usually obtained. This is caused by the effect of the crystallites in physically restraining the network from swelling and thereby reducing the rate of uptake of water in the early stages of swelling. In fact the swelling plot with the "kink" could be very approximately represented as a straight line instead of a curve as shown in Figure 10 .

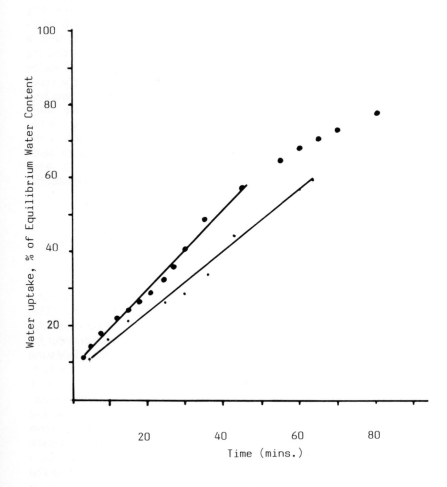

Figure 10 The swelling behaviour shown in Figure 8 represented as a straight-line for a significant proportion of the swelling curve.

It is thus clear that the presence of crystallinity would be expected to have some effect not only on the ingress of water but also on the egress of contained diffusate from the swelling xerogel. The time taken for a slice of xerogel to swell to the point where the last sign of crystallinity disappears from its centre depends on the thickness of the slice. If slices are compared and the times plotted against the initial thickness then a curve is produced in which the time increases rapidly with increasing thickness. It is readily calculated from a log/log plot that the time varies as the 2.6 power of the thickness. This means that a relatively small change in thickness results in a significant change in the time of swelling and indicates that thickness is a useful parameter to consider when wishing to control the time of diffusion processes occurring within hydrogels. The frequently desired constant or (Case 2) diffusion is the one that is abnormal and difficult to obtain. The rest of this paper deals with the obtention of drug release that is essentially Case 2 from swelling hydrogels and the conversion of this result into a commercial product.

3 THE PHARMACY

The Combination of the Hydrogel with a Drug

Any useful development of a programmed drug delivery system for a novel therapy must inevitably involve an interdisciplinary team. This team must include a medical specialist to have a chance of success. All members of the team must be able to communicate and have respect for each others skills and contributions. The dosage form under discussion here commenced with an introduction of the author to Mostyn Embrey, gynaecologist and obstetrician, of the John Radcliffe Hospital in Oxford. This meeting was mediated by the National Research and Development Corporation as it was then known. Mr. Embrey was working on the use of prostaglandin E_2 as an agent to assist mothers to be at full term when difficult labours were expected due to a so-called unripened cervix. It was his clinical judgement that 'the most convenient dosage form would be one which was administered vaginally' He also considered that the safest and best manner to administer the drug was at a controlled and sustained rate that effectively mimicked the natural process by the continuous delivery over several hours at a low level. As Prostaglandin E_2 is also a very unstable compound it was essential that an adequately stable dosage form was developed which would be able to be distributed through the normal channels and stored for use. The general specification was thus set and research commenced. From the outset the hydrogels based on the crosslinking of poly(ethylene glycols) were seen as very attractive carriers for the prostaglandin. This has been confirmed over the many years since the work commenced. Poly(ethylene glycols) have been used in foodstuffs and as components of pharmaceutical products for many years as well as being a major construction unit for many ubiquitous surface active agents. They can be turned into quite perfect and reproducible network polymers with almost no residual extractable material to provide toxicological and regulatory worries. Their most interesting characteristic was their ability to swell in water and organic solvents. This provided three most important properties.

The first was that by swelling the polymer slices in water their entire mass could be given a very high permeability thus allowing for a very thorough aqueous extraction of any small amount of residual reactants which allows a very clean material to be produced. This provides the minimum possible toxicological concern. The second property is that the polymer can be swollen by organic solvents which allows a very precise absorption of any given low molecular weight drug solution into the mass of the hydrogel thus guaranteeing the drug loading. This is of great importance, as a safety issue, the incorporation of the drug throughout the hydrogel polymer means that there can be no possibility of drug dumping. This removes another matter of concern to Regulatory Authorities. The third, and most surprising result of swelling was that when the drug was released from the dried down slice of hydrogel it was at a nearly constant rate for the first fifty percent of the release[10]. The typical release profile is shown in Figure 11 . Such a constant release from a device which cannot dump the dose was a novelty at the time and is still a highly desirable attribute of these systems. The drug loading and the half-life of drug-release could be readily changed to meet the clinician's specifications in the laboratory and so devices were prepared and evaluated in the clinic. These samples for human use had to be prepared and packaged in a purpose built clean room.

Finally two dosage levels containing 5 and 10mg of the prostaglandin were found to meet Mr. Embrey's clinical performance requirements and after writing up some papers on the work the project might have been thought of as completed. How wrong could one be. It was only just beginning! To turn the product concept into a real product which would perhaps help women all over the world required much, much more work than the chemical and clinical research. It required protection via patents, development programmes on the polymer chemistry, the engineering to produce the slices of polymer and the drug incorporation to produce the final dosage form, the establishment of a commercial company to make the speciality polymers, finance, the collaboration of a pharmaceutical company to manufacture the drug-containing dosage form, the obtention of a product license, agreements with marketing partners and finally the acceptance or otherwise of the product in the marketplace. Failure can occur at any one of these and it requires great persistence and fortitude to stay the course. The much discussed problems of Technology Transfer come from the very considerable problems of taking product concepts from Universities across these typically difficult hurdles.

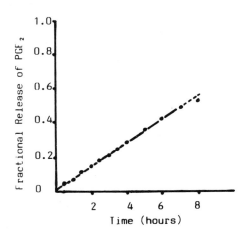

Figure 11 Results demonstrating constant release of PGE$_2$, for the first 50% of the loaded drug content, from an initially dry flat pessary of 2.6mm dry thickness. Initial PGE$_2$ content = 8.6mg. (Taken from J. Controlled Release, 1984, 1, 107, with permission from Elsevier).

4 THE ENGINEERING

To develop a commercial product will almost always demand a collaboration with engineers. Chemical scale up demands the skills of chemical engineers while the establishment of a piece of slicing equipment as in this case demands the skills of mechanical engineers. In this case assistance was provided by the University's Department of Process and Production Engineering, headed at the time by Professor Donald Ross. From the physical chemical point of view the delivery polymer would be a thin slab of given thickness and surface area. The precise shape of the cross section was of little importance as it did not alter the release profile significantly. Of course to an engineer who has to manufacture the moulds, cutting machines and production line it is of vital importance as it affects the tools to be used, the quality of finish and the cost. The shape is also of considerable relevance to the marketing image and to the convenience of use in the clinic. The shape that was finally chosen was a lozenge. This now had to be produced in a very precise cutting operation. A laboratory prototype machine was first assembled and the various problems slowly eliminated. This led to the design of the full scale manufacturing unit.

5 THE COMMERCE

The Establishment of the Commercial Enterprises

The only significant route to the establishment of a product in "The real world", as it is often called, is via a commercial entity and a commercial pattern of behaviour. In general investments are more readily obtained for products which are protected by a patent which provides a degree of monopoly for the product for a sufficient number of years to justify the considerable development time and cost. As this research was funded by the British Technology Group they filed and licensed the patent[11] which has been granted in many important countries including the United States of America. The filing of a patent and its completion in a wide range of countries costs many thousands of pounds and the "discussions" with the various patent examiners occupy much time over many years. Patents have to be backed by the ability to spend a very considerable sum of money in their defense and this is one of the strong assets and policies of the British Technology Group who have an excellent record in this respect. They are thus worthwhile allies in the prosecution of patents.

Having obtained the patent there is a need for a commercial company to take a licence and exploit it. If the licence is not taken by an already existing Company then if the technology is to be exploited a new Company must be financed and formed. Though it is beyond the scope of this paper to go into the detail of establishing a Company it is perhaps educational to consider some of the topics which must be addressed in reaching a successful product in the market place. The process is cyclical and is illustrated in Figure 12 for a pharmaceutical product. It is helpful to recognise these (and other) various steps and to use them in the progression of the programme towards the end point. Failure in any of the boxes can lead either to the project recycling back to an earlier box for more work or to abandonment of the project. The cycle commences with some appropriate (maybe inadequate!) capital injection which may perhaps be supplemented by some contract research. This funds the early and least expensive stages 1 to 5 which are also appropriately done in University laboratories. From box 6 onwards the project should move into a commercial operation. In the case of the product, which is the subject of this paper, in fact five different commercial entities were involved. The development work on the polymers was done under contract by the company Polysystems founded by the author, the manufacture of the polymers to FDA standards was done by a second Company, Polysystems Health Care Ltd. The Product Licence Holder is Controlled Therapeutics Corporation who have licensed the technology from the British Technology Group and manufacture the final dosage forms which was marketed in the UK by Roussel Laboratories. The final single hydrogel pessary containing 10mg of prostaglandin E_2 sold at a price of £26 in the UK. The polymer content by weight was a fraction of a gram. This certainly made it a high value polymer which is the theme of this volume. It is however clear that most of the price is from other than the polymeric component and it would not be possible to separate out the value of this specifically. Though it is clear that a very unusual, tightly controlled and high performance material is required, it is hoped that it has been clearly demonstrated that the polymer is only one of the many components which are essential to the successful attainment of the finished product. A considerable team of talented people with a wide range of skills is required for success and in the Company start-up situation one also wishes them to be continually blessed with a considerable measure of good fortune. However the norm is for Murphy's Law to apply.

It took some thirteen years to take this product from conception to sales in the marketplace and, sadly, Mr. Mostyn Embrey did not live to see it happen. The author no longer has any connection with Polysystems Ltd.

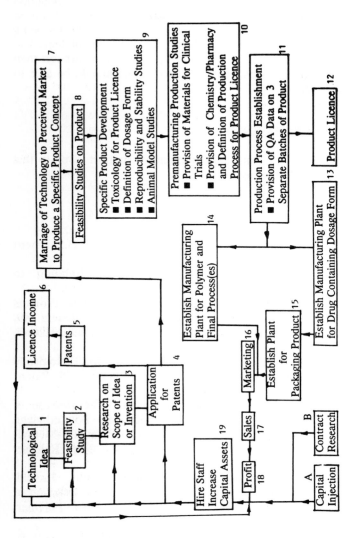

Figure 12 Outline of the Stages in the Commercial Development of a Pharmaceutical Product Involving a Polymeric Delivery System

6 CONCLUSION

It is clear that speciality polymers can produce highly desirable effects in drug delivery and make a contribution to health through the introduction of new or improved therapies. The commercial development of a new material is frought with difficulties and is a little like a horse being entered for the Grand National Steeplechase - it has a lot of fences to jump and it may fall at any one of them. Once it has won through then the material, like the horse, becomes very valuable and should be able to sire other related products.

7 ACKNOWLEDGEMENTS

The author acknowledges the considerable financial support given to this research by the British Technology Group.

REFERENCES

1.   Nan Shieh Chu, 'Water Swellable Poly(alkylene oxide)', U.S. Patent 3,963,805, 1976.
2.   W. R. Grace (assignee), 'Improvements in or Relating to Hydrophilic Foams', U.K. Patent 1,429,711, 1976.
3.   A. Abuchowski, 'Development of Therapeutically Active Proteins Through Modification with Poly(ethylene glycol)', Pacifichem '89, International Congress of Pacific Basin Societies, Honolulu, Hawaii, 1989.
4.   J. T. Fildes and F. G. Hutcheson, 'Delivery Means for Biologically Active Agents Comprising Hydrophilic Polyurethane', U.S. Patent 4,202,880, 1980.
5.   N. B. Graham, M. E. McNeill and A. Rashid, 'The Release of Prostaglandin E_2 from a Novel Crystalline-Rubbery Poly(ethylene oxide) Network Crosslinked by 3,4-Dihydro-2H-Pyranyl-2-Methyl-(3,4-Dihyrdo-2H-Pyran-2-Carboxylate). *J. Controlled Release*, 1985, 2, 231.
6.   N. B. Graham, M. Zulfiqar, N. E. Nwachuku and A. Rashid, 'Interaction of Poly(ethylene oxide) with Solvents : 4. Interaction of Water with Poly(ethylene oxide) Crosslinked Hydrogels', *Polymer*, 1990, 31, 909.
7.   N. B. Graham, M. Zulfiqar, (in part) N. E. Nwachuku and A. Rashid, 'Interaction of Poly(ethylene oxide) with Solvents : 2. Water-Poly(ethylene glycol)', *Polymer*, 1989, 30, 528.

8. N. B. Graham and M. E. McNeill, 'Morphology of Poly(ethylene oxide) Based Hydrogels in Relation to Controlled Drug Delivery', Die Makromolekulare Chemie, Macromolecular Symposia, 1988, 19, 255.

9. N. B. Graham, N. E. Nwachuku and D. J. Walsh, 'Interaction of Poly(ethylene oxide) with Solvents : 1. Preparation and Swelling of Crosslinked Poly(ethylene oxide) Hydrogel', Polymer, 1982, 23, 1345.

10. N. B. Graham, M. Zulfiqar, B. B. MacDonald and M. E. McNeill, 'Caffeine Release from Fully Swollen Poly(ethylene oxide) Hydrogels', J. Controlled Release, 1987, 8, 243.

Speciality Silicones as Building Blocks for Organic Polymer Modification

W. Gardiner and J. W. White

DOW CORNING LTD., CARDIFF ROAD, BARRY, SOUTH GLAMORGAN, CF6 7YL

1 INTRODUCTION

The Silicones are a unique class of materials which are often considered as high value polymers in themselves and this is especially true of silicones which are used in conjunction with organic plastics and resins. For much of the plastics industry rising costs have forced a shift in emphasis from creating new polymers to modifying and upgrading existing ones. In many applications a small proportion of a selected siloxane can dramatically influence the properties of the organic polymer. Silicones and silanes provide the means to enhance the value of plastic products by adding desirable properties and also by reducing production costs. The resulting materials combine improvements in processing, impact strength, wide temperature range, weatherability and chemical resistance of the siloxane with the mechanical properties and economic advantages of the organic.

Silicones are well known in the plastics industry[1] and have found a diversity of applications as internal release agents and lubricants, surfactants for foam production and antifoams for polymerisation processes, release agents and adhesion promoters and cross linking systems for thermoplastics. The combination of silicone and organic can be achieved in a number of ways from simple blending in of high molecular weight polydimethyl siloxane fluids and rubbers to custom synthesis of silicone organic copolymers using silicone resins, branched fluids or linear organofunctional siloxanes. In this paper we concentrate on the modification of organic plastics and resins using organofunctional siloxanes to form ABA or (AB)n block copolymers.

2 SYNTHESIS OF ORGANOFUNCTIONAL SILOXANES

Organofunctional silanes are often the primary building blocks in the synthesis of functional siloxanes. They have many uses in their own right, for example functional silanes are used extensively as coupling agents, providing the bond between organic resins and inorganic pigments and fillers, or as surface treatments to immobilise active species such as catalysts, enzymes, biocides, ion exchangers etc. and as adhesion promoters, primers and cross linkers. Their synthesis and applications have been extensively covered elsewhere [2-4] and will not be repeated here.

In principle the range of possible substituents at silicon is limited only by the ingenuity of the silicone chemist and a huge variety of such materials is described in the open literature and patents. In practice the commercialisation of organofunctional siloxanes has centred around a few selected groups of materials. Typical examples of the more useful reactive organic groups are:

$-CH=CH_2$, $-(CH_2)_3Cl$, $-(CH_2)_3NH(CH_2)_2NH_2$, $-(CH_2)_3(OCH_2CH_2)_nOH$

$-(CH_2)_3SH$, $-(CH_2)_nCOOH$, $-(CH_2)_3OCH_2CH\!-\!CH_2$ etc.
$\overset{\displaystyle \diagdown O \diagup}{}$

One of the best known synthetic routes to organofunctional siloxanes is ring opening copolymerisation of the appropriate organofunctional monomer with dimethylcyclo-siloxanes and a chain terminating agent or end blocker[5]. The process which is often abbreviated "equilibration" typically utilises strongly acidic or basic catalysts which open the cyclic monomers and redistribute the siloxane bonds. Suitable organofunctional silanes can be readily converted to the siloxane prepolymers in carefully controlled reaction with water, the resulting hydrolysates being mixtures containing low molecular weight linear oligomers and cyclic species. The general reaction scheme is:

$$X_2SiMeR + H_2O \longrightarrow (MeRSiO)_m + 2HX$$

$$(Me_3Si)_2O + (Me_2SiO)_n + (MeRSiO)_m \longrightarrow$$

$$Me_3SiO[Me_2SiO]_x[MeRSiO]_ySiMe_3$$

where X is a hydrolysable halide or alkoxy group and R is the organofunctional substituent. Functional siloxane monomers can also be synthesised from the appropriate

cyclic precursors by addition at unsaturated functional
precursors:

$$(MeHSiO)_4 + CH_2=CHCH_2Cl \xrightarrow{Pt} \underset{\underset{(CH_2)_3Cl}{|}}{(MeSiO)_4}$$

$$(MeVSiO)_4 + HSCH_2COOH \xrightarrow{h\nu} \underset{\underset{(CH_2)_2SCH_2COOH}{|}}{(MeSiO)_4}$$

The hydrolysis and equilibration steps are sometimes also
carried out in a one pot process where the water, silane,
cyclic monomers, end blocker and suitable catalyst are
combined at the beginning of the reaction. Clearly in such
reactions the type of catalyst is important and must be
chosen to minimise interaction with the reactive group on
the functional monomer. Thus if the final copolymer
contains amine groups then the catalyst would normally be a
strong base such as potassium silanolate whereas if thiol
groups were present then a strong organic acid would be
preferred:

$$(Me_3Si)_2O + \underset{\underset{(CH_2)_3NH_2}{|}}{(MeSiO)_m} + (Me_2SiO)_n \xrightarrow{\equiv SiOK}$$

$$Me_3SiO[MeSiO]_x[Me_2SiO]_ySiMe_3$$
$$\underset{(CH_2)_3NH_2}{|}$$

$$(Me_3Si)_2O + \underset{\underset{(CH_2)_3SH}{|}}{(MeSiO)_m} + (Me_2SiO)_n \xrightarrow{H^+}$$

$$Me_3SiO[MeSiO]_x[Me_2SiO]_ySiMe_3$$
$$\underset{(CH_2)_3SH}{|}$$

This method is not limited to polymers containing one type
of functionality nor are the structures necessarily of the
"rake"type. As long as the chemistry is compatible any
combination of functional groups and microstructure is
possible, and for organic polymer modification the
preferred structure is often the $\alpha,w-$ end blocked siloxane.
These materials have the added advantage that the end
groups are precisely defined (ie two per molecule) whereas
in the rake polymers the chemical formulae represent the
average structure and the materials will contain a normal
distribution of functional groups.

Hydrosilylation of unsaturated organics using ≡SiH containing siloxane prepolymers also serves as a useful route to functional polysiloxanes, eg.

$$HMe_2SiO[Me_2SiO]_xSiMe_2H + CH_2=CHCH_2OCH_2CH\overset{\displaystyle\diagup}{\underset{O}{-}}CH_2 \xrightarrow{\text{Pt}}$$

$$CH_2\overset{\diagup}{\underset{O}{-}}CHCH_2O(CH_2)_3Me_2SiO[Me_2SiO]_xSiMe_2(CH_2)_3OCH_2CH\overset{\diagup}{\underset{O}{-}}CH_2$$

One of the main advantages of this route is that the silicon hydride prepolymers are well characterised and readily available materials which in turn leads to well defined organofunctional products. A disadvantage is the low efficiency in the use of precious metal catalyst and the difficulty in recovering it. This is particularly true for 'dilute' systems ie. those containing a low level of functionality, but this can be partially overcome by using recyclable platinum catalysts on solid supports[6].

3 PROPERTIES OF ORGANOFUNCTIONAL SILOXANES

Although it is the intention of this paper to concentrate on the use of organofunctional siloxanes to modify organic polymers, it is worthwhile mentioning some applications of non-functional siloxanes to show how the uniqueness of these materials can be exploited in practice.

Release paper for pressure-sensitive adhesives

In this application, a cross-linked silicone film is formed on the backing paper, enabling labels, for example, to be removed from the backing paper and re-applied to another surface. The unique combination of properties which makes silicones suitable for this application are:

a) Incompatibility with the organic polymers used in pressure-sensitive adhesives,
b) Very little energy needed to activate viscous flow, making them easy to apply in very thin films and
c) Ideal surface energy to allow enough adhesion with the adhesive to make a laminate, but not too much to prevent easy parting of the silicone-adhesive laminate.

Extreme temperature elastomers

Polymers containing polyphenylmethylsiloxane, instead of just polydimethylsiloxane, are used as the base polymers for elastomers which can be used at both very high and very

low temperatures. Although these may not be the strongest
elastomers at room temperature, their usefulness lies in
being usable at temperatures where other rubbers would
either degrade rapidly, or lose their elasticity.

Personal Care Products

Silicone products are widely used in Personal Care products
such as skin and hair care preparations. The two
properties exploited in these applications are low toxicity
and unique surface properties, the latter imparting
aesthetic benefits such as "feel" and "appearance".

The value of silicones, especially polydimethylsiloxanes
(PDMS) is based on a unique combination of properties
rather than any single property. This unique feature has
been emphasised many times[5]. The properties most
important in organic polymer modification are their high
surface activity, low Tg ($\approx$ -120°C), thermal stability, low
solubility parameters and chemical and biological
inertness. Most of these characteristics are retained in
the organofunctional siloxanes and the surface properties
in particualar have been rationalised in terms of the
strength and flexibility of the siloxane bond, it's partial
ionic character and the low interactive forces between the
methyl groups[7,8]. The comparatively long Si-O and Si-C
bonds reduce steric conflicts between methyl groups on
neighbouring silicon atoms which would otherwise occur and
allow unusual freedom of rotation about the Si-O and Si-C
bonds. The partial ionic character (49%) of the Si-O bond
allows distortion of the bond angle at oxygen to relieve
such steric conflicts that do occur. This freedom of
rotation about the Si-O and Si-C bonds gives very effective
screening of the polar SiOSi backbone by the non-polar
methyl groups and enables them to adopt the lowest energy
configuration at interfaces. Thus the surface tension of
PDMS (σ = 20.4 dyn/cm at 20°C) is at least 10 dyn/cm lower
than most common organic polymers including that of the
isoelectronic polyisobutylene (σ = 33.9 dyn/cm).

PDMS has an extremely low solubility parameter which
approaches δ = 7.4 for the higher molecular weight
polymers, as the effects of end groups become
negligible[9,10]. This compares with δ = 8.5-14 for
typical organic polymers. Ordinarily this means that pure
PDMS is thermodynamically incompatible with most other
polymers and in simple blends the material will come to the
surface and leach out. However the presence of small
amounts of reactive organic

groups in the molecule permits new types of interactions to be exploited so that the organofunctional siloxanes generally have better solubility in organic polymers and solvents and their ability to co-react is enhanced. Once the copolymers have been synthesised, migration is prevented by permanent covalent bonding between the two dissimilar polymers. Copolymerisation of siloxanes with organic resins usually gives rise to microphase separation and the materials are composed of soft (siloxane) domains within a hard (organic) matrix. The bulk properties are naturally dependent on the ratio of the two phases but at levels of 10-40% siloxane modification the materials usually retain most of the mechanical strength of the continuous phase whilst their surface properties are dominated by the PDMS.

Experience has shown that groups attached to the silicone backbone display reactivity typical of the organic group. This is because in most examples[4], the functional group has a 3-carbon spacer between it and the silicone backbone which ensures that the reactive group is sufficiently far removed from silicon to confer solvolytic and thermal stability. If the substituent was attached to the carbon β to silicon then unexpectedly high reactivity would be observed resulting in anomalous properties and instability. This effect has been extensively studied[11,12] and it is now generally agreed to originate mainly from Si-C hyper-conjugation which tends to stabilise intermediate β-silyl carbenium ions, with lesser contributions from polarisation and inductive effects. Silicon compounds with substituents on the α-carbon are generally more stable than their β-substituted analogues however they are more difficult to synthesise than the propyl derivatives and represent a less practical alternative.

One of the most attractive properties associated with the chemical inertness of the dimethylsiloxane polymers is their low inherent toxicity. Silicones and some copolymers such as the silicone polyethers, are used in large volumes in topical applications in the personal care industry and the well documented[13] safety aspects of PDMS are often mirrored in it's functional counterparts[14].

4 APPLICATIONS OF ORGANOFUNCTIONAL SILOXANES IN PLASTICS

Organofunctional silicones have been extensively used to modify most classes of organic polymer[4,15,16] and some indication of the prolific activity can be gained from the scientific and patent literature. Publications have

recently been appearing at the rate of more than 1000 per
annum and are still increasing (this excludes references to
silanes and siloxanes used as coupling agents, surfactants,
in polyolefin catalyst manufacture and as low level
additives). Since the Plastics Industry is extremely
diverse and can be segmented in many ways we have chosen to
divide this discussion according to generic types and to
highlight some of the more recent and unusual aspects of
silicones used in high value materials. The references
cited are simply intended to give a flavour of the work in
silicone modified plastics and cannot be comprehensive.

Polycarbonates

Because of it's excellent combination of toughness, clarity
and thermal properties, polycarbonate (PC) is one of the
most widely used of the Engineering Resins and enjoys
growth rates of around 10% per year. Major applications are
found in the automotive, business machine, appliance and
protective screening markets. Silicone modified
polycarbonates have been known for more than 20 years[17].
Further improvements in impact strength and burning
characteristics were noted when siloxane polycarbonates
were blended into SAN compositions[18]. Several studies
have been made on the influence of micro structure on the
morphology and surface properties of PDMS-PC
copolymers[19,20].

Polyimides

Unmodified polyimides are well known for their high
temperature properties and consequent intractability. This
has encouraged researchers to modify their solubility and
melt flow characteristics to facilitate processing
operations. Amine and carboxy functional polysiloxanes
have been used to modify polyimides used in coating
electronic components. This is an area of intense activity
where improvements are also sought in impact strength,
thermal and flame resistance, moisture adsorption and
adhesion. Heat resistant siloxane polyimide copolymers
have been patented[21] with good flexibility and adhesion
to glass, in which the onset of thermal degradation occurs
at 423°C vs 340°C without the siloxane. A fourfold
improvement in notched Izod impact strength is claimed for
polyimide copolymers made with amine terminated PDMS[22].
Polyimide coatings for α-ray protection of liquid crystal
films and semiconductor devices were prepared using
bis(dicarboxyphenyl) tetramethyldisiloxane[23]. Diglyme
soluble polyimide-siloxanes have been prepared that have
high thermal stability and dielectric strength and do not

need primers for good adhesion[24]. The synthesis, properties and characterisation of various siloxane-polyimide copolymers are described in detail in several reports[25,26] and it has been shown that soluble polyimides can be formed with good adhesive and moisture resistant properties.

Polyurethanes

Silicone modified thermoplastic urethanes are important materials used in adhesives, medical devices, elastomers and coatings for textiles. Synthesis of block copolymers using a variety of diisocyanates and combinations of silicones with other soft blocks is known[27]. The use of a copolymer of aminopropyl end blocked PDMS, 1,3-diperidylpropane and isophorone diisocyanate was patented for pressure sensitive adhesive applications[28] and an electrically conductive adhesive was made from hydroxypropyl terminated PDMS, HMDI, 1,6-Hexandiol and silver leaf[29]. Silicone urethane copolymers made from hydroxybutyl end blocked PDMS, polyethylene oxide and toluene diisocyanate (TDI) were formulated with hydroxyethyl methacrylate and cured to give elastomers that are useful in the manufacture of contact lenses[30]. Silicone urethane rubber was prepared using TDI end blocked siloxane, the PDMS comprising 50% of the soft block. The rubber was claimed to have better abrasion resistance and friction properties than conventional urethane elastomers[31]. It also had less shrinkage and showed improved melt flow properties.

Epoxy Resins

Much of the recent work on modification of epoxy resins has been directed at the electronics and protective coatings industries. A huge number of patents and publications claim organofunctional siloxanes and rubbers for stress reduction in epoxy, cresol and phenol novolak resins[32-34] Silicone and silane modified epoxy coating materials have improved adhesion, weatherability, flow out, flexibility and impact resistance. Epoxy siloxane copolymer systems are capable of being designed to have one or two glass transition temperatures depending on the molecular weight of the oligomeric prepolymers[35,36].

Polyesters

Thermosetting silicone polyester resins have been used in coatings applications for many years[37]. They were amongst the first silicone organic copolymers to be commercialised and are still used in large volumes in coil,

appliance and cookware coatings today. These materials are
typically made by cooking silanol or alkoxy functional
silicone resins into the polyester process under carefully
controlled conditions. In some of the more recent examples
the use of linear siloxanes have been reported to improve
impact strength, hydrolytic stability and coefficient of
friction of thermoplastic polyesters[38] and surgical
threads have been prepared from hydroxy functional PDMS
reacted with dimethyl terephthalate and aliphatic
diols[39].

Miscellaneous Polymers

Silicone Acrylates: Silicone acrylic block copolymers have
been used in a variety high value applications including
antifouling coatings[40] and membranes for medical
devices[41]. Silicon chemistry has also been employed to
make moisture curable sealants and adhesives, to facilitate
Group Transfer Polymerisation of monodisperese acrylic
polymers and in the fabrication of contact lenses.
Polyamides: Aminofunctional polysiloxanes react with
caprolactam and adipic acid to form silicone polyamide
block copolymers which when blended with nylon 6 gave
improved low temperature properties[42]. Polysulphones:
Polysulphones are another important class of high
temperature materials which have been silicone modified.
Dialkylamine end blocked PDMS was reacted with carbinol
terminated polysulphones and the resulting block copolymers
characterised[43]. Liquid Crystal Polymers: Organofunctional silicones are important precursors in the
synthesis of polymeric liquid crystals. This is another
area of intense activity and a variety of mesogens have
been attached to the siloxane backbone. The low T_g of the
silicone allows high molecular weight LC polymers to be
constructed and yet still retain useful phase transition
temperatures. The high degree of flexibility of the
siloxane and it's high free volume also permit reasonably
fast switching times to be maintained. Polyglycols:
Although the largest uses of silicone EO and PO copolymers
are in surfactants for polyurethane foam production, EO
-PDMS copolymers treated with lithium salts have also been
studied as solid poly-electrolytes.

5 CONCLUSIONS

The preceding discussion indicates just a few of the many known examples in which siloxanes are employed to enhance the properties of organic polymers in high value applications. Some of the most convenient building blocks in copolymer synthesis are the organofunctional siloxanes, particularly the α,w-functional linear polymers and these are readily accessible using well established chemistry and common starting materials.

6 REFERENCES

1. "A Guide to Silicones in the Plastics Industry"; Dow Corning Corporation, Midland, MI 48686.
2. E.P. Plueddemann; "Silane Coupling Agents" Plenum, NY, 1982.
3. U. Deschler, P. Kleinschmit and P. Panster; Angew. Chem. Int. Ed. Engl., 1986, 25, 236.
4. T.C.Kendrick, B.Parbhoo and J.W.White; S.Patai and Z.Rappoport (eds.): The Chemistry of Organic Silicon Compounds Pt2, John Wiley, Chichester, 1989, Chapter 21, "Siloxane Polymers and Copolymers".
5. T.C.Kendrick, B.Parbhoo and J.W.White; G.Allen, J.C.Bevington, G.C.Eastmond (eds.): Comprehensive Polymer Science, Vol 4, Pergamon Press, Oxford, 1989, Chapter 25, "Polymerisation of Cyclosiloxanes".
6. G. Chandra, B.J.Griffiths; GB 1 527 598 (Dow Corning Ltd.)
7. M.J.Owen; Ind. Eng. Chem. Prod. Res. Dev., 1980, 19, 97.
8. M.J.Owen; Chemtech, 1981, 11, 288.
9. R.P.Gee; Dow Corning Corp., unpublished results.
10. A.F.M.Barton; Chem. Rev., 1975, 75, 731.
11. A.W.P. Jarvie; Organometal. Chem. Rev. A, 1970, 6, 153
12. S.G. Wierschke, J. Chandrasekhar and W.L. Jorgensen; J. Amer. Chem. Soc., 1985 107, 1496.
13. S.L.Cassidy; Manufacturing Chemist, 51, December 1989.
14. R.A. Parent; Drug Chem. Toxicol. 99, 1979, 2, 295.
15. I.Yilgor & J.E.McGrath; Adv. Polym. Sci., 1988, 86, 1.
16. R.Oda; Kagaku, (Kyoto Univ.); 43(5), 346, (1988).
17. H.A.Vaughn; J. Polym. Sci., B, 1969, 7, 569.
18. H.J.Kress, C.Linder, F.Mueller, H.Peters, D.Wittmann & J.Buekers; DE 3 628 904, (Bayer).
19. B.Hammouda; Makromol. Chem. Macromol. Symp., 1988, 20/21, 127.
20. E.R.Mittlefehldt and J.A.Gardella Jr.; Polym. Prep., 1988, 29(1), 305.
21. M.Kojima, et. al.; EP 260 833 (Hitachi).
22. E.N.Peters; EP 273,150 (General Electric).

23. Y.Yamada and N.Furukawa; JP 88 23 928 (Nippon Steel).
24. C.J.Lee; Electron. Mater. Processes, 1st Int. SAMPE
    Electron. Conf., 1987, 523.
25. N.Tsutsumi, A.Tsuji, C.Horie and T.Kiyotsukuri; Eur.
    Polym. J., 1988, 24(9), 837.
26. C.A.Arnold, J.D.Summers, D.H.Chen, Y.P.Chen and
    J.E.McGrath; Polym. Prepr., 1988, 29(1), 349.
27. G.A.Gornowicz & C.L.Lee; USP 4 631 329, (Dow Corning).
28. C.M.Leir, J.J.Hoffman, L.A.Tushaus and G.T.Wiederholt;
    EP 250 248, (3M).
29. C.Prud'Homme; EP 269 567, (Ciba-Geigy).
30. K.C.Su and R.J.Robertson; EP 267 158, (Ciba-Geigy).
31. H.Otani; JP 88 69 815, (Nippon Mectron).
32. Y.Okabe & H.Fujita; JP 88 161 014, (Sumitomo Bakelite)
33. J.S.Riffle, I.Yilgor, C.H.Tran, G.L.Wilkes,
    J.E.McGrath and A.K.Banthia; ACS Symp. Series 1983,
    No 221, 21.
34. Y.Morita and S.Shida; EP 258 900, (Toray Silicone).
35. C.H.Tran, J.S.Riffle and J.E.McGrath; Polym. Prepr.,
    1988, 29(1), 41.
36. J.L.Hendrick, D.C.Hofer, T.P.Russell, B.Haidar; Polym.
    Bull., 1988, 19(6), 573.
37. A.G.Short and J.W.White; Ullmans Encyclopedia of
    Industrial Chemistry, VCH Verlagsgesellschaft,
    Weinheim, A18, Chapter 2.11 Silicone Paints, in press.
38. T.Nakane, K.Hijikata and Y.Kageyama; EP 273 636,
    (Polyplastics).
39. H.Matsumoto et. al.; JP 88 40 560 (Kanegafuchi).
40. S.Matsuoka, H.Doi, Y.Honda; EP 273 457 (Nippon Oils).
41. A.Kishada, Y.Ikada; Koenshu-Kyoto Daigaku Nippon
    Kagaku Sen'i Kenkyushu, 1988, 44, 71.
42. D.Wittmann, U. Westeppe, O.Shlack, W.Paul and
    H.Brinkmeyer; DE 3 709 238, (Bayer).
43. N.M.Patel, D.W.Dwight, J.L.Hendrick, D.C.Webster and
    J.E.McGrath; Macromolecules, 1988, 21(9), 2689.

Aminoresins: New Uses for Old Polymers

J. R. Ebdon, B. J. Hunt, and M. Al-Kinany

THE POLYMER CENTRE, CHEMISTRY BUILDING, LANCASTER UNIVERSITY,
LANCASTER LA1 4YA

1 INTRODUCTION

Aminoresins are two-stage resins made by the reaction of
an amine or an amide with an aldehyde or, exceptionally,
with a ketone. The majority of commercially important
aminoresins are based upon either urea (1) or melamine
(2,4,6-triaminotriazine) (2), although resins based upon
other amines and amides have been described. Other amines
and amides from which resins can be made include thiourea,
formamide, acetamide, methylene diamine, hexamethylene
tetramine, guanamine, guanidine, cyanamide, dicyandiamide,
aniline, p-diaminobenzene, pyrroles, aminoacids, proteins
and urethanes.

Of the aldehydes which will react with amines or
amides to form resins, the most widely employed is
formaldehyde, HCHO. Formaldehyde is most commonly used in
the form of a 37–42% w/w solution in water, i.e., as
Formalin. In aqueous solution, formaldehyde exists mainly
in the form of methylene glycol (3) and is stabilised

NH_2CONH_2

(1)

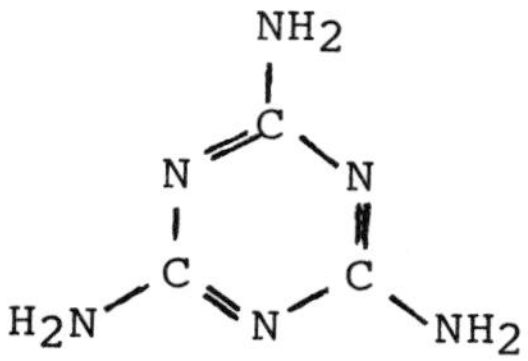

(2)

$HOCH_2OH$

(3)

against significant polymerisation through the addition of
up to 3% of methanol. Other aldehydes which have been

used to make aminoresins include acetaldehyde, malonic
dialdehyde, glyoxal, glyoxylic acid, chloral, hydroxy-
pivalaldehyde, acrolein, crotonaldehyde and a variety of
aldols. Acetone and other ketones have also been used
with amines to make resins.[1]

History of Aminoresins

The history of the two most important aminoresins,
urea-formaldehyde (U-F) and melamine-formaldehyde (M-F)
resins, is long but interesting and has been described in
detail.[2] Urea was first synthesised by Wohler in 1824 and
formaldehyde was identified by Butlerov in 1851. However,
resins formed by reaction of urea with formal- dehyde were
not investigated until 1884, by Tollens. The commercialisation of U-F resins could be said to date from 1899,
when Goldschmidt published the first important patent in
the area. Modern U-F resin technology, however, owes most
to the work of Hans John carried out in the first two
decades of this century. By 1930, U-F resins could be
said to have become a mature product with uses in surface
coatings, adhesives and as a moulding material.

Melamine was synthesised by Liebig in 1834 but
production of M-F resins did not begin until 1936. M-F
resins were used first as a moulding material; their most
important advantage over U-F resins in this application
lies in their comparatively better resistance to water and
staining.

2 CHEMISTRY OF AMINORESIN FORMATION

As mentioned above, aminoresins are two-stage resins,
i.e., their synthesis is normally accomplished in two main
stages. In the first stage, the amine (or amide) is
reacted with a controlled excess of the aldehyde in
aqueous solution under neutral or slightly alkaline
conditions and at moderate temperature to give a low
molecular weight resin. During this first stage a series
of addition and condensation reactions takes place to give
a mixture of short linear and branched chains with an
average molecular weight of the order of 10^3. The resin
may be concentrated to give a viscous liquid suitable for
use as an impregnant or as an adhesive, or may be spray-
dried to give a solid. At this stage, the resin is both
soluble and fusible although reactions will continue in
liquid resins during storage and shelf-lives of these
resins are consequently relatively short.

During the second stage, the resin is cured with the application of heat, pressure and, in the case of U-F resins, of an acid catalyst. During curing, further chain extension and also crosslinking takes place to give an insoluble, infusible product. Mouldings made from amino-resins are normally filled since the mechanical properties of unfilled cured resins are relatively poor.

Addition Reactions

The addition of an aldehyde to an amine can be cata-lysed by both acids and bases. Under acid conditions, however, addition is rapidly followed by condensation. Under neutral to basic conditions it is possible to study the kinetics of addition relatively free from inter-ference by condensation.

Addition of Formaldehyde to Urea. The chemistry of addition of formaldehyde to urea was elucidated first by de Jong and de Jonge in the 1950's using classical analytical methods.[3] The accepted mechanism for the base catalysed addition of formaldehyde to urea in aqueous solution is shown in Scheme 1 whilst that for the acid catalysed addition is shown in Scheme 2.

$$NH_2CONH_2 + \bar{B} \longrightarrow NH_2CO\bar{N}H + BH$$

$$NH_2CO\bar{N}H + HOCH_2OH \longrightarrow NH_2CONHCH_2OH + \bar{O}H$$

$$\bar{O}H + BH \longrightarrow \bar{B} + H_2O$$

Scheme 1

$$HOCH_2OH + HA \longrightarrow H_2\overset{+}{O}CH_2OH + \bar{A}$$

$$H_2\overset{+}{O}CH_2OH + H_2NCONH_2 \longrightarrow H_2NCO\overset{+}{N}H_2CH_2OH + H_2O$$

$$H_2NCO\overset{+}{N}H_2CH_2OH + \bar{A} \longrightarrow H_2NCONHCH_2OH + HA$$

Scheme 2

Urea is tetra-functional and so in all four molecules of formaldehyde may add to one of urea to give a variety of methylolmelamines. The addition reactions can be represented as a series of competitive and consecutive equilibria (Scheme 3).

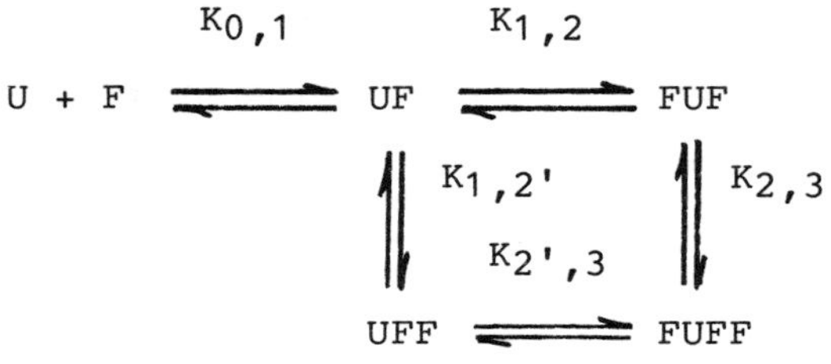

$$U + F \underset{}{\overset{K_{0,1}}{\rightleftharpoons}} UF \underset{}{\overset{K_{1,2}}{\rightleftharpoons}} FUF$$

$$\Big\Updownarrow {}^{K_{1,2'}}_{K_{2',3}} \qquad \Big\Updownarrow {}^{K_{2,3}}$$

$$UFF \rightleftharpoons FUFF$$

Scheme 3

The forward reactions are bimolecular and second order whilst the back reactions are unimolecular and first order. At low ratios of formaldehyde to urea, monomethylol- (UF, 4), N,N'-dimethylol- (FUF, 5) and N,N-dimethylolurea (FUF, 6) are the principal products whilst at higher formaldehyde to urea ratios, trimethylol-urea (FUFF, 7) is formed also. The formation of tetra-methylolurea has never been observed in practice but at high formaldehyde to urea ratios, and under strongly alkaline conditions, the cyclic product, dimethyloluron (8), has been identified.

$NH_2CONHCH_2OH$ $HOCH_2NHCONHCH_2OH$ $NH_2CON(CH_2OH)_2$

(4)   (5)   (6)

$HOCH_2NHCON(CH_2OH)_2$

(7)

$$\begin{array}{c} O \\ \| \\ C \\ HOCH_2N \diagup \quad \diagdown NCH_2OH \\ | \qquad\qquad | \\ H_2C \diagdown \quad \diagup CH_2 \\ O \end{array}$$

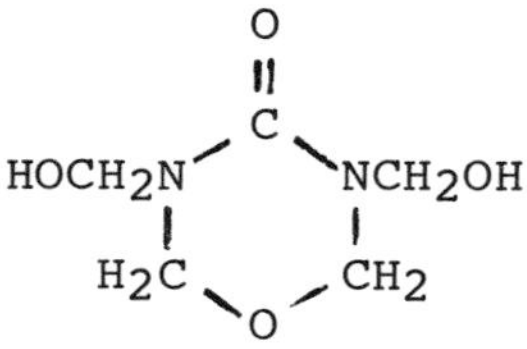

(8)

The kinetics of addition of formaldehyde to urea can be studied independently of condensation in dilute solutions and at relatively high pH. Rates of reaction and the approach to equilibrium can be conveniently studied by ^{1}H nuclear magnetic resonance (n.m.r.) in D_2O solution. In D_2O, only the protons of the methylene groups in formaldehyde and the methylolmelamines are visible; those of the primary and secondary amino groups exchange rapidly with the D_2O and are not seen. Reactions of urea with formaldehyde, or alternatively the re-equilibration of monomethylolurea or N,N'-dimethylolurea (both of which are obtainable as pure compounds), can thus be monitored continuously by [1]H n.m.r. and appropriate rate constants and equilibrium constants can be obtained. Figure 1 shows a set of typical [1]H n.m.r. spectra

corresponding to stages in the re-equilibration of a
sample of N,N'-dimethylolurea and Figure 2 shows
graphically the data obtained by integrating the peaks
within these spectra.[4]

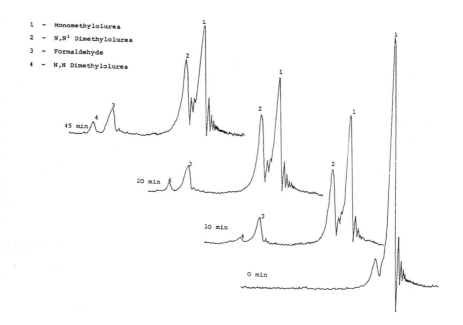

1 - Monomethylolurea
2 - N,N¹ Dimethylolurea
3 - Formaldehyde
4 - N,N Dimethylolurea

Figure 1   60 MHz ^{1}H n.m.r. spectra of re-equilibrating
monomethylolurea in D_2O at pH 9.8 and 60°C as a
function of reaction time. (Taken from
Reference 4)

Typical values for $K_{0,1}$, $K_{1,2}$, $K_{1,2'}$, $K_{2,3}$ and $K_{2',3}$
at pH 9.8 and 60°C are 32, 4, 0.5, 0.6 and 5 1 mol^{-1}
respectively.[4]   Thus, $K_{0,1} > K_{1,2} \approx K_{2',3} > K_{1,2'} \approx K_{2,3}$,
showing (i) that methylolation of a primary N takes place
more readily than of a secondary N and (ii) that methylolation of the first primary N of urea deactivates the
second.   Studies have shown that the equilibrium constants
depend upon pH, buffer concentration and temperature.   The
temperature dependences of the equilibrium constants give
$\Delta H_{0,1} = -17 \text{ kJ mol}^{-1}$, $\Delta H_{1,2} = -20 \text{ kJ mol}^{-1}$ and $\Delta H_{1,2'} =
-13 \text{ kJ mol}^{-1}$, i.e., methylolation is exothermic.[4]

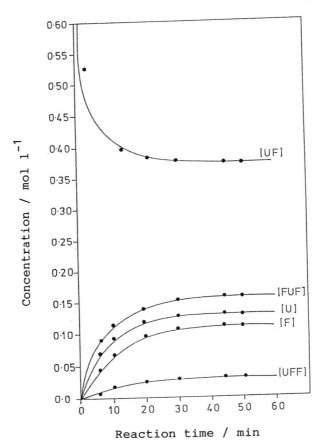

<u>Figure 2</u>   Concentrations of monomethylol- (UF),
               N,N'-dimethylol- (FUF), N,N-dimethylol- (UFF)
               and trimethylolurea (FUFF) as a function of time
               obtained by integrating the spectra shown in
               Figure 1.   (Taken from Reference 4)

        <u>Addition of Formaldehyde to Melamine</u>.   The addition
of formaldehyde to melamine takes place in a manner
similar to its addition to urea but more reactions are
possible because of the higher functionality of melamine.
In all, nine distinct methylolmelamines are formed in a
complex series of competitive and consecutive equilibria.
These addition reactions are depicted schematically in
Figure 3.

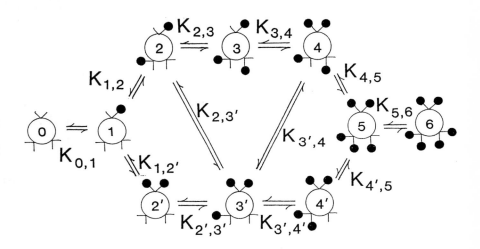

Figure 3 Addition reactions of melamine and formaldehyde depicted schematically. The larger open circles represent azine rings; the smaller filled circles represent methylol groups.

The complexity of the M-F system, coupled with the fact that only hexamethylolmelamine can be obtained as a pure compound by conventional means, makes it difficult to determine the kinetics of M-F addition by [1]H n.m.r. Instead, reverse phase high performance liquid chromatography (h.p.l.c.) has been used. Tomita obtained all the relevant equilibrium constants (at three different temperatures at pH 9.8) and also many of the constituent rate constants by monitoring by h.p.l.c. the appearance of the various methylolmelamines in (i) aqueous mixtures of melamine and formaldehyde and (ii) in aqueous solutions of hexamethylolmelamine which were allowed to re-equilibrate.[5] Additional rate constants have been obtained more recently by O'Rourke who separated from mixtures of methylolmelamines, by preparative h.p.l.c., pure samples of N,N'-dimethylolmelamine and N,N,N',N"-tetramethylolmelamine. These were re-equilibrated in water at pH 9 and at 48°C and the appearance of other methylolmelamines monitored by analytical scale h.p.l.c.[6] Equilibrium constants for M-F addition are compared, and some conclusions drawn from them, in Table 1.

<u>Table 1</u>   Comparisons of equilibrium constants for various
            stages of M-F addition at 48°C and pH 9 and
            conclusions drawn from them. (Data taken from
            Reference 6; all equilibrium constants are in
            l mol^{-1})

$$K_{0,1} > K_{1,2} > K_{2,3} > K_{3,4} > K_{4,5} > K_{5,6}$$

25   12   4.3   2.9   1.3   0.7

(Methylolation progessively becomes more difficult)

$$K_{0,1} \approx K_{2',3'} \approx K_{4',5} \approx 25$$

(Presence of tertiary N has little effect on reactivity of
neighbouring primary N)

$$K_{0,1} > K_{1,2} > K_{2,3} \text{ but } K_{1,2} \approx K_{3',4}$$

25   12   4.3   12   7

(Secondary N deactivates neighbouring primary N)

$$K_{1,2} > K_{1,2'}; \quad K_{2,3} > K_{2,3'}; \quad K_{3,4} > K_{3',4'}$$

12   0.9   4.3   1.8   2.9   0.5

(Formation of secondary N preferred over tertiary N where
possible)

Condensation and Crosslinking (Curing)

     Two types of condensation reactions are possible in
aminoresins: the mutual condensation of pairs of methylol
groups to give methylene ether links and the reaction of a
methylol group with an amino group to give a methylene
link. The former reaction has long been believed to be
favoured by high pH and by high formaldehyde to amine
ratios whilst the latter is believed to be favoured by low
pH, by relatively low formaldehyde to amine ratios and by
relatively high reaction temperatures. Reaction of a
methylol group with a secondary amino group is believed to
be the source of the crosslinking. These reactions are
depicted in Scheme 4.

$$-NHCH_2OH + HOCH_2NH- \quad --> \quad -NHCH_2OCH_2NH-$$

$$-NHCH_2OH + H_2N- \quad --> \quad -NHCH_2NH-$$

$$-NHCH_2OH + -NHCH_2NH- \quad --> \quad -NHCH_2NCH_2NH-$$

Scheme 4

Condensation and Cure in Urea-Formaldehyde Resins. A variety of techniques have been used to study condensation and curing in U-F systems. Differential scanning calorimetry (d.s.c.) has given information about the overall kinetics of cure and heats of reaction[7], whilst gel permeation chromatography (g.p.c.)[8] and h.p.l.c.[9] have allowed measurements of changes in molecular weight during condensation and even the appearance and disappearance of individual condensation products (e.g., dimers and trimers) to be monitored. [1]H, ^{13}C and [15]N n.m.r. have proved particularly useful for characterising the average structures of soluble resins.[10-12]

Cured resins are, because of their insolubility, more difficult to characterise from the structural point of view. However, Maciel and coworkers have followed cure in U-F resins by high-resolution solid-state (CPMAS) [13]C n.m.r. The spectra show the conversion of free methylol groups to methylene ether, and especially methylene, links as cure times increase and also show the development of a network structure.[13]

It is safe to say that most of the available evidence corroborates the long held views (stated above and depicted in Scheme 4) regarding the nature of condensation and curing in U-F systems and the effects upon these of urea to formaldehyde ratio, temperature and pH.

Condensation and Cure in Melamine-Formaldehyde Resins. Condensation and cure in M-F resins has been less extensively studied than in U-F systems. That M-F resins may be cured without the need for acid catalysts has led to some speculation that crosslinking may be less extensive in these resins. Of particular interest in these systems, therefore, are the extent and nature of chain extension and branching in both the soluble and the cured resins and the effects upon these of process variables.

As for U-F resins, useful information about the overall compositions and structures of soluble M-F resins can be obtained from g.p.c.,[14,15] h.p.l.c.[16] and [13]C n.m.r. analyses in particular.[17,18] [13]C n.m.r. spectra of soluble M-F resins show evidence of both methylene and methylene ether links whilst the chromatographic techniques show evidence for a range of oligomeric products.

Recently, four oligomeric fractions, X1, X2, X3 and X4, were isolated from a low molecular weight soluble M-F resin (precondensate) by preparative reverse-phase h.p.l.c. (Figure 4) and were characterised by high field ^{1}H and [13]C n.m.r. spectroscopy.[16] Fraction X1 was shown most probably to be a mixture of dimers containing mainly methylene links whilst X2 was shown most probably to be a mixture of dimers containing methylene ether links. It was concluded that X3 and X4 were most probably trimers and tetramers respectively containing both methylene and methylene ether linked species.

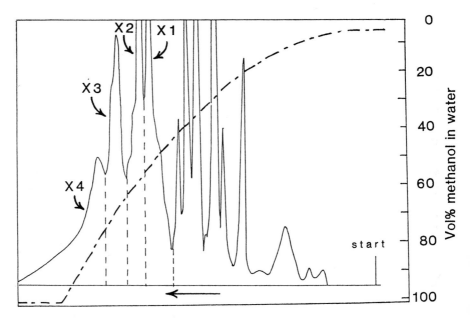

<u>Figure 4</u>   Reverse phase h.p.l.c. trace of an M-F precondensate showing peaks from oligomeric fractions (X1, X2, X3 and X4). (Taken from Reference 16)

The structures of cured M-F resins, like those of cured U-F resins, can be deduced in part by solid state (CPMAS) [13]C n.m.r. spectroscopy.[16] Figure 5 shows [13]C n.m.r. spectra of two cured M-F resins made with different melamine to formaldehyde ratios. The signals between 40 and 63 ppm arise from carbons in methylene links whilst those between 63 and 80 ppm arise mainly from methylene ether links.

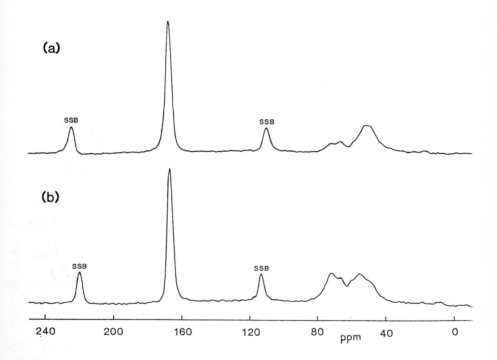

Figure 5 100 MHz solid-state (CPMAS) [13]C n.m.r. spectra of cured M-F resins made with M/F ratios of (a) 1:1.5 and (b) 1:3. SSB = spinning side-band. (Taken from Reference 16)

What is not clear from these spectra is whether or not there are many residual unreacted methylol groups. This question can be answered by studying the solid-state (CPMAS) [15]N n.m.r. spectra of the samples (Figure 6).[16]

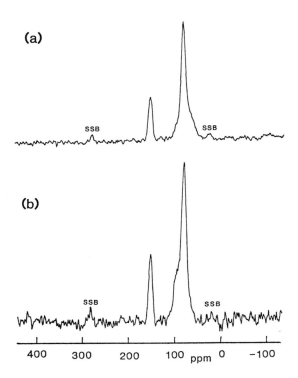

<u>Figure 6</u>   30 MHz solid-state (CPMAS) ^{15}N n.m.r. spectra of
            cured M-F resins made with M/F molar ratios of
            (a) 1:1.5 and (b) 1:3. SSB = spinning
            side-band. (Taken from Reference 16)

These spectra contain signals characteristic of azine ring
nitrogens (150 ppm) and secondary and tertiary nitrogens
in methylene and methylene ether links and at branch
points (77 and 93 ppm respectively), but no significant
signals are visible characteristic of nitrogen in -NHCH$_2$OH
and -N(CH$_2$OH)$_2$ groups (expected at 86 and 107 ppm respectively). Thus it may be concluded that M-F resins cure in
a manner similar to that of U-F resins with the formation
of methylene and methylene ether links and ultimately of a
crosslinked structure. Methylene to methylene ether
ratios appear to be controlled, like those in U-F resins,
partly at least by the amine to formaldehyde ratio in the
starting mixture.

3 TRADITIONAL USES OF AMINORESINS

Urea-Formaldehyde Resins

The major established use of U-F resins is as an adhesive, indeed roughly 70% of such resins are currently used for this purpose. Of this 70%, 60% is used for the construction of particle board in which wood flour, fibres or chips are mixed with a liquid resin and pressed into sheets and cured at high temperature. Such particle boards, because of their dimensional stability and relative cheapness compared with traditional timber, are finding increasing use in the building industry for the construction of roofs, flooring and partition walling, and by the furniture trade for the construction of flat-sided wall and floor units especially. Much of the remaining 10% of U-F resins used as adhesives is used in the construction of timber laminates of which plywood is still the most important and familiar example.[19]

Of the 30% of U-F resins not used as adhesives, 10% is used as a moulding compound (usually reinforced, for the construction especially of domestic electrical fittings), 8% is used in surface coatings (usually in butylated form to confer solubility in organic solvents), 6% is used as a paper treatment (to improve wet-strength) and 3% is used in textile treatments (especially of rayon and cotton to improve crease resistance). Of the remaining 3% not accounted for above, significant quantities are used in what might be regarded as relatively high value added applications such as in printing inks, in carriers for phosphorescent coding systems, and as encapsulants of inks in carbonless copying paper and of perfumes in 'scratch-and-sniff' test cards. A further minor use of U-F resins has been as foams for thermal insulation and as an agricultural mulch. The former application of U-F foams is however now rather discredited owing to perceived problems of water transmission when used as a cavity wall insulant in areas of high rainfall.

Melamine-Formaldehyde Resins

M-F resins are not as widely exploited as U-F resins and probably form no more than 7-8% of total aminoresin production. M-F resins are particularly useful in applications where better heat and/or water resistance are required.

The major use of M-F resins is currently for the production of decorative laminates. In such laminates, the decorative (uppermost) paper layer is laminated with a M-F resin whilst the lower layers (of Kraft paper) are laminated with a phenolic resin. M-F resins have also enjoyed uses as moulding materials but recently have been displaced from many of these uses by newer, more easily fabricated and no more expensive thermoplastics. M-F resins are often added in small quantities to U-F resins in many applications to upgrade the properties of the latter.

4 A POTENTIAL NEW HIGH VALUE USE FOR AMINORESINS

Thermosetting resins, because of their inherent cross-linkability, lend themselves, at least in theory, to any application where a network structure is desirable. Such an application is the use of polymers as chromatographic stationary phases and supports, especially for forms of liquid chromatography. The use of phenolic resins as an ion-exchange chromatographic medium is well established but aminoresins have received little attention for these or similar uses. Consequently, we have recently investigated the potential of aminoresins in macroporous bead form for use as a stationary phase in size exclusion chromatography (g.p.c.). Aminoresins seemed inherently attractive not only because they can be readily cross-linked but also because they have an affinity for water and therefore might be expected to have potential particularly for separations carried out on aqueous solutions, an area of separation science of increasing importance and where further developments of stationary phases would be timely.

Preparation of Macroporous Aminoresin Beads

We have made aminoresin beads in macroporous bead form by a reverse phase suspension polycondensation procedure starting either with a suspension of the primary reactants, amine and formaldehyde (a single step procedure), or with a suspension of a low molecular weight liquid precondensate (a two-step procedure).[20] Of the two procedures, the single step one has proved to be the more satisfactory; a typical 'recipe' for the production of beads from a U-F resin is given below.

A Typical Recipe for the Production of U-F Beads.
74 g of formaldehyde (37% solution in water) are added to 80 ml of water and the pH is adjusted to 7.5 by the

dropwise addition of sodium hydroxide solution. To the
formaldehyde solution are added 4 g of urea. This mixed
solution is then added to a suspending medium consisting
of 1,100 ml of liquid paraffin, 50 ml of petroleum ether
(boiling range 100-120°C) and 40 ml of SPAN 85 (non-ionic
surfactant) contained in a 200 ml round-bottomed flask
equipped with a 'Jiff' stirrer, nitrogen purge, temperature probe and water-cooled condenser. The mixture is
stirred at 200-220 rpm and heated at 82°C for 2 h, then
acidified to pH 5 with acetic acid. The stirring and
heating are continued for a further 16 h.

The beads are recovered by decantation or by
filtration on a Buchner apparatus. The beads are then
stirred with water, washed with acetone, and dried by
pumping in a vacuum oven at room temperature. The dried
beads are then extracted with diethyl ether in a Soxhlet
apparatus for 20 h and dried once more by pumping under
vacuum.

The beads are post-cured by intimately mixing them
with ammonium chloride (6% w/w) and heating the mixture at
105°C for 17 h. The beads are once more washed with water
and acetone and are finally dried by pumping in a vacuum
oven at room temperature.

Variations. The above recipe was found to be the
most satisfactory for the production of U-F beads of
reasonable sphericity, porosity, size distribution and
mechanical strength. However, a large number of
variations were tried in which different urea/formaldehyde
ratios, surfactant type, surfactant concentrations,
stirring speeds, reaction times, reaction volumes, pH
regimes, etc., were employed. Also, various water-
compatible substances were added to the reaction system in
attempts to change porosity: substances such as alcohols,
glymes, THF and DMSO. Also, recipes have been devised for
the production of beads from M-F resins, and also for
beads from phenol-formaldehyde, catechol-formaldehyde,
epoxy and other thermosetting resins.[21]

Characterisation of Beads

The sizes and size distributions of beads have been
assessed by light microscopy, surface characteristics of
beads by traditional and freeze-fracture scanning electron
microscopy (s.e.m.), surface areas by nitrogen adsorption
measurements, and degrees of crosslinking by solid-state
n.m.r.

Light microscope pictures of typical U-F and M-F beads are shown in Figure 7. Sizes of beads range from about 1 to 100 μm depending upon the exact conditions of preparation although always within a particular sample of beads there is a distribution of sizes, i.e., the beads are far from being monodisperse.

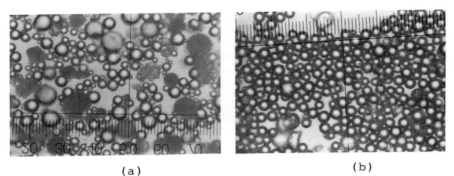

(a)   (b)

Figure 7   Light microscope pictures of typical (a) U-F and (b) M-F beads.   (Magnification x40)

Figure 3 shows s.e.m. pictures of a typical freeze-fractured U-F resin bead. The bead clearly has a complex morphology with numerous internal voids and fissures. Nitrogen adsorption measurements of such beads reveal surface areas of up to 300 m^2 g^{-1}, i.e., comparable with those found for other porous substances such as silica gel.

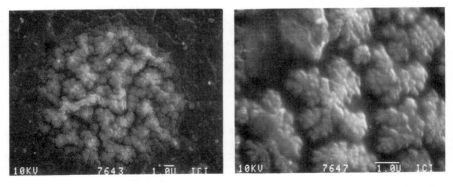

Figure 8   S.e.m. pictures of a freeze-fractured U-F bead

Performance of Beads in Size Exclusion Chromatography

To assess their potential as stationary phases for size exclusion chromatography, samples of U-F and M-F beads were packed (using a 14% w/v slurry of beads in methanol) into steel columns of internal diameter 4.6 mm and length 15 cm. These columns were end-capped and then attached in turn to a Waters Gel Permeation Chromatograph. Samples of polymer standards in various solvents were then passed through the columns at a rate of up to 1 ml min^{-1} and retention volumes were measured. From these data, calibration curves (plots of log$_{10}$ molecular weight vs. retention volume) were plotted and thus the abilities of the beads to act as size exclusion media were assessed.

Performance of U-F Beads. Figures 9 and 10 show the calibration curves for the separation of some poly-(ethylene glycol)s on columns packed with U-F resin beads and a commercial polyacrylamide gel (Polymer Laboratories 'Aquagel') respectively, using water as the mobile phase. It is clear that both packings are capable of resolving poly(ethylene glycol)s over the molecular weight range $10^2 - 10^4$. However, the calibration curve is slightly sigmoidal in the case of the column packed with the U-F beads indicating the presence within these beads of a non-uniform range of pore sizes. Also the range of volumes over which the poly(ethylene glycol)s elute is smaller in the case of the U-F packing compared with that for the Aquagel indicating that the overall pore volume is lower in the case of the former.

Figures 11 and 12 show calibration curves for the separation on the U-F packing of poly(ethylene oxide)s in water and of polystyrenes in THF. It is clear from these last two Figures that the U-F beads perform equally well in a polar organic solvent and in water. The reason for this probably lies in their macroporous nature, i.e, the beads are extensively crosslinked and therefore compara-tively rigid and so do not rely upon swelling to achiev? their porosity.

Performance of M-F and other Beads. The performances of M-F and other beads as size exclusion media when packed into columns have been disappointing when compared with the performance of the U-F packing. The porosities of beads other than those from U-F resins are seemingly rela-tively low. That U-F resin beads should apparently have high porosities we attribute to the hydrophilicity of U-F resins which is greater than those of M-F, phenolic and other resins. Thus in the suspension polycondensation of

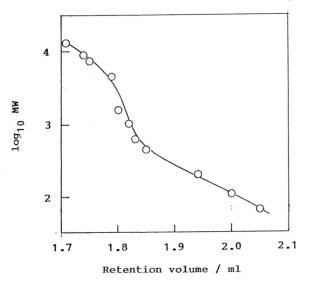

<u>Figure 9</u>   Calibration curve for the separation of poly-
              (ethylene glycol)s on a g.p.c. column packed
              with U-F resin beads.

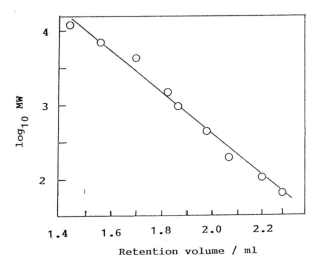

<u>Figure 10</u> Calibration curve for the separation of poly-
              (ethylene glycol)s on a g.p.c. column packed
              with Polymer Laboratories 'Aquagel'.

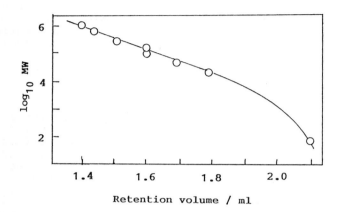

Figure 11 Calibration curve for the separation of poly-(ethylene oxide)s on a g.p.c. column packed with U-F resin beads.

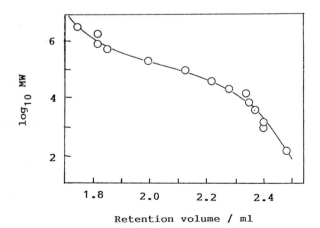

Figure 12 Calibration curve for the separation of poly-styrenes on a g.p.c. column packed with U-F resin beads.

U-F, the final beads are probably not significantly smaller than the initial suspended droplets since the resin retains much of its water-compatibility to beyond the point where it has crosslinked sufficiently to become rigid. However, in the cases of M-F and other resins, the resin beads probably become hydrophobic before cross-linking has advanced very far; thus the beads shrink, exuding water as they do so, and lose much of their porosity.

5 CONCLUSIONS

Aminoresins are well established materials with a range of valuable but essentially 'low tech' uses. Their inherent crosslinkability, however, and the ability to fashion them in porous bead form, renders them suitable for applications as chromatographic supports. So far, these applications have been explored only slightly but results to date encourage further work in this area. Not investigated so far is the potential of aminoresin beads to act as reactive supports; the beads can be expected to have reactive amino (and perhaps methylol) groups on their outer surfaces which could serve as anchor points for other species.

ACKNOWLEDGEMENTS

We thank the following for their work on various aspects of aminoresin preparation and characterisation at Lancaster: Mike Dawbarn, Steve Hewitt, Simon Bunce, Peter Stratford, Balraj Kalirai and Jim Bishop who, as undergraduate students, carried out valuable practical and literature projects in the area; Peter Heaton, John Parkin and Bill O'Rourke who, as postgraduate students, carried out considerable work on aminoresin chemistry for their PhDs; and Dr Tom Huckerby, a colleague on the staff at Lancaster, who has rendered valuable assistance with n.m.r. spectroscopy.

Thanks are also owed to the Science and Engineering Research Council, Formica Research, Blagden Chemicals and the Lancaster University Research Fund for financial support.

REFERENCES

1. B. Meyer, 'Urea-Formaldehyde Resins', Addison-Wesley, London, 1979.

2. Reference 1, Chapter 2.

3. J.I. de Jong and J. de Jonge, <u>Rec. Trav. Chim.</u>, 1953, <u>72</u>, 139 and earlier papers.

4. P.E. Heaton, PhD Thesis, University of Lancaster, 1977.

5. B. Tomita, <u>J. Polym. Sci., Polym. Chem. Edit.</u>, 1977, <u>15</u>, 2347.

6. W.T.S. O'Rourke, PhD Thesis, University of Lancaster, 1984.

7. A. Sebenik, U. Osredkar, M. Zigon and I. Visovisek, <u>Angew. Makromol. Chem.</u>, 1982, <u>102</u>, 81.

8. See for example, P. Hope, B.P. Stark and S.A. Zahir, <u>Brit. Polym. J.</u>, 1973, <u>5</u>, 363.

9. K. Kumlin and R. Simonson, <u>Angew. Makromol. Chem.</u>, 1981, <u>93</u>, 27 and earlier papers.

10. C. Duclairoir and J.-C. Brial, <u>J. Appl. Polym. Sci.</u>, 1976, <u>20</u>, 1371.

11. J.R. Ebdon and P.E. Heaton, <u>Polymer</u>, 1977, <u>18</u>, 971.

12. J.R. Ebdon, P.E. Heaton, T.N. Huckerby, W.T.S. O'Rourke and J. Parkin, <u>Polymer</u>, 1984, <u>25</u>, 821.

13. G.E. Maciel, N. Szeverenyi, T. Early and G. Myers, <u>Macromolecules</u>, 1983, <u>16</u>, 598.

14. D. Braun and V. Legradic, <u>Angew. Makromol. Chem.</u>, 1974, <u>36</u>, 41 and earlier papers.

15. T. Matsuzaki, V. Inoue, T. Odkuba and S. Mori, <u>J. Liq. Chrom.</u>, 1980, <u>3</u>, 353.

16. J.R. Ebdon, B.J. Hunt, W.T.S. O'Rourke and J. Parkin, <u>Brit. Polym. J.</u>, 1988, <u>20</u>, 327.

17. H. Schindlbauer and J. Anderer, <u>Angew. Makromol. Chem.</u>, 1979, <u>79</u>, 157.

18. B. Tomita and H. Ono,
    J. Polym. Sci., Polym. Chem. Edit., 1979, 17, 3205.

19. Reference 1, Chapter 11.

20. J.R. Ebdon, B.J. Hunt and M. Al-Kinany, Brit. Pat.
    Appl., 8715020, June 1987.

21. J.R. Ebdon, B.J. Hunt, M. Al-Kinany and J. Bishop,
    work to be published.

Liquid Crystalline Side-chain Polymers in Technology: Present Trends and Future Developments

G. S. Attard

DEPARTMENT OF CHEMISTRY, THE UNIVERSITY, SOUTHAMPTON SO9 5NH

1 INTRODUCTION

Organic and organometallic materials whose constituent molecules are highly non spherical (eg. rod-like, disc-like, or lath-like) often exhibit a complex melting/crystallization behaviour. The reason for this lies in the molecular shape and can be understood intuitively by considering the idealised situation in which the molecules are cigar-shaped. In the crystalline state the molecules are translationally ordered, forming a well-defined long-range lattice structure. In addition the molecules are orientationally ordered throughout the sample, lying essentially parallel to one another. As a consequence of the long-range spatial and orientational order the crystal phase is anisotropic, which means that measurements of physical properties, such as refractive index, will yield different values depending on the direction of the measurement relative to the ordering direction. By contrast, the liquid state has no long-range orientational or translational order and is therefore said to be isotropic. On cooling the liquid phase it is possible to envisage a phase transition from the isotropic liquid into a phase in which the molecules lie, on average, parallel with each other, but with a random distribution of their centres of mass. Such a phase would be both fluid and anisotropic (the latter due to the long-range orientational order). On further cooling a transition from this phase into one in which the orientationally ordered molecules tend to lie in planes with a random distribution of their centres of mass within each plane, and no positional correlations across planes, can be imagined. Cooling such phases further could produce phases with increasing degrees of spatial ordering until a final transition to the crystal phase. On heating the crystal the reverse sequence of transitions would occur. The fluid anisotropic phases that occur between the crystalline and the isotropic phases as a function of temperature have been observed in a large variety of organic and organometallic systems, and are collectively referred to as liquid crystalline phases or mesophases[1,2] The mesophase with the simplest ultrastructure is the one with orientational order and no translational order, and is called a nematic phase. Mesophases which have translational order in addition to the orientational order are collectively known as smectic phases. The molecular organisation in nematic and smectic A phases is illustrated in Figure 1. Smectic phases with differing degrees of translational order are distiguished by postscripts A, B, C, etc.; the progression down the alphabet denoting increasing degrees of translational order[3].

The unusual combination of fluidity and orientational anisotropy encountered in liquid crystalline phases is not only a source of fascination and challenge to condensed matter physicists, but is also the basis for their exploitation in technology.

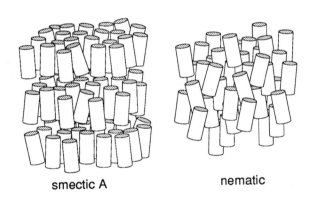

smectic A   nematic

FIGURE 1

The most commonly encountered use of low molar mass liquid crystals in technology is in display devices such as digital watches or the flat panel screens of lap-top computers. In this type of application the ability of the orientationally ordered phases to respond to electric fields by becoming uniformly aligned over macroscopic distances is exploited. Associated with such a realignment is a change in the optical properties of the materials due to the anisotropic nature of the phase. Although on a microscopic scale, say over 10^3 to 10^6 molecules, long range orientational order is present, the direction of preferred alignment, denoted by a unit vector called the director, varies randomly on a macroscopic scale. As a consequence of this, and of the defects present where there are discontinuous changes in the director, liquid crystalline phases are highly scattering and have an opaque appearance. If an external field (eg. a magnetic or electric field) is applied to a mesophase it induces a realignment of the directors to produce a macroscopically aligned sample characterised by a uniform director. When viewed along the director an aligned phase appears isotropic, ie. clear, such that if it is placed between crossed polarizers it will appear black. This type of macroscopic alignment is called homeotropic alignment. Conversely a sample whose director is aligned perpendicular to the viewing direction is said to be planarly (or sometimes homogeneously) aligned. To achieve a display device a mesophase having the required combination of viscosity and dielectric constants is sandwiched between two glass plates bearing patterned transparent electrodes, and which have been treated to induce a director alignment parallel to the surface (i.e. planar alignment). The top plate is rotated by 90° with respect to the bottom plate to achieve a uniform 90° twist in the director across the cell. This arrangement allows the transmission of light through the cell when the device is placed between crossed polarizers. Areas addressed by an a.c. electric field are realigned to become homeotropic to produce a black pattern against a bright background[4].

In addition to the very large number of low molar mass compounds which have been shown to exhibit liquid crystalline phases as a function of temperature, it is also possible to observe mesophases in polymeric systems. The same structural principles which account for the existence of the fluid anisotropic phases in monomeric materials apply to the polymeric materials, as illustrated in Figure 2. The required anisometry in molecular shape is achieved by incorporating rod-like or disc-like groups as an integral part of the polymer backbone (leading to so-called main-chain polymers), as pendant groups off the polymer (side-chain polymers), or in both the backbone and the side-chain (mixed main- and side-chain polymers)[5,6].

Main-chain polymers, which have found applications as structural materials due to their high tensile strength when in fibre form, will not be discussed here. Liquid crystalline side-chain polymers have considerable potential as high-value polymers because they are essentially hybrid systems, in that they exhibit the useful electro-optic properties of their monomeric counterparts, albeit on a much slower timescale, while at the same time showing many of the physical properties, such as mechanical integrity and glass-phase formation, of conventional plastics. This combination of properties can be exploited in a number of ways, for example, in more economic processing/enhanced performance over current materials, and in the realisation of new technologies such as ultrafast optical information processing based on non linear optical effects. The versatility of liquid crystalline polymers will be illustrated by their potential advantages in three areas of technology, namely optical information storage, non linear optics, and chromatography.

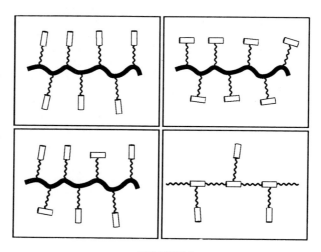

FIGURE 2

The market for materials that enable the storage or static display of optical information encompasses such diverse products as photographic film, microfiche, projector slides, and compact discs, all of which have been developed to a stage whereby they meet the standards set by highly demanding and critical consumers. A similar situation applies to stationary phases used in chromatographic separations, although in this case the market is more specialised. Thus in order for new materials to pose a serious challenge to materials that are currently available, they must satisfy one of two criteria. Either they must provide the same level of performance as existing systems but at a reduced processing cost, or they must enable a new line of products for which a realistic market niche has been identified. It is against these yardsticks that the commercial potential of liquid crystalline polymers must be measured.

2 LIQUID CRYSTALLINE POLYMERS FOR OPTICAL INFORMATION STORAGE

Given their unusual combination of properties and relatively low cost, liquid crystalline polymers could provide viable alternative solutions at the high technology end of the laser-addressed high-denisty information storage market, i.e. as novel media for microfiche or compact discs.

The earliest demonstrations of the use of liquid crystalline polymers as laser-addressed optical information storage media were based on a mechanism analogous to that used to display information in low molar mass liquid crystal devices, and is illustrated in Figure 3. Typically, a polymer, which contains dye molecules that absorb strongly at the laser wavelength, is sandwiched between two glass slides coated with a transparent conducting layer. A homeotropic monodomain is produced across the device by using an a.c. field. The blank medium thus appears optically clear (or dark when viewed between crossed polarizers). Irradiation of the material by a focused laser beam results in localised heating of the polymer, raising its temperature to above the clearing temperature. As the addressed area cools through the clearing temperature the liquid crystal phase forms with a random director distribution. As a result the addressed area appears opaque against a clear background[7].

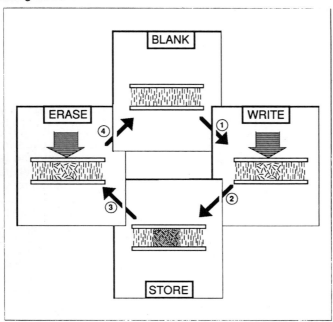

FIGURE 3. Write/Erase sequence in liquid crystalline polymer medium.

The contrast of the device is further enhanced if it is viewed between crossed polarizers, in which case the written data will appear white against a dark background. Selective erasure of information can be achieved by addressing a written area with the laser, and allowing the locally heated regions to cool in the presence of an aligning a.c. field to achieve the original homeotropic director distribution. Studies[8] have shown that liquid crystalline polymer devices of this type can sustain up to 10^4 to 10^5 erase-write duty cycles. Indeed it would appear that the fatigue characteristics are due to degradation of the dye rather than of the polymer itself.

Siloxane side-chain polymers for laser-addressed
optical information storage

A useful feature of many of these materials is that written information remains stored at temperatures well above the glass transition temperature, and can only be fully removed by heating into the isotropic phase. Thus the critical material parameters are a clearing temperature of between 50°C to 80°C, and a large positive dielectric anisotropy at a.c. frequencies of between 1kHz to 10kHz. Furthermore, the viscosity of the material must be such that cell filling times are realistic, and this entails selecting materials with low glass transition temperatures (<12°C). Because of the large a.c. fields (>50V) required to induce director alignment in these systems it is essential that trace ionic impurities are rigorously excluded from the polymer to minimise the probability of dielectric breakdown. Several prototype devices have been demonstrated using liquid crystalline poly(hydrogen methyl siloxane) derivatives and poly(hydrogen methyl siloxane/dimethyl siloxane) derivatives[7,8].

Although this type of device has many attractive features, namely high-contrast, high-denisity, and updatable storage, it has some serious disadvantages from the point of view of commercial exploitation. In particular the high voltages and laser powers required are incompatible with existing compact disc technology, and so would necessitate the development of dedicated hardware. Furthermore the use of polarizing films and the long cell-fill times (hours) add to the production costs.

An ingenious method of achieving information storage in liquid crystalline polymers has been demonstrated by workers at GEC Research[9], and relies on the contrast between the naturally highly scattering texture of the macroscopically disordered liquid crystalline phase and the clear appearance of the isotropic phase. Materials for use in this type of application are designed to exhibit a clearing temperature that is close to, but higher than, the glass transition temperature of the polymer. The area of material irradiated by a focused laser beam absorbs energy and its temperature increases to above the clearing point. Dye molecules which absorb strongly at the laser wavelength are usually added to the polymer or incorporated as pendant groups to enhance the energy transfer. When the radiation is removed rapid cooling of the addressed area occurs. Because of the proximity of the clearing temperature to the glass transition temperature the system requires time to achieve thermodynamic equilibrium, and providing the cooling rate is faster than some critical rate the irradiated spot will go directly into an amorphous isotropic glass without forming the liquid crystalline phase. Since in the amorphous glass the material is in a metastable state it is necessary for the glass transition temperature to be much higher than the working temperature of the medium. This prevents relaxation effects leading to the formation of a liquid crystalline glass. In order to erase information the written area is addressed by the laser and allowed to cool sufficiently slowly for the liquid crystalline phase to form prior to the onset of the glass transition. From a practical point of view selective erasure of information is not feasible since it is difficult to achieve sufficient control over the cooling rate under normal operating conditions. Thus this type of device is essentially a WORM (write once read many times) device which, however, has the advantage of being potentially reusable.

The GEC device has several attractive features. Because of the recording method it does not require electrodes, and so its embodiment can take several forms, for example, a free-standing film, or a supported thin film. This enables the media to be designed for compatibility with existing read-out systems. The processing of thin films or free-standing films relies on available and reliable technology which would make production costs attractive. The achievable contrast and information density compare favourably with those of commercially available microfiche products. Furthermore, in contrast with currently available media, liquid crystalline side-chain polymers do not require wet processing, thereby affording a further and potentially significant reduction in production costs.

3 LIQUID CRYSTALLINE POLYMERS FOR NON LINEAR OPTICS

The interaction of light with matter can be described in terms of the polarization induced by the electric field component of the radiation. For low intensity light this polarization is proportional to the electric field vector, and the proportionality constant is a second rank tensor called the linear susceptibility $\chi^{(1)}$

$$\mathbf{P} = \chi^{(1)}.\mathbf{E} \tag{1}$$

$\chi^{(1)}$ is related to the refractive index of the material. This simple relationship is no longer valid for high-intensity radiation, such as laser radiation, and the induced polarization contains contributions from higher powers in $\mathbf{E}$

$$\mathbf{P} = \chi^{(1)}.\mathbf{E} + \chi^{(2)}.\mathbf{E}.\mathbf{E} + \chi^{(3)}.\mathbf{E}.\mathbf{E}.\mathbf{E} + \ldots \tag{2}$$

The coefficients $\chi^{(2)}$ and $\chi^{(3)}$ are hypersusceptibility coefficients (they are third and fourth rank tensorial quantities respectively). Since $\mathbf{E}$ is a periodic function of frequency ω, then the optical field in the z direction can be written as

$$E_z = E_o \cos\alpha \tag{3}$$

where $\alpha = (-k_z\omega t)$. Substitution into equation (2) yields an expression for the frequency dependence of the induced polarization in the z direction

$$P = \chi^{(1)}E_o\cos\alpha + (1/2)\chi^{(2)}E_o^2(\cos 2\alpha + 1)$$
$$+ (1/4)\chi^{(3)}E_o^3(3\cos\alpha + \cos 3\alpha) + \ldots \tag{4}$$

From equation (4) it can be seen that the effect of the hypersusceptibility coefficients leads to the polarization having frequency components at ω, 2ω, and 3ω. As a consequence of this, radiation whose frequency is the second and third harmonic of the incident frequency can be obtained from an incident beam with a fundamental frequency of ω.

The ability to generate harmonics at twice or three times the frequency of the incident beam is of great interest in technology[10]. For example, the information density that can be achieved in laser addressed storage devices is proportional to $(1/\lambda^2)$[11], where λ is the wavelength of the laser. Thus an information read/write system operating at wavelengths in the visible region of the spectrum would enable a greater storage density than a system operating in the near infra red. However lasers emitting in the visible region of the spectrum are bulky, expensive, and not readily amenable to the miniaturisation demanded by a mass market. By contrast solid-state lasers, which emit in the near infra red are compact, cheap to manufacture and readily available. Indeed they form the basis for the readout system of currently available compact disc players. Materials capable of converting efficiently near infra red radiation into visible radiation, say via second harmonic generation, would thus offer an attractive and profitable route to achieving commercially viable high density systems. A similar situation exists in telecommunications, where the amount of information that can be launched down an optical fibre is limited by wavelength.

Because of the technological importance of achieving efficient second harmonic generation a wide range of materials have been evaluated as second order non linear optical media. Of these, inorganic crystals such as lithium niobate are commercially available. They suffer from two major drawbacks. Firstly their second order hypersusceptibilities are too small to be of practical use in the frequency conversion of solid state laser radiation. Secondly they have low radiation damage thresholds, which severely limits their reliability and operational lifetime. By contrast organic materials have been shown to exhibit large second order coefficients[12], arising from a combination of electron delocalisation in conjugated aromatic systems, and intramolecular charge-transfer states due to electron donating and electron accepting groups being located at either end of the delocalised system.

Mesogenic polymers with non linear optical chromophores

For organic systems an equation analogous to (2) can be written which describes the polarization induced in a molecule by the incident radiation

$$P = \alpha.E + \beta.E.E + \gamma.E.E.E + \ldots \quad (5)$$

where the coefficients β and γ, which are molecular parameters, are termed hyperpolarizabilities. The coefficient α is the familiar (linear) polarizability. Because of their hybrid properties, side-chain polymers with n.l.o. chromophores are in principle amenable to processing into the miniaturised geometries desired for integrated optical elements. This is a feature not shared by organic single crystals.

The link between the macroscopic non linear optical coefficients and the molecular hyperpolarizabilities is given by statistical averaging over the orientations of the molecules in the sample

$$\chi^{(n)} = <\mathbf{R}_{[n+1]}>\zeta_n \qquad\qquad (6)$$

where n denotes the order of the coefficient (ie. n= 1, 2, 3,...), ζ_n is the molecular hyperpolarizability (ie $\zeta_n \equiv \beta$ for n=2 etc.), $\mathbf{R}$ is the rank (n+1) rotation matrix which defines the transformation of the components of an (n+1)th rank tensor from its molecular co-ordinate system to the components of an (n+1)th rank tensor in a laboratory reference co-ordinate system. The angular brackets denote an ensemble average over all orientations. It is the relationship shown in equation (6) which lies at the heart of the potential of liquid crystalline materials as non linear optical media, since their inherent long range orientational order can lead to an enhancement of the macroscopic hypersusceptibilities through anisotropic averaging.

Since the contribution of the hypersusceptibilities to the optical-field induced polarization decreases with increasing rank of the susceptibility, $\chi^{(2)}$-active materials are of primary interest for applications involving harmonic generation. However $\chi^{(2)}$ is a third rank tensor and is therefore not symmetric with respect to a mirror plane perpendicular to its principal axis. Since liquid crystalline and isotropic phases are centrosymmetric media they are $\chi^{(2)}$ inactive. They can, however, be processed into non-centrosymmetric phases by the action of a d.c. field. The effect of a d.c. field is to bias the centrosymmetric distribution of dipoles to give an effective macroscopic dipole moment in opposition to the applied field. In the case of polymeric materials their ability to form glassy phases can be exploited to advantage. Typically a liquid crystalline side-chain polymer is cooled from the isotropic phase in the presence of an a.c. field in order to produce a monodomain (i.e. a phase with a uniformly aligned director). Then, at temperatures close to the glass transition temperature a large d.c. field is applied to the material to bias the dipole distribution, and the material is cooled to below T_g with the field applied. The co-operative cessation of motion which occurs at the glass transition temperature results in the biased dipole orientations being locked into the glassy phase.

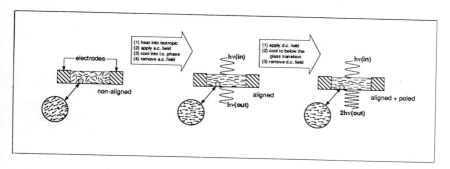

FIGURE 3. Poling of liquid crystalline medium

The acentrosymmetry created in a liquid crystalline phase will be proportional to the magnitude of the d.c. field employed and to the magnitude of the molecular dipole moments. Consequently, the magnitude of $\chi^{(2)}$ will also depend on these quantities. The relationship between the principal component of the macroscopic tensor and the principal component of the molecular tensor is given by[13]

$$\chi_{zzz}^{(2)} = \{(3/5)<P_1>^{(E)} + (2/5)<P_3>^{(E)}\} N\beta_{zzz} \qquad (7)$$

The quantities $<P_L>^{(E)}$ are ensemble averages over Legendre polynomials, and are the field-induced odd-rank orientational order parameters which define the extent of polar ordering in the system. If we assume that the electric field does not significantly affect the apolar orientational order, characteristic of liquid crystalline phases, then the orientational averages in equation (7) can be related to the even rank (field free) order parameters of the system by[13]

$$\chi_{zzz}^{(2)} = \{(1/5) + (4/7)<P_2> + (8/35)<P_4>\} N\beta_{zzz}(\mu E/kT) \qquad (8)$$

In practice the assumption of linear response appears to be reasonable for the poling fields typically used ($\sim 10^7$-$10^8 Vm^{-1}$)[14].

For a perfectly ordered liquid crystal $<P_2>$ and $<P_4>$ both equal one and equation (8) reduces to

$$\chi_{zzz}^{(2)} = N\beta_{zzz}(\mu E/kT) \qquad (9)$$

More realistic values for the order parameters ($<P_2>=0.65$ and $<P_4>=0.1$) give

$$\chi_{zzz}^{(2)} \sim (3/5)N\beta_{zzz}(\mu E/kT) \tag{10}$$

By contrast, in the isotropic phase both order parameters vanish and equation (8) reduces to

$$\chi_{zzz}^{(2)} = (1/5)N\beta_{zzz}(\mu E/kT) \tag{11}$$

A comparison of equations (10) and (11) shows that the use of perfectly ordered liquid crystalline phases can result in a fivefold enhancement of the second order macroscopic hypersusceptibility. More realistically, a threefold enhancement can be expected.

Based on the insights provided by equations (10) and (11) one may derive guidelines for the design of liquid crystalline systems showing maximal enhancement of the second order effects. Thus materials with large values of $\mu\beta_{zzz}$ and values of $<P_2>$ approaching 1 would be expected to have the best performance. Values of $<P_2>$ >0.8 are typically found in smectic phases. So the ability of a polymer to form one or more smectic phases would appear to be advantageous. However it should be noted that obtaining monodomains in smectogenic systems through the use of a.c. electric fields is not trivial, and often requires either cooling from the isotropic phase, or repeated cycling through the isotropic phase, with a large field (50-200V r.m.s.) applied.

The theoretical treatment outlined above is based on a very important assumption, namely that the dipoles are independent of each other. It is well known from dielectric studies of polar liquids that extensive dipole-dipole correlations occur in condensed phases, with the dipoles prefering an antiparallel arrangement. This type of association will clearly hinder efficient poling of highly dipolar materials and consequently lead to lower performance than theory might lead us to expect. The tendency for dipole-dipole associations among the highly polar moieties that are typically used as n.l.o. chromophores must be minimised in order to achieve high-performance materials. One route to minimising dipole-dipole associations is to introduce bulky lateral substituents (e.g. methyl groups) in the mesogenic core. However this has the undesireable effect of decreasing the degree of ordering, since it also decreases the anisotropic interactions that stabilise the mesophase. Another solution to the problem involves the use of copolymers in which the n.l.o. chromophores are physically separated from each other. Since $\chi^{(2)}$ is related to the concentration of chromophores this route is also of limited value. A more subtle approach to the problem is to design chromophores with a smaller permanent dipole moment than is found in conventional n.l.o. chromophores (e.g. by substituting a terminal nitro group by a fluorine atom). For smectogenic systems this leads to the possibility of inducing a bilayer-type arrangement of the molecules instead of the interdigitated arrangement favoured by highly polar cores. A decrease in μ would thus be compensated by retaining the high order characteristic of smectic phases.

In addition to the design restrictions imposed by the requirement to minimise dipole-dipole interactions, further restrictions are imposed by operating conditions and by the processing conditions involved in achieving monodomains and polar ordering. Since a.c. field induced alignment frequently requires large voltages, materials should have a low clearing temperature in order to minimise the probability of dielectric breakdown, and hence increase production efficiency. Excluding thermally or oxidatively unstable materials, and assuming that stringent precautions have been taken to exclude ionic impurities, in itself a major task, materials with clearing temperatures <130°C are desireable. This requirement has important implications on the type of n.l.o. chromophore that can be employed, since, as a rule of thumb, the longer the mesogenic core the higher the clearing temperature. Cores consisting of more than two aromatic units linked by an alkene bond are thus likely to be of little practical value. Furthermore, it is essential from a device reliability point of view for the normal operating temperature to be well below the glass transition temperature, particularly when the possibility of radiation-induced heating is taken into consideration. This ensures that the polar ordering does not relax significantly over the operational lifetime of the device (10 years). For devices operating at room temperature a glass transition of >80°C is desirable.

A more serious drawback of virtually all side-chain polymers with non linear optical chromophores (including non mesogenic systems) is their absorbance in the visible region of the spectrum; a consequence of the extended conjugation characteristic of most chromophores. This would lead to the second harmonic radiation being absorbed by the material. In principle the λ_{max} of a chromophore can be shifted to lower wavelengths by a suitable choice of substituents on the conjugated system while still retaining a high β. Realistically however, a sufficiently low extinction coefficient (in the condensed phase) cannot be achieved below ~500nm. This means that the materials can only be used for harmonic generation from fundamental radiation with wavelengths >1.0μm.

Ultimately it should be recognised that the best liquid crystalline side-chain polymer for non linear optics will represent a compromise between the various conflicting requirements outlined above. In spite of these limitations it is likely that liquid crystalline side-chain polymers with values of $\chi^{(2)}$ of the order of 10^{-6} esu will be achievable. The commercial viability of such a material or group of materials, either for second harmonic generation or for electro-optic modulation applications, will then depend on the level of optical attenuation in waveguides fabricated from them. Unfortunately the few results reported in the literature suggest that liquid crystalline materials have unacceptably high optical losses (>8dB cm^{-1}). These losses arise primarily from fluctuations in the director orientations which are frozen into the glassy state leading to optical inhomogeneity. Some improvement can be expected for highly ordered smectogenic materials. However it is unclear to what extent these loss levels can be reduced.

4 LIQUID CRYSTAL POLYMERS FOR CHROMATOGRAPHY

The basis of chromatographic separation is the partitioning of an analyte between a mobile phase (which is an inert carrier gas in the case of gas chromatography) and a stationary phase. The stationary phase can be a liquid adsorbed on a solid substrate, an organic species bonded to a solid surface, or a solid. The supporting substrate is usually an inert material (e.g. a diatomaceous earth) with particle size in the range of 5 to 10μm[16]. As the anaylte molecules are carried along a column packed with particles coated with the stationary phase an equilibrium is established between the concentration of molecules in the gaseous state and in the stationary phase. Differences in the free energies of mixing arising from entropic effects, enthalpic effects, specific chemical interactions, or a combination of these, lead to materials with different chemical structures having dissimilar partition coefficients into the stationary phase. This is reflected in their retention times, and their sequential elution out of the column. Since the existence of the liquid crystalline state of matter is intimately linked to molecular anisometry, it is not surprising that the thermodynamic factors which govern the operation of liquid crystalline stationary phases are dominated by shape selectivity.

Solute-solvent interactions in low molar mass liquid crystalline materials have been the subject of extensive studies over the past twenty years or so[17,18]. The orientational energy of a solute molecule (B) in a solvent (A) can be written, within a simple molecular field approach, as

$$U_B(\beta) = \{(1-x)u^{BA}<P_2>^{(A)} + xu^{BB}<P_2>^{(B)}\} \, P_2(\cos\beta_B) \qquad (12a)$$

where x is the mole fraction of solute, u^{AB} is the strength of the anisotropic orientational interactions of a solute molecule with the molecular field of the solvent, u^{AB} is the strength of the interaction between a solute molecule and the moleculer field due to other solute molecules, $<P_2>^{(B)}$ and $<P_2>^{(A)}$ are the second rank orientational order parameters of the solute and solvent respectively, and $P_2(\cos\beta_B)$ is the second rank Legendre polynomial for a solute molecule. More sophisticated models make use of volume fractions rather than mole fractions. Analogously to equation (12a), the orientational energy of a solvent molecule can be written as

$$U_A(\beta) = \{(1-x)u^{AA}<P_2>^{(A)} + xu^{AB}<P_2>^{(B)}\} P_2(\cos\beta_A) \qquad (12b)$$

The coefficients u^{AA} and u^{BB} can be related to the molecular shape anisotropies of the solute and the solvent, and so encode structural information. The coefficient u^{AB} contains information about solute solvent interactions, and so is sensitive to factors such as specific chemical associations etc. The interaction strengths are related by

$$u^{AA}u^{BB}/(u^{AB})^2 = \delta \qquad (13)$$

The quantity δ provides a measure of specific interactions within a mixutre, and in the absence of specific interactions $\delta=1$.

The partitioning of a solute into a stationary phase can be described in terms of the activity coefficient of the solute γ_B^∞ at infinite dilution

$$RT \ln\gamma_B^\infty = -\partial/\partial x(\Delta A) \mid_{x\to 0} \qquad (14)$$

where ΔA is the free energy of mixing. The activity coefficient is related to the retention time, τ, of the solute in the stationary phase by

$$\gamma_B^\infty = RT/(\tau F M_A \, p_B^\circ) \qquad (15)$$

where F is the volume that passes the column per unit time, M_A is the molecular weight of the solvent, and p_B° is the vapour pressure of the pure solute. The contribution to the actvity coeffiecient from the orientational interactions in a liquid crystalline solvent can be calculated from the Helmholtz free energy of mixing ΔA^{aniso}, which in turn can be constructed from equations (12a) and (12b)

$$RT \ln\gamma_B^\infty = -\partial/\partial x(\Delta A^{aniso}) \mid_{x\to 0} \qquad (16)$$

For low solute concentrations this gives

$$\ln \gamma_B^\infty = (u^{AA}/2kTV^2)[<P_2>^{(A)}]^2 - \ln Z_B \tag{17}$$

where Z_B is the solute partition function

$$Z_B = \int \exp\{- U_B(\beta)/kT\} \, d\cos\beta \tag{18}$$

and V is the molar volume of the solvent. From equation (17) it can be seen that the solute activity coefficient is composed of a term that is independent of solute (but which is related to the liquid crystalline solvent) and a negative term arising from solute solvent interactions. The retention time for a particular solute will thus depend on its partition function. To a first approximation, and in the absence of specific chemical interactions, this can be related to the polarizability anisotropy of the solute and it can be shown that τ is directly proportional to $\Delta\alpha$ ($\Delta\alpha = \alpha_\parallel - \alpha_\perp$). Thus the elution sequence of a series of aromatic geometrical isomers can be predicted from a knowledge of the polarizability anisotropies and/or molecular shape. A cautionary note is appropriate at this stage. The thermodynamic models described above are, strictly speaking, only valid for bulk mesophases and therefore a more complete understanding of the thermodynamics of thin liquid crystal fims adsorbed on surfaces must await the developement of suitable mathematical models.

Early studies of liquid crystals as stationary phases involved the use of low molar mass mesogens such as N,N'-bis(p-methoxybenzylidene)-α,α'-bi(p-toluidine) (BMBT)[19]:

$$CH_3O-\langle O \rangle-CH=N-\langle O \rangle-CH_2CH_2-\langle O \rangle-N=CH-\langle O \rangle-OCH_3$$

BMBT melts at 181°C and is nematic up to 337°C. Although BMBT and related systems were shown to be highly efficient in the separation of isomeric polycyclic aromatic hydrocarbons, they have poor thermal stability, an unacceptably high volatility at typical operating temperatures, and do not wet the substrate, thus precluding their commercial exploitation.

The ability to promote liquid crystal phase formation in polymers bearing pendant mesogenic groups, together with our understanding of the molecular physics of solute-solvent interactions in liquid crystalline phase, offers a significant tool in the rational design of liquid crystalline stationary phases with low volatility, good thermal stability, and high selectivity. The tendency of many side-chain polymers to form glassy phases, which effectively suppresses crystallisation, could also be exploited to achieve materials with wider mesomorphic ranges and hence, at least in principle, to a broader range of operating temperatures. That systems with this combination of properties are achievable in practice has been demonstrated for liquid crystalline poly(siloxane) and poly(acrylate) derivatives which exhibit a superior performance than conventional stationary phases in the separation of cis and trans fatty acid methyl esters, heterocyclic aromatics, and polycyclic aromatic hydrocarbons such as the isomers of methylchrysene, methylphenanthrene, and benzopyrene[20-21] .

Side-chain polymers used as stationary phases
in gas chromatography

The incorporation of chiral centres in the mesogenic groups, leading to polymers with chiral nematic phases, has also been reported. These materials are of interest because they could be used as stationary phases for the separation of optically active molecules. Although attempts to achieve separation of optically active isomers in chiral nematic phases have been disappointing, it should be noted that promising results have been reoprted for analogous, though non mesogenic, side-chain polymers. In view of the various structural elements that are amenable to molecular engineering it would appear likely that mesogenic stationary phases for chiral resolutions will also be achieved.

The ability to achieve an efficient and unambiguous separation of the various isomers of polycyclic aromatic hydrocarbons (several of which are potent carcinogens) is of considerable importance in environmental monitoring. Furthermore, the ability to separate optical isomers is of importance to the pharmaceutical industry as well as being of general analytical applicability. In view of this, the prospects for the commercial exploitation of liquid crystalline stationary phases appear to be particularly good.

The physico-chemical principles underlying the operation of liquid crystalline materials as stationary phases could also be exploited in the design of coatings for sensor applications. Although this is a much more speculative and longer term application, it offers the exciting prospect of being able to design materials which through specific chemical interactions (e.g. charge transfer interactions) exhibit semi-specific recognition of materials containing particular chemical groupings. It is thus possible to envisage an array of sensor elements, each with a coating that binds strongly with particular types of molecules, which, through a suitable transducing system (e.g. a piezoelectric or surface acoustic wave device), and image processing software, could be used in the detection of a broad range of organic vapours, effectively mimicking the process of olfaction.

5. CONCLUSION

It has been shown that the unique combination of physical properties characteristic of liquid crystalline side-chain polymers can be exploited in a variety of technological contexts. The design of new materials with properties optimised for a particular application is dependent on a sound understanding of the molecular physics underlying orientationally anisotropic phases. By drawing on the extensive body of mathematical models developed for low molar mass mesogenic systems we can establish guidelines for the rational design of materials. However, it should be stressed that although current mathematical theories do provide insights into some aspects of the fundamental physics of these complex systems, it is desirable for models to be developed which specifically take into account the conformational and dynamic aspects of polymeric systems. Furthermore, successful commercial exploitation requires more than just molecular engineering, and in particular the importance of careful market research, and a detailed analysis of production and development costs should be stressed. When all these factors are taken into account it is realsitic to expect that liquid crystalline stationary media for gas chromatography will reach the market place in the next few years. The use of mesogenic side-chain polymers as optical information storage media is also likely to find a small but significant market niche in the near future. By contrast, liquid crystalline non linear optical media still represent a highly speculative proposal.

REFERENCES

1.   G.W. Gray, in "The Molecular Physics of Liquid Crystals" (G.R. Luckhurst and G.W. Gray eds.), Academic Press,1979, Chapter 1, p1.
2.   G.W. Gray, ibid, Chapter 12, p263
3.   G.W. Gray and J.W. Goodby, "Smectic Liquid Crystals", Leonard Hill (Blakie), 1984.
4.   I. Sage, in "Thermotropic Liquid Crystals", (G.W. Gray ed.), John Wiley and Sons, (1987), Chapter 3, p 64.
5.   H. Finkelman and G. Rehage, Adv. Polym. Sci., 1984, 60/61, 173.
6.   M. Engel, B. Hisgen, R. Keller, W. Kreuder, B. Reck, H. Ringsdorf, H-W. Schmidt, and P. Tschirner, Pure and Appl. Chem., 1985, 57, 1009.
7.   C.B. McArdle, M.G. Clark, C.M. Haws, M.C.K. Wiltshire, A. Parker, G. Nestor, G.W. Gray, D. Lacey, and K.J. Toyne, Liquid Crystals, 1987, 5, 573.
8.   C.B. McArdle, in "Side Chain Liquid Crystal Polymers" (C.B. McArdle ed.),Blakie, 1989, Chapter 13, p357.
9.   C. Bowry, and P. Bonnett, "Optical Data Storage Mechanisms using Liquid Crystal Polymers", presented at the British Liquid Crystal Society Annual Conference, 1990.
10. R. Lytel, G.F. Lipscomb, J. Thackara, J. Altman, P. Elizondo, M. Stiller, and B. Sullivan, in "Non Linear Optical and Electroactive Polymers" (P.N. Prasad, and D.R. Ulrich editors), Plenum Press, 1987, p415.
11. J.P. Huignard, F. Micheron, and E. Spitz, in "Optical Properties of Solids: New Developments", (B.O. Seraphin editor), North Holland Publ. Co.,1976, Chapter 16, p847.
12. D.J. Williams, Angew. Chem. Int. Ed., Engl., 1984, 23, 690.
13. M.G. Kuzyk, K.D. Singer, H.E. Zahn, and L.A. King, J. Opt. Soc. Am., 1989, B6, 742.
14. C.P.J.M. van der Vorst and S.J. Picken, J. Opt. Soc. Am., 1990, B7, 320.
15. R.L. Grob, "Modern Practice of Gas Chromatography", Wiley Interscience, 1985.
16. D.E. Martire, in "The Molecular Physics of Liquid Crystals" (G.R. Luckhurst and G.W. Gray eds), Academic Press,1979, Chapter 10, p221.
17. D.E. Martire, ibid, Chapter 11, p239.
18. G.M. Janini, K. Johnston, and W.L. Zielinski, Analytical Chemistry, 1975, 47, 670.
19. G.M. Janini, G.M. Muschik, H.J. Issaq, and R.J. Laub, Analytical Chemistry, 1988, 60, 1119.

20.   G.M. Janini, R.J. Laub, J.H. Purnell, and O.S. Tyagi, in "Side Chain
       Liquid Crystal Polymers" (C.B. McArdle ed.), Blakie, 1989, Chapter
       14, p395.
21.   S.K. Aggarwal, J.S. Bradshaw, M. Eguchi, S. Parry, B.E. Rossiter,
       K.E. Markides, and M.L. Lee, <u>Tetrahedron</u>, 1987, <u>43</u>, 451.

Semi-conducting Polymers as Electro-rheology Substrates

H. Block, J. P. Kelly, and T. Watson

CENTRE FOR MOLECULAR ELECTRONICS, SCHOOL OF INDUSTRIAL
SCIENCE, CRANFIELD INSTITUTE OF TECHNOLOGY, CRANFIELD, BEDFORD
MK43 0AL

1 INTRODUCTION

It has been known since 1947[1] that certain dispersions
when subjected to electric fields undergo dramatic
increases in their resistance to flow. The phenomenon,
termed electrorheology (ER), can result in the complete
cessation of flow, in which event the material acts as
a Bingham body when in an electric field E. In the
early years after its discovery the phenomenon was
largely investigated by fluid engineers, with the
primary motive of increasing the field induced
resistance to flow, or more importantly, the
discrimination of this resistance relative to the flow
in the absence of electric fields. This electric field
enhancement of shear stress can be in excess of kPa
under E $\approx$kV mm^{-1} and that makes ER of interest for use in
a variety of devices in the area of fluid mechanics,
hydraulics and robotics. However, for some of the
potential applications ER fluids still require
improvements and this need, together with some
fundamental interest in an unusual fluid phenomenon, are
the motivations for much current research into ER. The
phenomenon involves a number of disciplines: rheology,
dielectric behaviour, colloid science as well as general
aspects of physics and chemistry.

The present paper describes some of our
investigations which have been targeted to improve our
understanding of the mechanism for ER. Before doing so
however, it is desirable to outline the salient features
of ER and ER fluids and to indicate in which respects
present ER fluids require improvement, for improvements

are necessary before ER as a scientific phenomenon
becomes of large scale practical use. We conclude the
paper with some brief comments about the applicability
of ER fluids and their future. More extensive
information about the ER phenomenon and its application
can be found in the general literature, including a
number of reviews[2-6].

2 THE ELECTRO-RHEOLOGY PHENOMENON

All current ER fluids comprise dispersions of solid
particulates within insulating oils and act under the
application of electric fields thereby changing the
developed shear stress $\tau(\dot{\gamma},E)$ at shear rate $\dot{\gamma}$. The
phenomenon is dependent on E rather than voltage[7].
Unless very dilute, most fluids under field show a
static yield at a level of stress, termed the static
yield stress $\tau(0,E)$ but this is often larger than the
enhancement of stress $\Delta\tau(\dot{\gamma},E) = \tau(\dot{\gamma},E) - \tau(\dot{\gamma},0)$
occurring under flow. Various types of Bingham type
behaviour can occur, some of which are illustrated in
Figure 1. The simplest, shown as curve 2, is when
$\tau(0,E)$ is maintained as $\Delta\tau(\dot{\gamma},E)$ in flow[2,7]. Other fluids
we have examined[3] are not so straightforward. Many ER
fluids follow curve 3 behaviour with a region following
yield where steady flow is not present and the fluid is
accelerating, after which steady flow is maintained
under moderately climbing stress. However there is a
decline in the excess stress needed for the flow
vis-a-vis the yield level required to start the flow.
On relieving the stress, the return curve does not
follow curve 3 for some fluids but diverges, following
curve 4 and ending with an initially weaker fluid.
Given a sufficiency of time such a fluid hardens back to
its original state. The system shows hysteresis.
Finally, some of our poorer fluids do not even manage to
maintain any significant level of excess stress under
field when in flow but collapse (curve 5) to the
non-field flow line[3]. Possible causes for these effects
are discussed in section (3).

For most investigations and applications of ER the
causative field has been dc although there have been a
few investigations of the effect under ac fields[7,8]. The
rheological response of ER fluids is only attenuated at
field frequencies (f) > kHz and this factor has led to
the often expressed view that the response of ER fluids
is ≈ms. Although it is becoming apparent that the time
dependence of the phenomenon may be complicated by the

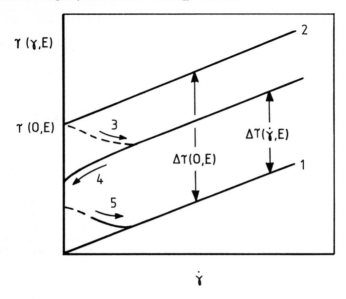

<u>Figure 1</u> Examples of the shear stress $\tau(\gamma,E)$ - shear rate ($\dot{\gamma}$) dependence shown by ER fluids. Curve 1, flow when E = 0 and assumed Newtonian for simplicity. For the description of other flows when E $\neq$ 0 see the text.

possible participation of several stages, resulting for example in the hysteresis, it is true that the switching between field and no-field response in flow is rapid and approaches that time scale; the full hardening in the absence of flow probably takes longer. Rapid switching in flow is an important property for many potential applications of ER fluids[6].

At moderate levels of dc fields, $\tau(0,E)$ and $\Delta\tau(\dot{\gamma},E)$ are often found to depend upon E^2 but at high enough fields there is either dielectric breakdown or a saturation in the developed stresses[2,4,10,11].

The active components for an ER fluid are the solid particulates which are dispersed at volume fractions between 0.2 and 0.4. The ER response is very concentration dependent with little effect at low

<u>Table 1</u>   Examples of components for promoted ER fluids[2].

<u>Dispersed Phase</u>	<u>Promoters</u>	<u>Dispersing Fluids</u>
Starch	Water	Hydrocarbons (eg
Barium titanate	Salt	kerosene, oils)
Cellulose	solutions	Silicone fluids
Diatomites	Alcohols	Chlorinated
Silica	Diethylamine	hydrocarbons
Polyacrylate		
salts		
Polymethacrylate		
salts		
Ion exchange		
resins		
Titanium dioxide		

concentrations and poor discrimination between on and off states at higher concentrations. These factors and the propensity for the fluids to have higher conductances at higher concentrations sets practical limits to the concentration range. The dependence of ER on concentration is a consequence of the mechanism and although the details of this are not clear there are a number of suggested relations in the literature. The reader is referred to the reviews mentioned above for further information.

A large number of powdered materials have been patented or otherwise reported for ER fluids. Many but not all of these materials require the addition of small quantities of water or other polar additives to be effective as ER substrates, and distinction is often made in the modern literature between anhydrous and promoted (most commonly water) fluids. The need for an additive reflects upon the mechanism of ER which is more fully dealt with in section (3) but it can be seen by the examples quoted in table 1 that the promoted fluids contain ions, either directly or from the promoter or surfactant. As described below, there are modern fluids which do not require water or other promoters. Although the list of active materials is diverse it is incorrect to deduce that any dispersed powder will provide an ER fluid. The other components for an ER fluid are a dispersing liquid, which must of necessity be electrically insulating, plus, in many cases, surfactants whose primary purpose is to provide stability to the dispersion. It should be stressed that when promoters are present, they are bound within or on the surface of the particles and not free as droplets or dissolved in the dispersing fluid, conditions to be avoided if the resistivity of the fluid is to be high.

The conductance of ER fluids under the high fields used presents one of the major problems in the application of the ER effect. Whether and to what degree the currents dissipated in use relate to the mechanism of the rheological changes or are extraneous is not known at present and must await a clearer, quantitative understanding of the mechanism of ER. Certainly, extraneous conductance can occur and elimination of impurity conductance is an important part of the technology of ER. It is often stated that the current density ≤ 1 A m^{-2} for an ER fluid when used in a heavy duty role [section (4)].

Temperature is another critical parameter in ER technology. Since most ER fluids have a large positive temperature coefficient of conductance, the difficulties associated with conductivity get worse with temperature and further, the power dissipation, which is not trivial under the voltages employed, can cause runaway problems. Water promoted systems often fail at elevated temperatures (>70°C) due to drying, particularly when used over protracted periods. At low temperatures ($\lesssim 20$°C) such fluids also fail, presumably because ice structures are ineffective as promoters. However, some of the anhydrous systems also have temperature limitations and we believe that other temperature sensitive mechanistic factors described in Section (3) may play a key role in ER.

There are a number of other potential variables relating to ER fluids including particle shape, size and size distribution. Little is known about the influence of these on the phenomenon and this suggests that they do not have any sensitive or critical role in the effect since, with the many fluids known, such effects would have been more manifest. Size is limited on the one hand by the electrode spacing which is often of the order of mm or fractions of mm. Particles approaching such dimensions can cause mechanical jamming or shorting of the electrodes because they are conductive. For very small, sub-micron particles it may be expected that Brownian motion would interfere with the effect but the lower limit is not known. As far as the authors are aware there are no known homogeneous fluids which show field induced Bingham behaviour suggesting that this

<u>Table 2</u>   Requirements for ER fluids excluding those
            arising from the phenomenon which are described
            in the text.

   *   Absence of sedimentation, including long
            periods of storage, and absence of
            centrifugation if used in devices such as ER
            clutches.

   *   Fluids should be non-corrosive and
            non-abrasive, the latter being a problem with
            some dispersed materials.

   *   Materials should be non-toxic, environmentally
            friendly and of low flammability.

   *   ER materials should have good oxidative and
            thermal stability.

   *   Low cost which implies a high volume demand to
            recover development costs.

aspect of ER cannot occur with entities of molecular
dimensions. Most ER fluids operate with particles in
the range 5 - 50 μm. Since most powders for ER fluids
are mechanically ground prior to dispersion, the
distribution of size and the shape of particles are not
well controlled or characterized in practical systems.
As far as the authors are aware there have been no
reported studies targeted to these parameters.

   Table 2 lists a number of other practical
requirements for ER fluids but which are not inherent to
the mechanism of the phenomenon.

   3.   THE MECHANISM OF ELECTRO-RHEOLOGY.

   It is generally agreed that the ER effect follows
from the polarization induced by the electric field.
What is less clear is by what means the polarization
induces the rheological changes or what changes at the
molecular level are involved in the polarization. It is
not appropriate here to deal in detail with the various
processes which can lead to polarization in materials.
Two or three types of charge re-arrangements have been
suggested as possible causes for the polarization in ER.
They comprise dipolar orientation, interfacial
polarization and double layer distortion, of which the

last two are the most likely contenders, because they can provide large polarizations and because the phenomenon occurs with dispersions where interfaces between particle and oil are present. It is arguable that on occasion, either or both of the latter mechanisms are involved in the ER effect and indeed they have common features, although in detail there can be conceptual differences between double layer and interfacial polarization. Mobility of charge carrier is a requirement in both, but for double layer distortions the carriers are ionic and reside within the double layer region between particle and bulk dispersing fluid. Mobile carriers are also necessary for interfacial polarization but they can be ionic or electronic and originate within the bulk of the particles, near the surface, or even in the dispersing fluid. Obviously, the distinction between double layer and interfacial polarization involving ions near a surface becomes blurred, and double layer polarization can be considered as one specific example of interfacial polarization. It is the interface that restrains the charge migration in all these mechanisms, causing charge separation and a field induced dipole to appear within each particle as is shown diagramatically in Figure 2.

Following dipole development it is then to be expected that interactions between particles occur. This is a possible route for the fibrillation of particles which has been observed in ER fluids under field[1,2,4,5,10-13] and may account for the Bingham behaviour. A level of structuring could also be maintained in flow but one would expect shorter chains or clusters with a more transient and dynamic character depending on the shear rate. In consequence there could be differences in the development or permanence of such structuring during flow and when the fluid is left stationary, leading to complex rheology.

The above widely drawn mechanism does not explain some apparent limitations for the dispersed materials nor the influence of promoters such as water or diethylamine. Until relatively recently it was presumed that promoters were essential and this led to a view that either water had a very specific role[13] or that the promoter was required to form a double layer, particularly when surfactants were also present as is often the case. There was also the question why, if interfacial polarization could operate, dry dispersed electronic conductors did not appear to be effective. Although Winslow[1] had listed carbon as an ER material,

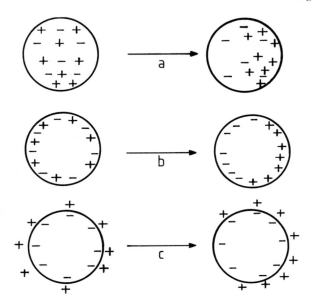

Figure 2 Polarization mechanism consequent upon a
particle - oil interface. (a) Interfacial
polarization through bulk charge migration, (b)
the same process involving surface charge and
(c) double layer distortion where the counter
ions are in the liquid phase.

many investigators found that carbon was ineffective.
Electronic conductors isolated in an insulating matrix
are the classical Maxwell-Wagner-Sillars[14] interfacial
polarizing systems and provide very large polarizations
and in consequence potentially powerful interactive
forces between particles, but many like metals and good
semi-conductors, are ineffective as ER substrates. We
have hypothesized[3], on the basis of both dielectric and
rheological evidence, that the rate as well as the
magnitude of the polarization plays an important role
in ER, and that with this extra consideration many
aspects of the ER phenomenon can be explained in terms
of dipole interactions. The role of promoters such as
water is in mediating the rate and extent of
polarization by influencing both the number and mobility
of ionic carriers, rather than any specific role,
although it is true that in many water promoted systems
double layers are present and the extent and rate of

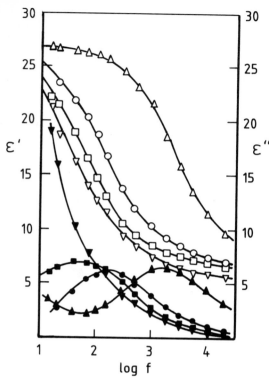

<u>Figure 3</u> Relative permittivity (ϵ') and dielectric loss
(ϵ'') as a function of frequency (f/Hz) for a
selection of ER fluids based on
poly(methacrylate) salts including water as
promoter (content based on weight fraction
polymer). o, lithium salt in pyralane 3010
with 15% water; $\triangle$, potassium salt in pyrelene
3010; $\square$, lithium salt in pyrelene 3010 and
∇, lithium salt in dibromo-diphenylmethane.

their polarizations are then likely to be important in
ER.

Before discussing the influence of polarization rate
on ER mention should be made of the one rather specific
water promoted mechanism which has been proposed[13]. In
this mechanism, ionic migration is the first effect of
the field which is accompanied by electro-osmosis of the
water. This is thereby transferred to one or both
charged poles of the particle. There follows

interparticle capillary attraction between particles
causing water bridges and structure formation. If this
were the only mechanism for ER then all fluids require
water and anhydrous fluids could not exist.

The alternative view that water just mediates
polarization in ionic systems and that polarization rate
plays a role stemmed from the observation that the
dielectric spectra of promoted and now other known ER
fluids show a dielectric relaxation in the region ≈ 10
to 10^4Hz, which reflects the effective rate of charge
migration to provide interfacial - double layer
polarization for ER. As examples, Figure 3 shows the
dielectric spectrum of some ER fluids based on dispersed
moist lithium or potassium poly(methacrylate) and Figure
4 their $\tau(0,E)$ dependence on E^2. This and similar data
indicated that polarization alone was not the only

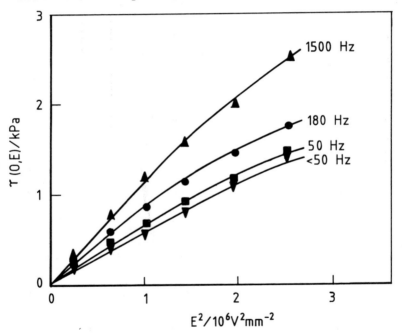

__Figure 4__ The field dependence of the static yield stress
of the poly(methacrylate) salt based ER fluids
described in Figure 3. Frequencies for maximum
dielectric loss are indicated on the left of
the curves.

factor in ER strength. Thus, in the examples shown, the level to which the low frequency limited relative permittivities are tending do not differ very much and the correlation to ER strength is better reflected by the dielectric relaxation frequency. This is not to imply that polarization strength is irrelevant, only that relaxation frequency also has an influence and that ER fluids outside an optimum rate for polarization are less effective in static situations and more particularly in flow. Many electronic conductors have carriers of much higher mobilities than the ionic systems of which poly(methacrylate) salts are typical examples. Such electronic conductors would, if dispersed as particles in an insulating matrix, relax their interfacial polarization at frequencies in the mega to giga Hz regions, well outside the range of effectiveness in ER. To simulate relaxations at the lower frequencies which appear to be necessary for ER, rather poor semi-conductors in the range 10^{-3}-10^{-5} S m^{-1} are needed and these have recently been shown to be effective as ER substrates without the need for promoters[3,15,16].

The reasons for the existence of a time limitation of polarization may originate in the following way. In shear flows, particles or clusters of particles are involved in rotation forced by the shear gradient and these particles or clusters 'see' the electric field in rotation. Within any framework local to these entities, E is constantly changing direction with a rate governed by the shape of the particle or cluster and $\dot{\gamma}$. There is experimental evidence for such an effect of flow on the polarization of particles using the technique of flow modified permittivity (FMP) which monitors the complex permittivity of a liquid as a function of $\dot{\gamma}$ and f. Figure 5 shows the relative permittivity of a lithium poly(methacrylate) dispersion as a function of both shear rate and field frequency but using only a small sensing field. Thus, there are no strong dipolar interactions and the particles move independently. The example selected shows two important aspects of FMP: the development of resonance and the trend for the low frequency relative permittivity to decline with shear rate. The resonances are known[17] to arise from a coupling between the rate of rotation of the particles and their rate of polarization, and the decline in relative permittivity stems from the rather slow polarization in the example shown (relaxation frequency < 10^2Hz) preventing its achievement at high $\dot{\gamma}$ as well as high f. The conclusion to be drawn, which is of

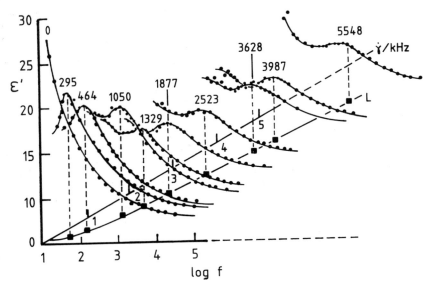

<u>Figure 5</u> Flow modified permittivity (FMP) of a
lithium poly(methacrylate) based ER fluid.
Relative permittivity (ϵ') shown as a function
of shear rate ($\dot{\gamma}$) and field frequency (f) with
individual shear rates indicated on the $\dot{\gamma}$
constant curves. The shown locus L of the
resonance maxima in the $\dot{\gamma}$ - f plane is
calculated from theory for spherical
particles[17].

relevance to ER, is that since polarization is not
instantaneous, the polarization will fluctuate under
such circumstances even if the external field is dc.
Particles which polarize too slowly relative to the
rotational motion will not polarize in such flows and
there would then be no dipole-dipole interactions even
in large fields and in consequence no ER effect. That
case involving as a simplified illustration just two
particles interacting is shown in Figure 6a. If the
relaxation frequency for the fluid is high relative to
the scale of $\dot{\gamma}$ or f then FMP shows little change because
the polarization vector is always in the field
direction. That fact does however have a bearing on ER
fluids in flow if such ER effects are due to
dipole-dipole interactions in the manner illustrated in
Figure 6c. Then dipole forces will constantly tend to
chain particles in the field direction but the flow will

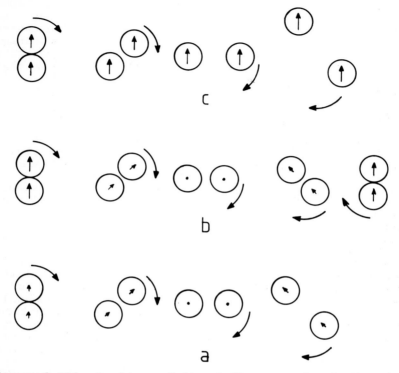

Figure 6 Illustration of the influence of polarization rate on a simplified example of the interaction of two polarizable particles in shear flow. Case a: rotation in shear >> polarization rate. Case b: the situation for optimum retention of dipole - dipole interaction due to matching of polarization and rotation rates. Case c: polarization rate >> rate of rotation in shear leading to the development interparticle forces of repulsion.

rotate these incipient clusters so that the major axis is rotated across the field direction[17]. In this configuration and with rapid polarization following the orientation, the dipole-dipole forces become repulsive and will disrupt the clusters which will tend to reform with the major axis again along the field direction. The net result is a dynamic component of motion of the particles rather than clustering of any permanence. This may well lead to extra viscous dissipation over the

no field case but probably not as much as if some
permanence of association was achieved. Then the
situation exemplified in Figure 6b would occur which
should only arise at some optimum rate of rotation
relative to polarization, when polarization can occur
but not relax to the field direction so quickly that
dipole-dipole repulsions would dominate. A
characteristic polarization rate is then a consequence
for optimising the ER effect in flow with this
mechanism.

This dynamic picture of the ER phenomenon is not so
easily applied to the static situation when ER fluids
form Bingham bodies. Although in such cases there is
fibrillation, there is no large scale motion and it
would seem that polarization magnitude alone should
determine ER effectiveness. It does appear however that
the same general trends in optimum relaxation time as
well as polarization magnitude have influence. That
sluggish and poorly polarizing particles are not good ER
substrates might well be expected but what of highly
interfacially polarizing particles in an insulating oil?
Firstly, these tend not to give practical ER fluids
because fluids based on them pass too much current,
especially in the absence of flow, when they can easily
fail because of the chain formation shorting the
electrodes. This problem of particles which are too
conductive (including those containing too much water)
leading to high and in the end unachievable current
demands, is a well known practical difficulty. However,
some of our investigations which are described below and
which used semi-conducting polymers covering a range of
conductances suggests that those having high as well as
low conductances are less effective than those of an
intermediate conductance. Thus the indications are
that, even for the development of static yield stress,
the ER phenomenon is dependent for effect upon a
combination of polarization magnitude and rate of its
achievement.

The polymer systems referred to are:
  i)   poly(acenequinone radicals) (PAQRs) of the type
       shown in Figure 7 and,
  ii) poly(anilines) in the "emaraldine" form but
       partially neutralised to reduce their
       conductance.

Finely ground and dispersed in base fluids such as
silicone oils or chlorinated paraffins, some of these
provide dispersions which are ER active. They also

show dielectric relaxation processes believed to arise from interfacial polarization involving electronic transport. As investigated by us, the base fluids and powdered polymers were thoroughly dried before mixing

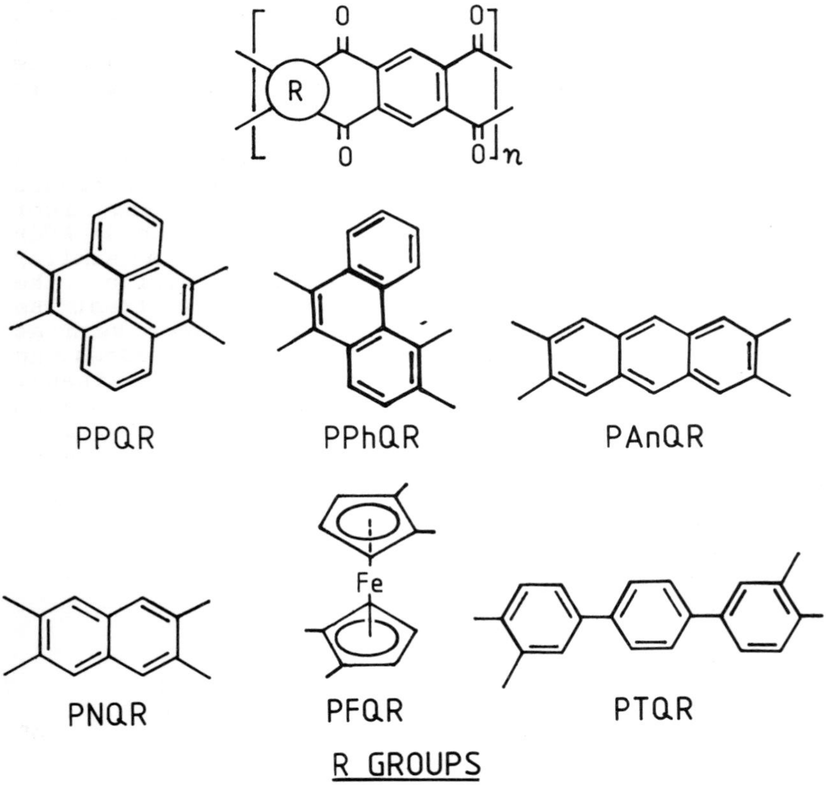

PPQR   PPhQR   PAnQR

PNQR   PFQR   PTQR

R GROUPS

Figure 7 The poly(acene quinones) (PAQRs) studied as ER substrates. Codes (conductances/$S\,m^{-1}$):- PPQR, poly(pyrenequinone radical) (10^{-2}); PPhQR, poly(phenanthrenequinone radical) (10^{-3}); PAnQR, poly(anthracenequinone radical) (10^{-4}); PNQR, poly(naphthalenequinone radical) (10^{-5}); PFQR, poly(ferrocenequinone radical) (10^{-7}); PTQR, poly(p-terphenylquinone radical) (5×10^{-9}). Materials were synthesized following the method of Pohl et al[18] whose estimates of conductance are quoted.

and no water, surfactants or other promoters were added: when ER effective, they are examples of the anhydrous fluids referred to above.

Figure 8 shows the dielectric loss behaviour of members of the PAQR series. Their relative positions follow the general tenet that the critical frequencies in interfacial polarizations depend upon the conductances (the values as quoted by Pohl et al[18] are given in the legend to Figure 7) and permittivities of the component phases[14]. As figure 9 shows, their effectiveness in terms of static yield stress is a maximum for conductances of $\approx 10^{-5}$ S m^{-1} which corresponds to a loss frequency in silicone oil of $\approx 2 \times 10^3$ Hz. Poor conductors such as PTQR or good conductors such as PPQR do not provide good substrates for ER. A similar finding has recently been obtained with the poly(aniline) system[16] where it is necessary to treat the emeraldine salt form as prepared, with bases such as ammonia to reduce the conductance[19]. Figure 10 shows the dielectric spectrum of various dispersed poly(anilines). The acid form has its interfacial loss process beyond 10^5 Hz as can be deduced from the retention of

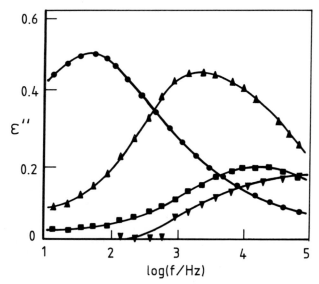

Figure 8 The dielectric loss of PAQRs dispersed in silicone oil at a volume fraction of 0.3 and at ambient temperature. ●,PTQR; ▲,PNQR; ■,PAnQR; ▼,PPhQR.

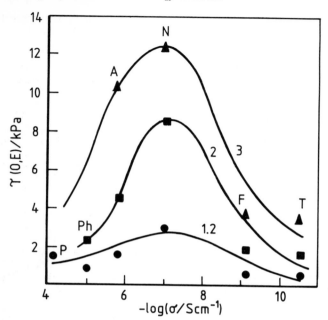

Figure 9 The variation of τ(0,E) of PAQR based ER fluids [volume fractions 0.35 in a chlorinated hydrocarbon (ICI, Cereclor 50 LV)] with the solid state conductance (σ) of PAQR's. ● ,E = 1.2; ■ ,E = 2; ▲,E = 3 kV mm⁻¹ dc.

considerable extra relative permittivity above solvent (≈3) at this frequency. There is also very large bulk conductance of the fluid as shown by ε'' at low frequency and this is causing the further increase in low frequency ε' by the development of space charge due to the 'blockiness' of the electrodes. This dielectric behaviour of the acid is entirely in keeping with a reported level of the solid state conductance of poly(aniline)[19] of 100 S m⁻¹. Measurements of any ER effect were found impossible due to the excessive current demand under even very moderate fields. In contrast the free base form obtained by equilibrating the acid form with excess aqueous ammonia and then working up and drying gave a dielectric response indicating a low frequency relaxation, again consistent with a reported[19] level of solid state conductance of 10⁻⁸S m⁻¹. By using a shorter reaction time for

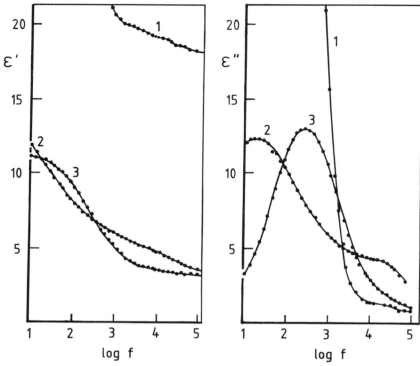

<u>Figure 10</u>   The dielectric spectrum of various
poly(anilines)in silicone fluid at 0.3
volume fraction and ambient temperature.
Type 1 - acid 'emeraldine' form; type 2 -
free base 'emeraldine' form and type 3 -
partially neutralised 'emeraldine' form.

neutralization with aqueous ammonia a partially
neutralised form of poly(aniline) was obtained and this
showed a relaxation frequency of 500 Hz. Both the
fully and partially neutralised samples provided ER
effects when dispersed in silicone oil at 0.3 volume
fraction. Values of $\tau(0,E)$ of 2.1 and 2.7 kPa and
current densities of 0.16 and 0.43 A m^{-2} were achieved at
$E = 4$ kV mm^{-1}, values which have promise for application.

The dynamic response of semi-conducting polymer
based ER fluids also reflects the dynamic consequences
of polarization. Figures 11 and 12 respectively show
the $\tau(\dot{\gamma},E)$ - $\dot{\gamma}$ behaviour of PAnQR and PTQR based

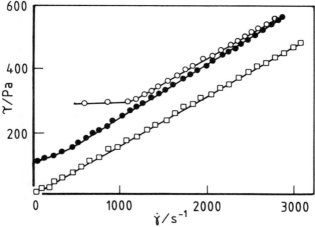

Figure 11 The shear stress [τ($\dot{\gamma}$,E)] - shear rate ($\dot{\gamma}$) dependence of an ER fluid based on PAnQR [0.15 volume fraction in a chlorinated paraffin (ICI, Cereclor 50LV)]. Open squares E = 0; open circles under increasing shear and filled points with decreasing shear both with E = 1.5 kV mm^{-1}.

fluids. Both require a static yield level to be exceeded before flow, following which there is accelerating rather than uniform flow until a state of dynamic equilibrium is reached. Different responses then occur with the two fluids. With PAnQR the τ($\dot{\gamma}$,E) - $\dot{\gamma}$ locus now parallels the zero field curve but at an enhanced shear stress and does so for the total range of $\dot{\gamma}$. This is not the case for PTQR where the curve rapidly collapses to the zero field level. Also shown in these figs are the 'down curves' which result when the stress is progressively relieved. Initially at high $\dot{\gamma}$ the return curves follow the 'up curves' but as $\dot{\gamma}$ decreases further the behaviour between increasing or decreasing τ($\dot{\gamma}$,E) diverges with the fluids showing an inability under flow conditions to quickly recover the modulus shown when flow originated from a fully 'solidified' state. Thus there is hysteresis in both cases with the fluid not initially recovering the static yield stress level after flow. Given a longer, but at this stage not accurately known time for recovery, the fluids harden fully in the absence of flow and the cycle can be repeated. PTQR and PAnQR based fluids only

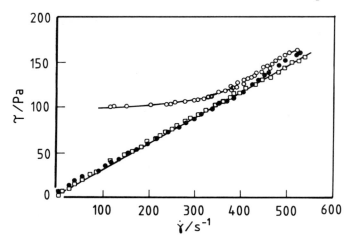

<u>Figure 12</u>   The shear stress $[\tau(\dot{\gamma},E)]$ - shear rate $(\dot{\gamma})$
                dependence of an ER fluid based on PTQR
                [0.2 volume fraction in a chlorinated
                paraffin (ICI, Cereclor 50LV)]. Open
                squares E = 0; open circles under
                increasing shear and filled points with
                decreasing shear both with E = 2 kV mm^{-1}.

differ in that the former, once having failed in an ER
role, does not recover in flow whilst the latter
maintains a level of excess stress under flow. Neither
fluid is able to transmit the total static yield stress
as an excess stress in flow (type 2 behaviour of Figure
1) but rather follow type 3 or 5 behaviour of Figure 1.
We have found this inability of ER fluids in flow to
retain the total $\tau(0,E)$ as an excess stress is shown by
some other, including poly(electrolyte) based fluids.

    The various dynamic responses are readily explained
in terms of time dependent polarizations and the
mechanism explained above. Flow disorganizes what may
be termed the fully field hardened state of the Bingham
body but after the onset of flow there is, at least at
low enough $\dot{\gamma}$, sufficient permanence to the interactions
or clusters to provide some enhancement of $\tau(\dot{\gamma},E)$. This
process is controlled by the ability to induce dipoles
and to retain cluster maintaining interactions. Slowly
polarizing particles (PTQR) fail in the former respect
whilst we have suggested that there is an optimum in the
rate of polarization for effective clustering. The
hysteresis and extra time requirement for full hardening

is probably due to the need to diffuse particles to their fully fibrillated positions.

4.   THE POTENTIAL FOR THE APPLICATION OF ELECTRO-RHEOLOGY.

The phenomenon of ER has the scope for application to many engineering problems and yet at the present time there is no ER device which is commercial! The reasons are several.

i)   ER fluids still demand rather large currents at big voltages to provide some of the large stress transmissions which have the biggest commercial potential; for example in the automotive industry for use in active dampers, engine mounts, in electric clutches and differentials.

ii)   In these applications they would need to operate over a large temperature range (circa -30° to 120°C) which is too wide for most current ER fluids and particularly for water promoted ones. Furthermore, for many fluids an increase of temperature causes too large a decrease in the fluids resistivity.

iii)   For many ER fluids the stability against sedimentation and centrifugation (a particular problem in clutch application) needs improving.   ER fluids should be such that, if they do sediment over long time storage (shelf-life), they readily redisperse on moderate shaking.

iv)   If used in moving machinery, ER fluids must not be abrasive. Many dispersed powdery materials, particularly inorganic ones are likely to be so.

v)   An ER fluid is a complex system with many variables influencing its properties. It is multi-component and can to some extent be tailored to the application. This does mean however that commercial decisions relating to application and diversification are more difficult, particularly as fluids are not available 'off the shelf' nor even some of their components.   Extensive developing technology is needed for quality assurance and control, and the technology involves the expertise of the chemical and engineering industries.

vi)   Current ER fluids could now very successfully be used in very many situations where low force transmission is involved and extremes of temperature are not. The field of robotics is one. One reason that this is not happening appears to lie in this difficulty of multiple industrial expertise and inaccessibility of fluids or fluid components. The costs for developing a commercial ER fluid appear only attractive to large industrial companies when there is a large potential market. This seems to be in heavy engineering, particularly the automotive industry. It is probable that if an ER fluid is commercially developed for such large scale use then it, or other variants, will find applicability in a wider market.

Research into ER fluids has of recent years become much more active in both Universities and Industry and involving all manners of disciplines. This is an encouraging sign that the technology will follow, since research into how ER fluids operate, how they can be improved and the design criteria for their use is a prerequisite for their commercial success.

5.   ACKNOWLEDGMENTS.

We wish to thank the Electro-Rheology Research Syndicate, the British Technology Group, Air Log plc and the SERC for support of our research programmes in ER. Thanks are due to Air Log plc for providing samples of the polyelectrolyte based fluids discussed above.

REFERENCES

1. W.M. Winslow, US Patent 2417850, 1947.
2. H. Block and J.P. Kelly, J.Phys. D Appl.Phys., 1988, 21, 1661.
3. H. Block, J.P. Kelly, A. Qin and T. Watson, Langmuir, 1990, 6, 6.
4. A.P. Gast and C.F. Zukoski, Adv. Colloid and Interface Sci., 1989, 30, 203.

5. Z.P. Shulman, R.G. Gorodkin, E.V. Korobko and V.K. Gleb, J. of Non Newtonian Fluid Mechanics, 1981, 8, 29.

6. W.A. Bullough and D.J. Peel, J.Japan.Hydraul.and Pneumatics Soc., 1986, 17, 520; D.J. Veasy and A. Woodruff, British Technology Group Prospect, 1987-8, 1, 14; D. Brooks, Physics World, 1989, 2, 35; W.A. Bullough, Hong Kong Engineer, 1989, 17, 10;14.

7. D.L. Klass and T.W. Martinek, J.Appl.Phys., 1967, 38, 67.

8. D.L. Klass and T.W. Martinek, J.Appl.Phys., 1967, 38 75; H. Uejima, Japan.J.App.Phys., 1972, 11, 319.

9. G.G. Petrzhik, O.A. Chertkova and A.A. Trapeznikov, Dokl. Akad. Nauk SSSR, 1980, 253, 173.

10. W.M. Winslow, J.App.Phys., 1949, 20, 1137.

11. L.Marchall, C.F.Zukoski and J.W. Goodwin, J.Chem.Soc.Faraday Trans.I, 1989, 85, 2785.

12. T. Sasada, T. Kishi and K. Kamijo, Proc.17th Japan Congr. Mater.Res., 1974,228; N Sugimoto, Bull JSME,1977, 20,1476; Z.P. Shulman, B.M. Khusid and A.D.M Matsepuro, Vesti.Akad. Navuk BSSR Ser.Fiz.Energ.Navuk, 1977, 3, 116, 122.

13. J.E. Stangroom, Phys.Technol., 1983, 14, 290.

14. J.C. Maxwell, 'Electricity and Magnetism' Vol 1, Clarendon Press, Oxford, 1982, p 452; K.W. Wagner, Arch. Elektrotech., 1914, 2, 371; 'Die Isolierstoffe der Elektrotechnik', Ed: H. Shering, Springer, Berlin, 1924; R.W.Sillars, J.Inst Elec.Engrs.(London), 1937, 80, 378.

15. H. Block and J.P Kelly, Proc.IEE Colloq., 1985, 14, 1; U.S. Patent, 4687589, 1987.

16. H. Block, J. Chapples and T. Watson, U.K. Patent App., 8908825.6, 1990.

17. H. Block, 'Polymers in Solution, Ed: W.C. Forsman, Plenum Press, New York, 1986; H. Block, E. Kluk, J. McConnell, B.K.P. Scaife, J.Colloid Interface Sci., 1984, 101, 320.18. H.A. Pohl and E.H. Engelhart, .Phys.Chem., 1962, 66 2085; H.A. Pohl and D. Opp, J.Phys.Chem., 1962, 66, 2121; R. Rosen and H. A. Pohl, J.Polymer Sci.Polymer Chem.Ed., 1966, 4, 113.

19. A.G. MacDiarmid, J.C.Chiang, A.F. Richter and A.J. Epstein, Synthetic Metals, 1987, 18, 105.

Some Applications of Conducting Polymers

P. Kathirgamanathan

COOKSON GROUP PLC, COOKSON TECHNOLOGY CENTRE, SANDY LANE,
YARNTON, OXFORD, OX5 1PF

1 INTRODUCTION

Conducting polymers and adhesives play a major role in our everyday life.[1] For example, conducting polymers (e.g. carbon loaded polymers) are used for charge dissipation in industries, whether electronics or otherwise, hospitals and military establishments. They are also used for electromagnetic shielding, in self-heating regulators, heating panels, etc. Conductive adhesives (for example silver loaded epoxy) are used in electrical circuitry and to bond (or "solder") two electrical wires.

Nearly 15 years ago, a new class of conducting polymers was discovered and these were called "inherently conducting polymers" (ICP) because the manner in which they are formed renders them conductive. However, in conventional semiconductor terminology, they have to be classified, in general, as extrinsic semiconductors although their conductivities are in the metallic regime. Conducting polymers can be broadly classified into two types as shown in Figure 1.

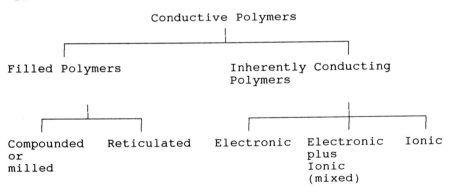

Figure 1

This paper is dedicated to Dr V. Arkley

Traditional conductive polymers fall into the class of filled polymers, where an inorganic or organic conductors are compounded or milled with a base polymer of interest. A typical example of a filler is a conductive grade carbon black compounded into PVC, PP or PE whose compounds are used to make charge dissipative tubings, charge dissipative containers and charge dissipative bags respectively. However, fibres and flakes of aluminium, brass, stainless steel, silver coated glass beads, nickel-coated mica and carbon fibres are also used. Other types of filled conductive polymers are classified as reticulated polymers[2]. Although, these are little known at present, they have enormous potential because of the very high conductivity (up to $1-10$ S cm^{-1}) that can be achieved with a very low loading (1-2% by wt.) of a conductive filler. This type is normally formed by crystallising a needle-shaped conductive adduct (e.g. Tetrathiafulvlene-Tetracyanoquinodimethane (TTF-TCNQ).

Inherently conducting polymers can be further divided into three types based on the mechanism of their conduction. Polyacetylene, polypyrrole(s), polythiophen(s), polyphthalocyanine(s), polyphenylene(s), polyphenylvinylene(s), polyphenylsulphide(s) are all electronic conductors. Polyethylene oxide(s) alkylammonium salts, polyphosphates, polyphosphonate(s), polytungstate(s) are all examples of ionic conductors. However, polyaniline(s) are curious examples of a mixed conductor, i.e. electronic and ionic conductor, although the electronic contribution is small in comparison to the ionic contribution. However, a new electronically conducting polyaniline has been isolated[3] and is an inherently conducting polymer with a conductivity of 10^{-1} S cm^{-1}.

A conductivity chart in Table 1 compares some selected inorganic and organic materials that are currently available.

Figure 2 shows chemical structures of some selected inherently conducting polymers.

Inherently conducting polymers have suffered from various problems in the past such as instability to air and moisture, insolubility in common organic solvents, non-processability using conventional techniques, and control of shape and size. For example, polypyrrole, though unstable at high temperatures in the presence of

Table 1   Conductivity Chart

	Material	Conductivity /S cm^{-1}	ICP
	Silver	10^6	Oriented doped (CH)$_x$
	Copper	10^4	
Metals	(SN)$_x$		Polyphenylene(s) Polythiophen(s) TTF-TCNQ
		10^2	Polypyrrole(s)
	Carbon Fibre		
			Polyaniline(s) Poly(aniline)coated talc
	Carbon Black	1	Reticulated polymers
	Germanium	10^{-2}	
	Doped Silicon		Undoped polypyrrole(s)
			Undoped polythiophen(s)
Semi-conductors		10^{-4}	Trans undoped (CH)$_x$
		10^{-6}	
			Undoped polyaniline(s)
		10^{-8}	
			Cis undoped (CH)$_x$
	Glass	10^{-10}	
	PVC		
Insulators		10^{-12}	
	Diamond	10^{-14}	
	Nylon		
	Polyethylene	10^{-16}	
	Polyimide		
	Polytetra-fluoroethylene	10^{-18}	

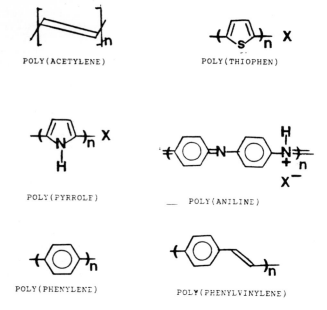

POLY(ACETYLENE)

POLY(THIOPHEN)

POLY(PYRROLE)

POLY(ANILINE)

POLY(PHENYLENE)

POLY(PHENYLVINYLENE)

FIGURE 2. SOME SELECTED INHERENTLY CONDUCTING POLYMERS

moisture under situations of either high temperature or
moisture alone is stable enough to be useful. Polypyrrole
is insoluble (although swellable) in many common solvents
and therefore cannot be solvent processed to produce clear
films. Because of thermal instability, this cannot be
hot-pressed, milled, compounded or extruded, which are the
main conventional techniques of processing.

Why such improvements in a number of properties are
desirable is summarised below.

Property	Application
1. Solubility	Conductive paints and production of transparent conductors.
2. Flexibility	Non-rigid substrates.
3. Swelling polymers	Conductive dispersions.
4. High thermal stability (200-300°C)	Processability made easy using conventional techniques.

It has now been found that the stability of ICP's can
be improved significantly by choosing the right stable
counterion, and the solubility and flexibility can be
enhanced by choosing the right organic substituent.[4]

Inherently conducting polymers can be used in a
variety of applications:

Antistatics (Charge dissipation)
EMI/RFI shielding
Gas sensors
Ion Sensors/Biosensors
Optoelectronic devices, e.g. photovoltaic cells
Display devices
Conducting circuits/Resistive circuits
Conducting adhesives
Electrode protective materials
Self-heating regulators
Pressure sensors
Strain gauges
Electroprinting
Electrorheology
Fibre optics

2   CURRENT STATUS OF INHERENTLY CONDUCTING POLYMERS

The current status of inherently conducting polymers along with the past and the extrapolated future is shown in Table 2. The realisation that a homogeneous conducting polymer would bring a lot of benefits to the filled system made the scientific community very optimistic. However, their instability made the general mood pessimistic. But, in the following years (namely 1984-1989), what makes an ICP stable has been mastered and the future looks bright.

Again as shown in Table 1, a conductivity as high as 10^5 S cm^{-1} has been achieved for oriented polyacetylene. Processing using conventional extruders and moulders is now a reality. Processability using contemporary methods (e.g. electrochemical composite formation[10,12] has been in place since 1984). These ICPs have already found applications with superior properties to conventional materials.

2a   Traditional Conducting Polymers (TCP)

Almost all the traditional conducting polymers are filled polymers where a conductive filler is compounded at elevated temperatures into a base polymer (e.g. with a twin screw compounder) with one or several processing aids (e.g. plasticisers, stabilisers, anti-oxidants, lubricants, etc.) to produce "compounds". These are then compression, injection or blow moulded to produce the desired shapes.

2b   Current Commercial Conductive Fillers

Some selected commercial fillers are tabulated in Tables 3 and 4.

Table 2 Status of inherently conducting polymers

	1980	1984	1989	1995
General mood	Optimistic	Pessimistic	Bright and Optimistic	Bright and Optimistic
Conductivity S/cm	1	10-100	$10-10^5$	$>10^5$
Method of Production	Electro-chemical	E.C. + Chemical	E.C. + Chemical Photochemical	E.C. + Chemical Photo-chemical Sono-chemical Magneto-Electro-chemical
Stability of the polymer	Poor	Poor	Good-Excellent	Excellent
Processability using conventional methods	No	No	Yes	Yes
Processability using contemporary methods	No	Yes	Yes	Yes
Commercial Applications to date	-	-	Antistatics (Polaroid) Shielding I. R. camera (BASF) Batteries (Verta/BASF) (Lockheed) Conductive circuitry Electro-chromic Display (IBM)	Radar Absorbers I. R. Sensors Conductive Circuits Resistive Circuits Gas/Ion Sensors Electro-chromic Display Device
Commercial Applications to date	-	-	-	Electro-chromic printing, Pressure sensors, Stain gauge, Electrode protection, Conductive adhesives, Self heating regulators, Opto-electronic devices, Electrorheology and Molecular Devices

__Table 3__   Conductive Fillers, Ionic Conductors and Mixed Conductors

Filler Type	Conductivity $(S\ cm^{-1})$ range obtainable in PVC at various loading levels	Comments
a) Quaternary ammonium salts b) Triethanol amine based salts c) Surfactants d) Amides	$10^{-8}-10^{-10}$	Susceptible to moisture, some require moisture for conduction, not permanent but light in colour, requires 5-10% by wt.
Mixed Conductors a) Metal oxides b) Doped metal oxides c) Doped metal oxides coated on minerals e.g. $Sb_2O_3/$ $SnO_2/Mica$, $Sb_2O_3/SnO_2/$ TiO_2, ZnO/ZnS, $In_2O_3/Sb_2O_3/$ Mica	$10^{-6}-10^{-10}$	Susceptible to moisture, expensive £25-£50/kg, 40-50% by wt loading required.
Mixed Conductor Polyaniline(s)	$10^{-2}-10^{-10}$	30-50% by wt loading required.

__Table 4__   Electronic Conductors As Fillers

Filler Type	Conductivity $(S\ cm^{-1})$ achievable in PVC	Comments
Carbon black several grades, e.g. Akzo EC300 CABOT (Vulcan XC72) and (BP 2000)	$10^{-1}-10^{-3}$	Intermediate conductivities (e.g. 10^{-6} to 10^{-9} $S\ cm^{-1}$) not easily achievable at all. Colour black; 10-20% by weight loading required Price: £1.5-£7/kg
Cookson's ICP based fillers (__not__ commercially available)	$10^{-2}-10^{-9}$	Various conductivities including $10^{-6}-10^{-9}$ $S\ cm^{-1}$ easily achievable, processing easier. Colour: blue-black
Metals (fibres, flakes, spheres)	$10^{+1}-10^{-1}$	Price: £25-50/kg
Metals coated on minerals, glass	$1-10^{-1}$	Price: £50/kg Colour: grey

3 THEORETICAL AND PRACTICAL CONSIDERATIONS ON COMPOUNDING CONDUCTIVE FILLERS

Loading Levels

The concentration (volume percent, in particular) of a filler is the overriding factor determining the conductivity of the composite obtained.

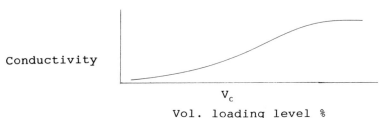

Conductivity

V_c

Vol. loading level %

Figure 3

The above Figure 3 represents the shape of a conductivity versus loading curve for virtually any filler material. Up to a certain loading, the conductivity of the composite is roughly the same as that of the bulk resin. At the critical loading V_c, a network formation commences and conductivity increases. A small increase in loading beyond the critical loading volume (V_c) or the corresponding weight (W_c) will result in a dramatic increase in conductivity and the asymptote of the curve should approach the conductivity of the pure filler. In practice, the maximum conductivity achievable is in the range 1/10-1/100 of that of the pure filler owing to incomplete dispersion and other processing factors.

The critical composition has nothing to do with the conductivity of the filler, but rather with its size, geometry and density. For two fillers of equal particle size and shape, the one with the lower density will have a lower critical composition by weight. The higher the aspect ratio, the lower is the critical composition (or volume) of any filler. For example, needle-shaped materials are better than flakes, and flakes are, in turn, better than spheres. This can be further illustrated by the following computer simulation (Figure 4). As can be seen, at 20% loading levels, needles are making very

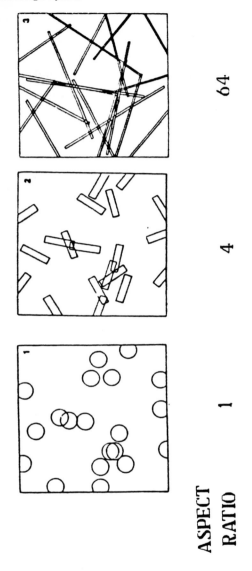

**ASPECT
RATIO**   1   4   64

FIGURE 4. COMPUTER SIMULATION OF ASPECT RATIO ON NETWORK FORMATION. EACH
FRAME CONTAINS 20 AREA PERCENT FILLER. THE FILLER PARTICLES HAVE THE SAME
POSITION AND ORIENTATION. THE ONLY DIFFERENCE BETWEEN THE FRAMES IS THE
FILLER ASPECT RATIO. (FROM REFERENCE 5.)

efficient conducting pathways, flakes make a few contacts
and spheres none at all[5].

   It should be mentioned, however, that particle-
particle contact is not required for conduction if they
are all less than 100 nm in diameter when tunnelling
conduction will operate and high conductivity may be
obtained well below the critical loading level.

3a   Carbon Black

   Various conductive grades of carbon black are available from companies such as CABOT Corporation (USA),
Degussa (W. Germany), and Akzo Chemie, American Division
(U.S.A). These are also available as polymer compounds in
polypropylene, EVA rubbers, polyethylene, high density
polyethylene, linear low density polyethylene, and polystyrene from CABOT Corp.

   One of the major problems with carbon black is that
the conductivity of the compounded material cannot be controlled or fine tuned as a result of sharp percolation[6,18]
behaviour as shown in Figure 5. In other words, the
conductivity of the composite is normally at the extremes,
either insulating or highly conducting[6]. In many cases, it
is preferable[6] to have volume conductivities in the range
10^{-6}-10^{-8} S cm^{-1}.

   It is worth summarising the various parameters that
govern the final conductivity of a composite[7,13].

1)   The conductivity of the filler itself. This deter-
     mines the maximum conductivity achievable in a com-
     posite.

2)   Surface area, structure, particle size and surface
     chemistry of the filler.

3)   Type of polymer system

     a)   polymer construction: co-polymer, blends, etc.

     b)   crystallinity

     c)   wetability

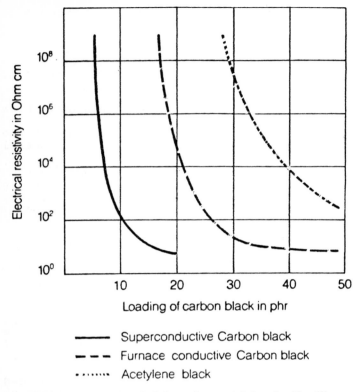

- ———— Superconductive Carbon black
- — — — Furnace conductive Carbon black
- •••••••• Acetylene black

FIGURE 5. Typical resistivity values obtained with different carbon blacks.

d)   rheological behaviour

e)   conductivity of the pure polymer

4) Processing method

a)   mixing

b)   moulding - whether injection, compression or blow

c)   screw speed, jet dimension, cooling rate of the melt, temperature, time (these factors control the dispersion and orientation)

5) Other additives

a)   nature of the stabiliser system

b)   nature of the plasticiser

c)   nature of the antioxidant

d)   nature of dispersants

e)   nature of lubricants

All the above factors have to be considered before deciding on the optimum conditions for the formulation of a conductive composite.

BP 2000, Akzo EC300 and poly(aniline) were compounded into polypropylene and the results are shown in Figure 6. It is clear that poly(aniline) can be compounded into PP and intermediate conductivities can be obtained consistent with what has been observed by Wessling[6].

3b Metallic Fillers

They are generally flakes of silver, copper, nickel, aluminium, silver-coated copper or silver-coated nickel, or fibres of stainless steel. Alloys like brass flakes are also used. Low-melting alloys are reputed to give very high conductivity at low loading levels if the melting point is tailored to match the processing temperature of the particular polymer concerned. There has been

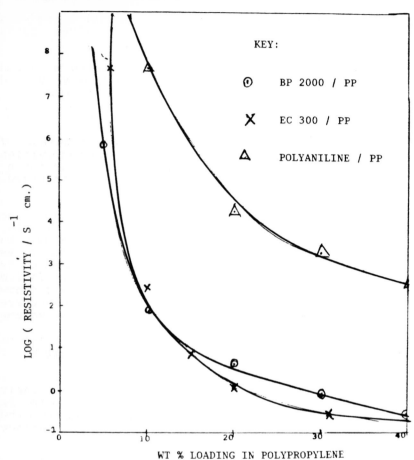

FIGURE 6. RESISTIVITY VERSUS LOADING LEVELS

an improvement in the quality of stainless fibres produced by Baekert. Also of interest are nickel- or silver-coated mica.

Metallised materials such as nickel-coated mica and metallised glass spheres and fibres are extensively used to produce composites for electromagnetic shielding.

3c   Semiconducting Oxide Fillers

Semiconducting oxide fillers have the advantage of being lightly coloured. However, if higher conductivities are required, they need to be doped and the colour intensifies or darkens as the doping level is increased, counteracting the advantages.

Doped or undoped tin oxides, titanium dioxides, zinc oxide and zinc sulphide are generally used. For example, indium oxide doped tin oxide, titanium dioxide coated with tin oxide doped antimony oxide are now available as commercial samples. Indium oxide is very expensive to be used as a filler. However, coating a mineral with indium tin oxide might be useful in some instances; these are now available: Mitsubishi (Japan) produces three conductive grades based on Sb_2O_3 and SnO_2, which require a volume loading of 20-35% to impart any conductivity.

3d   Quaternary Ammonium Salts As Antistatic Fillers

Quaternary ammonium compounds have been used for antistatic applications for over three decades. Their mechanism of conduction is by diffusion of the ions to the surface of the polymer and by absorption of water from the atmosphere. The absorbed water molecules form a surface film.

The conduction is therefore via proton hopping and is ionic in nature. As the diffusion of the quaternary ammonium salt is a prerequisite, this is the rate determining step and the induction period could be up to six months. High or reasonably high humidity is required, so that the abstraction of moisture is easy. Atmer 163, a quaternary ammonium compound (liquid) marketed by ICI, when compounded into PP at 10 weight percent (or 9.97% vol. loading) did not show any conductivity (i.e. $<10^{-13}$ S cm^{-1}). However, when the samples were immersed in water, for three months, they gained a conductivity of 1.5×10^{-9} S cm^{-1}.

The quaternary ammonium salts have clearly the disadvantage of attracting water and also requiring water for conduction, but they do produce white composites which can be pigmented.

4 RETICULATED POLYMERS

The preparation of these polymers involve (1) dissolving a traditional polymer (e.g. PVC) and a desired amount of charge transfer complex (e.g. TTT-TCNQ$_2$) in a hot solvent (e.g. chlorobenzene), (2) casting the film onto a glass (or any other inert substrate) kept at a predetermined temperature and (3) evaporating off the solvent. Normally, a dendritic type crystalline network is formed (see Figure 7a) and conductivities as high as 10^{-4} S cm^{-1} achievable at a 0.5 % wt loading level producing a transparent film as shown in Figure 7b.

Although this method is very attractive, the process is of limited industrial use as the organic solvents used may present safety problems.

5 ICP BASED CONDUCTIVE FILLERS FOR CHARGE DISSIPATION

Static charge generation is a phenomenon that happens frequently in daily life. This problem has been appreciated for at least a century and is becoming a major problem in the electronic age in relation to various semiconductor devices and their miniaturization. Billions of pounds are lost owing to static charges and the loss is growing[15]. The damage can be in the form of failures of electronic equipment as a whole or in parts like printed circuit boards, integrated circuits, transistors, MOS-FET devices, etc.

The damage can also be caused by static charge induced ignition and consequent fire.

Table 5 summarizes the minimum voltage required to cause damage to various devices and is a good indication of how sensitive modern electronic devices are, particularly

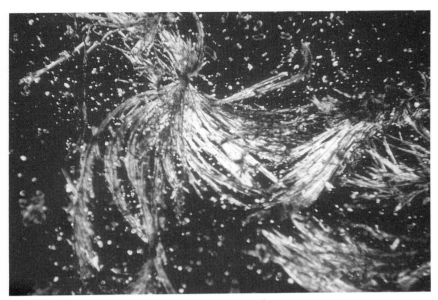

FIGURE 7 .a. PVC FILM CONTAINING TTT(TCNQ)$_2$
SHOWING DENTRITIC CRYSTALS(700 x MAGNIFICATION,
POLARISED LIGHT).

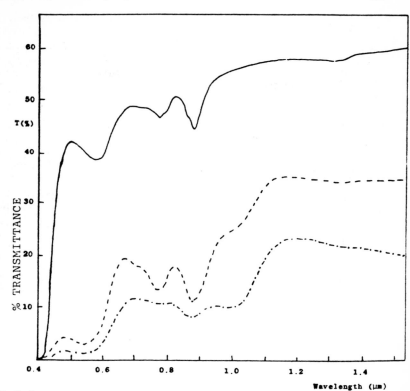

FIGURE . 7 . b . Transmittance spectra of PVC films containing TTT(TCNQ)$_2$: (———) 0.5 wt.% (50 μm); (— — —) 3 wt.% (50 μm); (— · — · —) 5 wt.% (20 μm).

Table 5   Minimum Voltage Required to Damage Various Devices

Type	Voltage/Volts
V-Metal oxide semiconductor (VMOS) device.	30
Metal oxide semiconductor field effect transistor (MOSFET).	100
Erasable programmable read only memory (EPROM).	100
Junction gate field effect transistor (JUGFET).	140
Surface acoustic wave devices.	150

Table 6   Typical electrostatic voltage

Means of Static Generation	Electrostatic Voltages/Volts	
	10 to 20 percent Relative Humidity	65 to 90 percent Relative Humidity
Walking across carpet	35,000	1,500
Walking over vinyl floor	12,000	250
Worker at bench	6,000	100
Vinyl envelopes for work instructions	7,000	600

when it is compared with Table 6 showing the various levels of voltages developed in the normal course of life.

Charge build-up can be avoided by making the surface in question sufficiently conductive so that the charge is dissipated at a controlled rate. Just as too slow a decay of charge is hazardous, too fast a decay is equally hazardous by causing ignition and fire or emitting a secondary radiofrequency wave which necessitates the protection of devices from radiofrequency interference.

The surfaces can be made conducting by either compounding a conductive filler (e.g. carbon black) into a polymer (making it volume conductive) or coating the substrate with a conductive paint (e.g. carbon loaded acrylate systems). However, carbon black has two serious disadvantages in that the percolation curve is so sharp (see Figure 5) that the intermediate conductivities required for safe charge dissipation (see Tables 7 and 8) are not possible, and the colour is black and therefore pigmentation is not possible.

Metal-loaded systems are prone to corrosion and have unreliable thermal characteristics, i.e. the conductivity changes on thermal cycling.

The inherently conducting polymers do offer the possibility of fine-tuning the conductivity and producing lightly coloured materials for example, poly-(isothianaphthene). The charge dissipation time depends not only on the conductivity but also on the dielectric constant of the material. Normally, a surface resistivity in the range 10^8-10^{10} ohms per square is required to give a charge decay time constant between 10 ms and 500 ms required for safe charge dissipation[15].

Future specifications will essentially specify the charge decay time rather than the conductivity requirements[14].

5a Using Inherently Conducting Polymers As Fillers

Inherently conducting polymers can be advantageously used just like carbon black and metal-based fillers, but offering superior electrical properties. One such polymer is poly(aniline) para-toluenesulphonate (PA).

Table 7   Electrical requirements for industrial products

Product	Electrical resistance/Ω min.	max.
Flooring for antistatic purposes	5×10^4	10^8
Hose	3×10^3 per m	10^6 per m
Hose with conducting cover only	3×10^3 per m	10^6 per m
Non-wire-reinforced hose with permanently attached metal fittings	3×10^3 per m	10^6 per m
Antistatic tyres	10^4	10^7
Footwear, antistatic soles and heels	7.5×10^4	5×10^7

Note: Electrical resistance as measured according to the procedures given in BS 2050 (1978).

Table 8   Electrical requirements for hospital products

Product	Electrical resistance/Ω min.	max.
Anaesthetic tubing	3×10^4 per m	10^6 per m
Flooring material	5×10^4	2×10^6
Footwear	7.5×10^4	5×10^7
Mattresses and pads	10^4	10^6

Note: Electrical resistance as measured according to the procedures given in BS 2050 (1978).

However, the material can be made cheaper by coating PA on to cheap mineral fillers (see Table 9) such as talc (PAT).

The choice of talc was not entirely based on its low cost. Amongst the minerals, talc, mica, etc. have platelet type of structure and Wollastonite, fibrous structure (see Figure 8). From Figure 4 it is clear that the higher the aspect ratio, the lower would be the threshold volume loading required to make particle-particle contact. As Wollastonite (a fibrous material) is attacked by acid, a reagent which is required in the sythesis of poly-(aniline), talc was employed.

5b   The Advantages Of PAT

PAT when compounded into PVC, gives green transparent films at or below a thickness of 50 μm. However, above this thickness, the composite is dark blue in colour. It certainly allows easy access to intermediate conductivities as shown in Figure 9 with various charge decay times. This way, controlled charge dissipation is possible. Thus, at a 25% volume loading of PAT, a charge decay constant of ~10 ms is achievable.

PAT can be compounded into other polymers such as PP, LDPE, PETG and PC and conductive composites with conductivity in the range 10^{-4} S cm^{-1} can readily be obtained.

5c   Other Conductive Polymer Coated Minerals

Other conducting polymers such as poly pyrrole(s), and polythiophen(s) can be coated on to various minerals such as Wollastonite, aluminium trihydrate, calcium carbonate, mica, etc.[16]. Table 10 summarises a selected number of such coated minerals.

5d   Stability Of Poly(aniline) Coated Talc (PAT)

The conductivity of PAT has been monitored over a 20 month period both in the dark and under UV, and no significant decay in conductivity has been observed. The thermal stability of PAT has been compared with that of nickel powder and pure polyaniline at 200°C and the relative decay constants are tabulated in Table 11 based on the gradient of the $\ln(\sigma_t - \sigma_\alpha)$ vs time plot.

Table 9   Typical cost of selected minerals

Mineral	Price (£/1000 kg)
Talc (Luzenac 1445)	221
Mica	410
Aminosilane coated Wollastonite	470
Aluminium trihydroxide	
- (Baco FRF 10)	297
Calcium carbonate (Polcarb S)	95
Clay	213

Table 10   Physical Properties of Some Selected Coated Minerals

	Mineral		
Property	PAT	PATH	PAPyrLS
Room Temperature Conductivity/S cm^{-1}	2.0 ± 0.5	1.0 ± 0.1	0.4 ± 0.1
Average Particle Size (μm) by SEM	17.8	10.0	15.0
Surface Area (m^2 g^{-1})	5.5	7.02	-
Density (g cm^{-3})	1.97	1.88	1.72

PAT: Poly(aniline) coated talc, PATH: Poly(aniline) coated on Aluminium trihydrate (Baco FRF-10), PAPyrLS (Polyaniline coated on polypyrrole coated calcium carbonate).

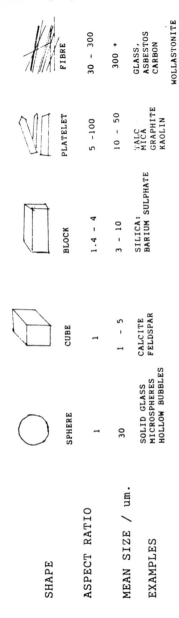

SHAPE	SPHERE	CUBE	BLOCK	PLATELET	FIBRE
ASPECT RATIO	1	1	1.4 - 4	5 -100	30 - 300
MEAN SIZE / um.	30	1 - 5	3 - 10	10 - 50	300 +
EXAMPLES	SOLID GLASS MICROSPHERES HOLLOW BUBBLES	CALCITE FELDSPAR	SILICA: BARIUM SULPHATE	TALC MICA GRAPHITE KAOLIN	GLASS, ASBESTOS CARBON WOLLASTONITE

FIGURE 8. VARIOUS SHAPES OF SOME COMMONLY USED FILLERS

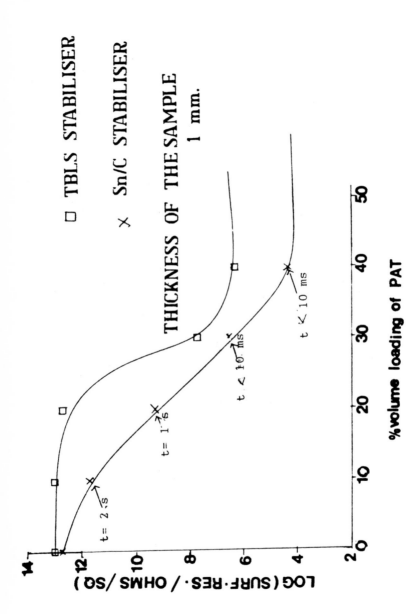

FIGURE 9. SURFACE RESISTIVITY OF PAT / PVC COMPOSITES

WITH DIFFERENT STABILISERS.

Table 11

Material	Relative Conductivity Decay Constant/Arbitrary Unit
PA	0.83
PAT	0.13
Ni (flakes)	0.05

The higher the decay constant, the more unstable the product is. It is clear from the above that the stability of PAT is comparable to that of nickel and the stability of PAT is much greater than that of PA. TGA/DTA study indicatesthat PAT is stable up to 300°C.

Conductivities of PA and PAT have been measured between 250 K and 480 K and semiconductor behaviour with activation energies 0.04 eV and 0.05 eV respectively have been obtained.

6   USE OF INHERENTLY CONDUCTING POLYMERS FOR ELECTRO-
    MAGNETIC SHIELDING

Electromagnetic interference is of major concern as it causes malfunctioning of electronic and electrical systems. Electromagnetic field arises from virtually any electrical source either DC or AC. For example, a high tension cable in a car engine can interfere with the in-car radio if the radio is not protected from EMI/RFI. An AC source can be any piece of electrical equipment which runs off the AC-mains; a transmission cable; a short pulse of electrical discharge (EMP) between two objects; switches or relay contact. Other day-to-day experiences are the interference of TV reception by a microwave oven. The latter can interfere with a telephone conversation as well.

A calculator which has not been shielded could cause problems with the reception of a radio placed nearby. A portable computer, if not properly shielded, could cause interference with the radar signal or other radio signal to and from aircraft. At worst, it could result in malfunctioning. The implications are severe in the case of military equipment including aircraft. All the equipment has to be protected for the efficacious functioning of the equipment and to prevent electronic espionage.

6a   Choice Of Shielding Materials

An electromagnetic shielding material has to be so chosen to shield both the electrical and magnetic component of the wave[20]. In other words, the conductivity as well as magnetic permeability have to be matched and their frequency dependence must also be considered. At low frequencies (<100 MHz), the magnetic component is important. As the ASTM-ES7083 (emergency standard) deals only with frequencies above 30 MHz, most manufacturers do not have to concern themselves with the lower frequencies.

Both electrical and magnetic shielding are now required by the Federal German regulations. Other countries are expected to follow suit.

6b   Advantages Of Inherently Conducting Polymers

As shown[9,11] in Figure 10, polypyrrole (curve a) shows a constant spectral response, which does not change on thermal cycling between 20°C and 80°C. Even a mixture of magnetic iron powder (20% by wt) and conductive aluminium flakes (30% by wt) in a thermoplastic is no match to poly-(pyrrole) film (cf. curve a with b). On thermal cycling between 20°C and 80°C, the S.E. has dropped very significantly in the case of Al and Fe mixture (cf. curve b and c). This is because the ohmic contacts between particles have become capacitive on thermal cycling. A good quality nickel paint coating of the same thickness (viz. 100 μm) shows a dip (curve d) at 300 MHz and the S.E. in the region 250-500 MHz is not acceptable. Similar behaviour is exhibited by brass flakes in ABS (curve e). It is therefore suggested that poly(pyrrole) is a good material for EMI/RFI shielding throughout the frequency range 10 kHz to 1 GHz.

ICPs can be produced in a variety of forms; viz. thin films, dispersion, (for example, acrylate or PVC/PVA based), aerosols and composites. Unlike carbon black, ICP's can be compounded with minimum environmental pollution and some polymers possess light colours.

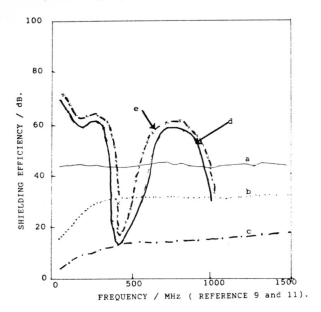

FREQUENCY / MHz (REFERENCE 9 and 11).

a: polypyrrole(100 μm)
b: 30 % (wt) Aluminium flakes and 20 % (wt) Iron powder
   in a thermoplastics.
c: b,after 3 cycles between 20^0 C and 80^0 C
d; good quality nickel paint (100 μm).
e: 58 % (wt) brass flakes in ABS.

FIGURE 10. SHIELDING EFFICIENCY VERSUS FREQUENCY
OF SELECTED MATERIALS.

Furthermore, ICPs' thermal characteristic, are better than carbon-filled or metal-filled polymers on repetitive thermal cycling.

In this era of information transport, low-frequency shielding of data transmission cables (10 kHz-30 MHz) becomes very important and ICP's may offer applications in this area[19]. With the new EEC Directive on EMC coming into effect on 1st January 1992, shielding of components, equipment and cables in the frequency range 10 kHz-18GHz becomes important. Therefore, there is an urgent need for novel materials for this application.

7   SYNTHESES OF INHERENTLY CONDUCTING POLYMERS

Inherently conducting polymers can be synthesised chemically (generally by oxidative polymerisation), electrochemically (constant current or constant potential), photochemically or photoelectrochemically. These are well documented in references 1,12,16,17.

8   CONCLUSION

Conducting polymers, whether traditional or inherently conducting (ICP) are expected to find a lot of applications with the growing awareness of EMC and ESD.[14,15,19,20].

ACKNOWLEDGEMENT

I would like to thank Dr J. S. Campbell (Research Director, Cookson Group plc), Mr R. Clabburn (Chief Executive, Cookson Devleopment Division), Dr J. D. Coyle and Dr J. B. Jackson for their strong support and my family for their patience and understanding. I would like to thank Angela Timberlake for her patience during several revisions of this manuscript.

REFERENCES

1. a)   Conducting Polymers and Plastic, ed. by J. M. Margolis, Chapman and Hall, London (1989).

    b)   Inherently Conducting Polymers, M. Aldissi, Noyes Data Corporation, Park Ridge, New Jersey, USA (1989).

c) Electroactive Polymer Materials, Anders Wirsen Technomic Publishing Company Inc., Pennsylvannia (1990).

d) P. Kathirgamanathan, D. Shah, A. S. Bhuiyan, R. Hill and R. W. Miles, in Surface Engineering Practice, p. 377-386, Ellis Horwood (London) 1990.

2. K. Quill, A. E. Underhill and P. Kathirgamanathan, Synthetic Metals, *32*, 329 (1989).

3. P. Kathirgamanathan, University of Southampton (unpublished work) (1985).

4. a) M. R. Bryce, A Chissel, P. Kathirgamanathan, D. Parker and N. R. M. Smith, J. C. S. Chem. Comm., 466 (1987).

b) M. R. Bryce, A. D. Chissel, N. R. M. Smith, D. Parker and P. Kathirgamanathan, Synthetic Metals, *26*, 153 (1988).

c) D. Parker, P. Kathirgamanathan, A. C. Chissel, M. R. Bryce and N. R. M. Smith, E. Pat. Appl. No. 87 306133.7 (1988).

d) P. Kathirgamanathan and N. R. M. Smith, E. Pat. Appl. No. 873061 34. 5 (1988).

e) P. Kathirgamanathan, P. N. Adams, A. Marsh and D. Shah, E. Pat. Appl. 88306224, 2 (1989).

f) P. Kathirgamanathan, P. N. Adams, K. Quill and A. E. Underhill, E. Pat. Appl. 883062 88. 7 (1988).

g) P. Kathirgamanathan, K. Quill and A. E. Underhill, E. Pat. Appl. 88306489. 1 (1988).

5. N. T. Kortschot, B.Sc. Project Report, University of Toronto, Canada (1984).

6. B. Wessling, Kunststoffe German Plastics, *76* (10), 69 (1986).

7. Degussa, Technical Bulletin: Pigments No. 69.

8.   J. D. Van Dumpt, Plastic Compounding, 37 (1988).

9.   a)   K. H. Mobius, Kunstoffe German Plastics, **78**, 53 (1988).

     b)   V. Krause, ibid **78**, 17 (1988).

10. a)   P. Kathirgamanathan and D. R. Rosseinsky, unpublished work (1981).

     b)   J. Roncali, A. Mastar and F. Garnier, Synthetic Metals, **18**, 857 (1983).

     c)   J. Roncali and F. Garnier, J.C.S. Chem. Comm., 783 (1986).

11. "Filimet" Internal Report.

12. P. Kathirgamanathan, J. Electroanal Chem., **247**, 351 (1988).

13. Cabot Corporation, Technical Report, S. 39

14. W. B. Lord, Conference Report, ESD '90, 2.3 (1990). Polygon Hotel, Southampton, UK, 25-26 April 90.

15. a)   Electrostatics, Summer School Report, University College of North Wales, Bangor, Gt. Britain (1985).

     b)   P. Kathirgamanathan, Conference Report, ESD '90, 3.4 (1990), Polygon Hotel, Southampton, UK, 25-26 April 1990.

     c)   Static Electrification, Fundamental Concepts, Hazards and Applications, University College of North Wales, Bangor, Gt. Britain, (1986).

16. P. Kathirgamanathan, P. N. Adams, A. M. Marsh and D. Shah, UK Pat. Appl. No. 8728149 (1987).

17. a)   Handbook of Conducting Polymers, Vol I and Vol II, ed. by T. A. Skotheim, Marcel Dekker Inc. (1986).

b) P. Kathirgamanathan, ISE 38th Meeting Extended Abstracts Vol II, Maastricht, Netherlands, Sept. 13-18 (1987).

c) Electronic Properties of Conjugated Polymers, Proceeding of the International Winter School, Kirchberg, Tirol, March 14-21 (1987), ed. by H. Kuzonany, M. Mehring and S.Roth, Springer Verlag, (Berlin).

d) E. E. Havinga, W. Ten Hoert, E. W. Merger and H. Wynberg, Chemistry of Materials, 1, 650 (1989).

e) S. Bask, K. Nayak, D. S. Marynick and K. Rajesh-war, Chemisty of Materials, 1, 611 (1989).

f) K. Yoshino, S. Hayashi and R. Sugimoto, Japanese J. of Appl. Phys., 23, L899 (1984).

g) T. Ohasaura, K. Kaneto and K. Yoshino, Japanese Journal of Appl. Phys., 23, L663 (1984).

h) J. Roncali, H. K. Youssoufi, R. Garreau, F. Garnier and M. Lemaire, J. C. S. Chem. Comm., 414 (1990).

i) I. Rodriguez and J. G. Valasco, J. C. S. Chem. Comm., 387 (1990).

18. Data Sheet "Pre-Elec", Premix Oy, P. O. Box 12, 05201 Rajamaki, Finland.

19. B. Wessling, Kunststoffe German Plastics, 80 (3), 21 (1990).

20. IEEE Transactions on Electromagnetic Compatibility, 30 (3), (1988) and all the papers therein.

A Review of Recent Developments Concerning Applications of the "Durham" Route to Polyacetylene

C. S. Brown

BP RESEARCH CENTRE, SUNBURY-ON-THAMES, MIDDLESEX TW16 7LN

1 ABSTRACT

The synthesis, kinetics, structure, electronic and optical properties of the "Durham" route to polyacetylene have been reviewed previously. Examples of particular electronic devices such as diodes, FETs, and electro-optic modulators employing the novel physics of polyacetylene have also been published. However, it is less well known that the "Durham" route can be modified to make passive optical devices such as gratings, holographic optical elements and waveguides. In this review the processing steps required to produce these elements from a single film of poly[[5,6-bis(trifluoromethyl)-bicyclo [2,2,2,] octa-5,7-diene-2,3-diyl]-1,2-ethenediyl] (PFX) is outlined. The optical properties of the three materials formed by processing PFX in three different ways are summarised and examples of prototype devices are illustrated. The possibility of combining these passive optical elements with active third order non-linear optical elements or other active devices in a fully integrated fashion is also discussed briefly.

2   INTRODUCTION

Polyacetylene has been the subject of considerable scientific and technological interest because of its particular electronic and non-linear optical properties [1,2]. For example an electro-optic modulator has been fabricated which makes use of the energy state in the middle of the band gap of polyacetylene [3,4]. Field effect transistors and diodes have also been demonstrated [5,6].

Polyacetylene is also well known for its high third order susceptibility ($\chi^{(3)}$ in excess of 10^{-8} esu at 1907nm [7]), its very fast response time 10^{-13}s [8] and its broad band width [9]. This has led to interest in producing non-linear optical devices with polyacetylene as the active element. The first of these devices to be reported was a simple phase conjugate mirror [10]. A particular feature of the Durham route to polyacetylene [11,12] is that it allows optical quality coherent films to be prepared without diluting the non-linear optical effects. Thickness variation below 3% over a 20 cm^2 area has been measured [13] and samples prepared from 20 nm to 5 microns thick. This is achieved by first spin coating a precursor polymer from solution onto a substrate and then heating the film in vacuum. See Fig 1. A fluoroxylene molecule is eliminated leaving polyacetylene. By this means the intractability of polyacetylene is neatly side-stepped.

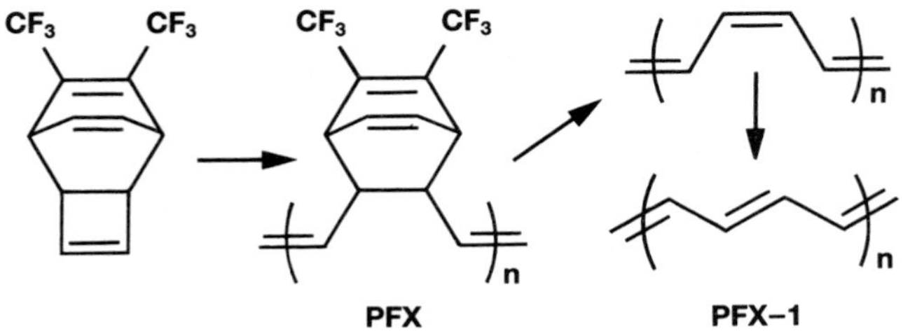

Figure 1(a)   The "Durham" route to polyacetylene.
The processable precursor polymer (PFX) is heated to
form predominantly cis polyacetylene which on further
heating isomerises to trans polyacetylene (PFX-1).

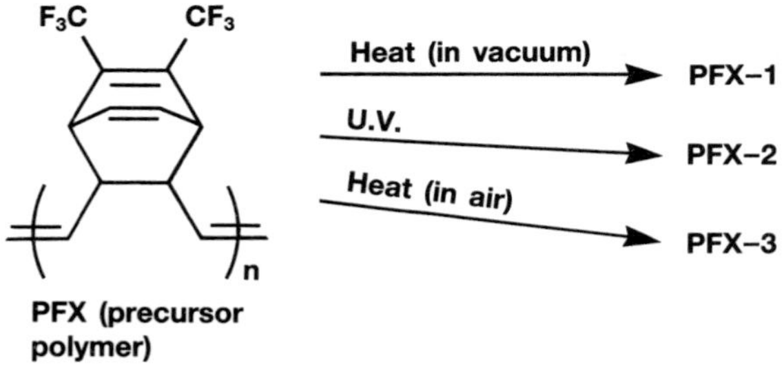

Figure 1(b)   Diagram illustrating the three PFX
derivatives that can be formed from a single PFX
film.   Combinations of these materials have led to
the fabrication of gratings, H.O.Es and waveguides.
Further work may lead to the integration of active
and passive devices.

This precursor route also provides a second advantage which has received very little attention in the literature. It has been found possible to photo-oxidatively cross-link the precursor polymer with ultra violet radiation [14] (This material has been named PFX-2 for convenience, see Fig 1b. PFX-1 is polyacetylene). The elimination reaction of Fig 1a will then not occur. Furthermore if PFX is heated in an oxidising environment, instead of in vacuum, a third material results which has been named PFX-3, (see Fig 1b). The implications of these effects have been investigated [13,15,16]. A range of linear optical devices have been demonstrated including gratings, waveguides, transmissive and reflective holographic optical elements operating in the visible and near infra-red. In principle these devices could be integrated with the active functions associated with polyacetylene itself on a single film without resorting to wet processing steps. It is the purpose of this article to review these linear optical effects and their device applications.

The Linear Optical Properties of Polyacetylene and its Precursor Polymer

The precursor polymer shown in Fig 1 is a transparent insulator (band gap >2.5 eV) without conjugation. Rotation is allowed about the carbon carbon single bonds which renders the polymer soluble in organic solvents, typically ketones. Molecular weights up to 800,000 have been measured by low temperature gel permeation chromatography [17]. In many respects it is a typical fluoropolymer. However, on heating, each 1,2-bis(trifluoromethyl) benzene molecule lost gives rise to one triene unit in the polymer. If two adjacent units are lost a sequence containing five double bonds is formed.

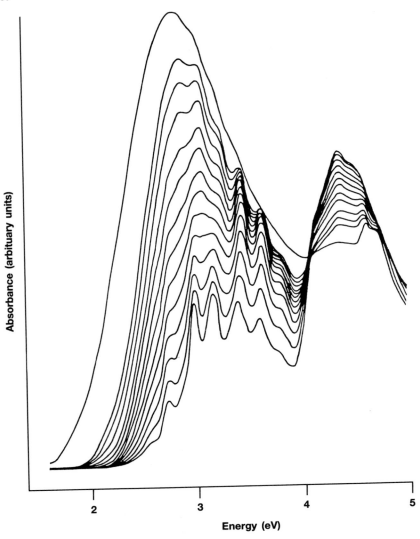

Figure 2.   Optical absorption spectra of the
precursor polymer (PFX) during transformation to
polyacetylene.

Gradually the average conjugated sequence length increases until polyacetylene itself is formed. The material transforms from insulator to semiconductor, the band gap narrows and the optical absorption spectra alter accordingly (see Fig 2). These "ene" sequences absorb light efficiently which in the limit gives rise to polyacetylene's high absorption coefficient of up to $1.6 \ 10^5$ cm^{-1} (see Fig 3).

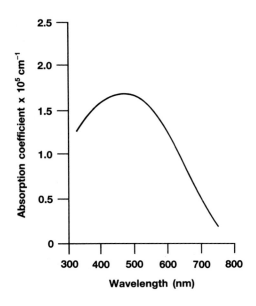

Figure 3. The absorption spectrum of unoriented trans polyacetylene between 325 and 750 nm.

Polyacetylene's absorption spectra can be further altered by doping. Fig 4 shows the effect of chemical doping with sodium. It may be possible to use chemical doping to tailor polyacetylene's absorption to a wavelength of particular interest.

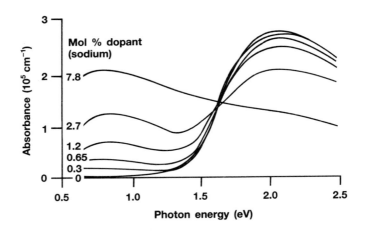

Figure 4.   Optical absorption spectrum of polyacetylene (PFX-1) as a function of dopant concentration.

Considerable insight is gained into polyacetylene's intrinsic optical properties by examining oriented samples. If a sample of the precursor is stretched during its transformation to polyacetylene, using an apparatus such as that shown in Fig 5, an oriented but not perfectly crystalline material results. The azimuthal full width at half maximum of the (200/ 110) x-ray diffraction peak is 6-8° (see Fig 6 and [18,19]). The optical properties of these oriented films have been measured[20] and are shown in Figs 7 and 8.

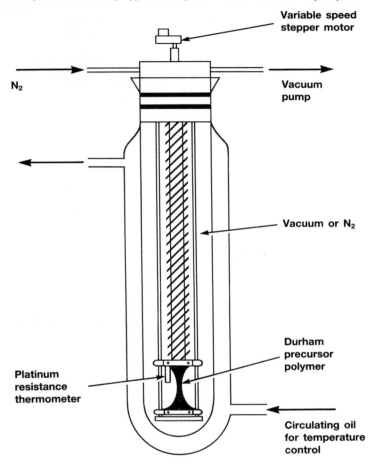

Variable speed stepper motor

N₂

Vacuum pump

Vacuum or N₂

Durham precursor polymer

Platinum resistance thermometer

Circulating oil for temperature control

Figure 5. Diagram of apparatus used to orient a film of polyacetylene by sketching it as it transforms.

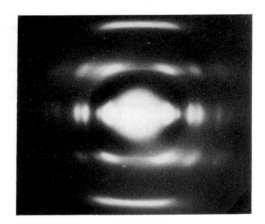

Figure 6.   Electron diffraction pattern recorded
from oriented Durham polyacetylene.

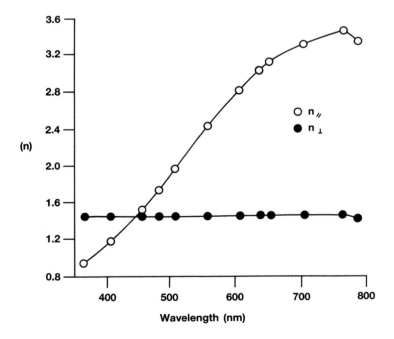

Figure 7.   Refractive index of oriented
polyacetylene (PFX-1).

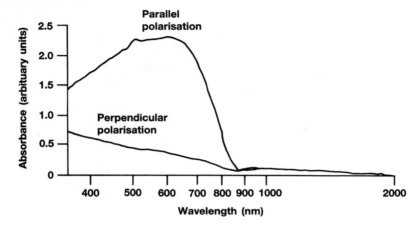

Figure 8.   Optical absorbance of oriented
polyacetylene.

With light polarised parallel to the conjugated
chains strong dispersion of the refractive index is
observed which is similar to observations of
semiconductors or metals. However, perpendicular to
the chains the refractive index has a constant value
of 1.4 from 400 through to 800nm consistent with a
non-conjugated polymer. These figures also show that
samples can be highly birefringent (Δn max = 2.0) and
have a very high anisotropy of absorbance ($\alpha_{//} / \alpha_{\perp}$
greater than 27 [21]). The optical properties of
unoriented polyacetylene have also been measured in
detail by ellipsometry [22]. Fig 9 shows the
refractive index and extinction coefficient as a
function of wavelength measured by this technique.

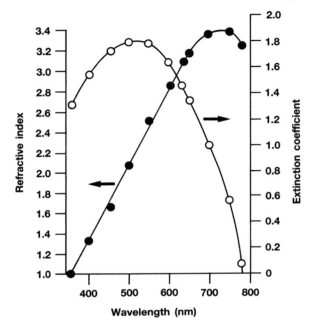

Figure 9.   Optical constants of unoriented
polyacetylene.

Gratings and Holographic Optical Elements

If the precursor polymer of Fig 1 is heated slightly (ie one hour at 40°C) it will form some short conjugated sequences, as illustrated in Fig 10, it will absorb ultra violet light very efficiently and a complicated photo-oxidative crosslinking reaction can occur [15,16]. Further heating of this cross-linked material <u>does</u> <u>not</u> result in polyacetylene. Instead a stable, transparent material with a refractive index of 1.5 is produced, see Fig 11. We have named this polymer PFX-2.

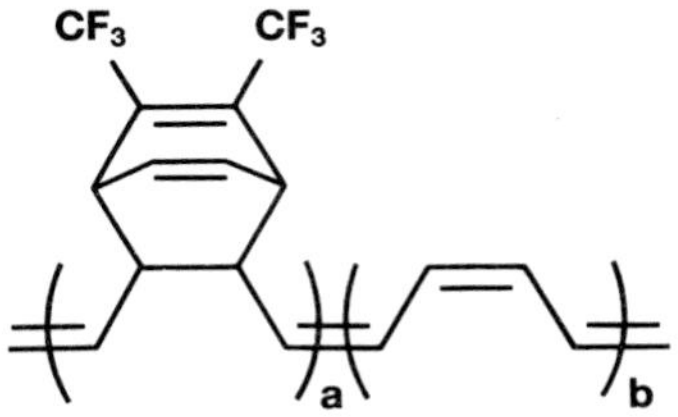

Figure 10. The structure of the precursor polymer after partial heating. Blocks of conjugated sequences (b) start to build up which are separated by uneliminated units (a). The conjugated sequences absorb u.v light very efficiently and are thought to initiate photo-oxidative cross-linking reactions.

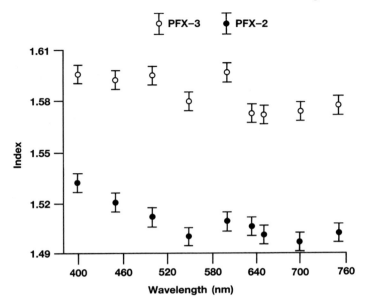

Figure 11.   Comparison of refractive index of PFX-2
and PFX-3.

Gratings and holographic optical elements can be
made using this effect. A film of PFX is irradiated
with ultra violet light through a mask, in air, as
shown in Fig 12. The irradiated areas crosslink to
PFX-2. On heating in vacuum the unirradiated areas
transform to polyacetylene (PFX-1). There is now a
difference in height (Δd), refractive index (Δn) and
absorption coefficient (Δα) between the two areas as
illustrated schematically in Fig 13.

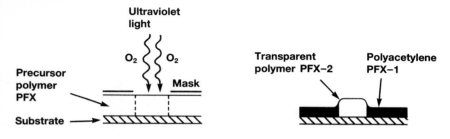

Figure 12.   Schematic for the image generation
process in precursor   polymer PFX.

$\Delta \propto$ absorption

Δ n refractive index

Δ h height

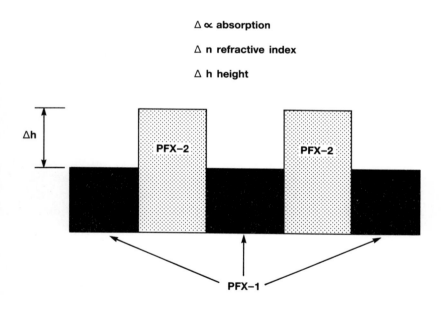

Figure 13.   Illustration of the type of grating
obtained by the process shown in Figure 12.   Note
that contrast arises from three sources; differences
in height (Δh), absorption ($\Delta\alpha$) and refractive index
(Δn).

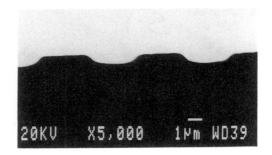

Figure 14.   Scanning electron micrograph of the edge
of a diffraction grating produced by the method
illustrated in Figure 12.

If this film is coated with a thin layer of metal the
difference in thickness (Δd) gives rise to an
efficient reflection grating (Fig 14). The height of
the relief can be controlled by varying the exposure.
Fig 15 shows the power law relationship between
relief height and exposure time.   The mask of Fig 13
can also be a computer generated holographic optical
element (HOE).   Such a mask will be limited in
diffraction efficiency to 6%. However, when copied
into PFX a diffraction efficiency of $\approx$30% can be
achieved close to the theoretical maximum for this
type of reflection hologram (34%).   Fig 16 shows such
a HOE in operation.   A helium neon laser beam is
being coupled into a fiber optic via the HOE.   This
is a primitive example of interconnection between
optical components using a polymer HOE.   Direct
recording of holograms by interfering two coherent
ultra violet laser beams has also been demonstrated.

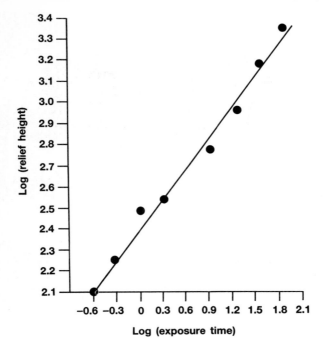

Figure 15. Graph showing the relationship between relief height and time of exposure at a fixed UV intensity.

Figure 16.   A reflection HOE made from PFX.   A
helium neon laser beam can be seen to be diffracted
into several orders.   The first order is coupled into
a fibre optic.

   Transmission gratings and HOEs can also be made
from PFX.   The difference in absorption coefficient
Da between PFX-1 and PFX-2 (Fig 17) is large in the
visible and this fact could be used to provide a
transmission HOE.   However, the diffraction
efficiency would not be enhanced with respect to the
original mask.   Nevertheless,   if the PFX film is
heated in air (eg 100°C for 20 minutes) instead of in
vacuum, a transparent stable material (PFX-3) results
14,16.   The relief height difference can be used to
induce a phase difference in transmitted light and

hence the production of gratings and HOEs up to a theoretical diffraction efficiency of 34%. A demonstration is illustrated in Fig 18. This sample is now air stable because it contains no polyacetylene. There is also a refractive index difference between PFX-2 and PFX-3 of about 0.08 as shown in Fig 11. This effect could also be used to form transmission gratings and HOEs. However, this effect counteracts the phase difference provided by the surface relief height difference.

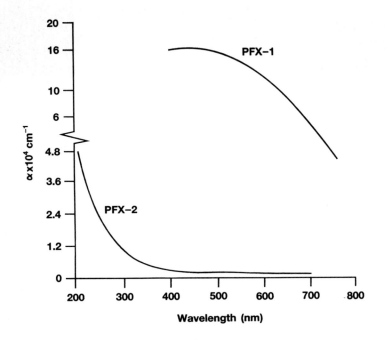

Figure 17.   Comparison of absorption spectra of PFX-1 and PFX-2.

Figure 18.   Photograph of a transmission HOE made
from PFX.   A helium neon laser beam can be seen to be
diffracted into several orders.   The first order is
focused and coupled into a fibre optic.

Waveguides

The refractive index difference referred to above
can be used to form waveguides. Guiding over 4cm has
been demonstrated as shown by Fig 19.   The light
travels in the higher refractive index PFX-3 which is
surrounded by PFX-2.   The large refractive index
difference (0.08), which extends into the near infra-
red, may allow waveguide structures with tight bends

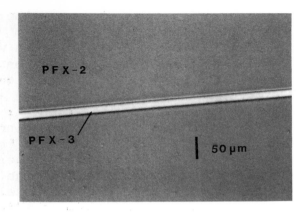

Figure 19.   PFX-3 stripe waveguide, with PFX-2
cladding.

to be constructed.   Polyacetylene itself has a high
enough refractive index for guiding but its high
absorption coefficient ($\approx 10^5$ cm^{-1} at 550nm) prevents
guiding in the visible.   However, its absorption
coefficient falls off in the infra-red leading to the
possibility of guiding over small distances (<mm) at
infra-red wavelengths.

Integration Potential

   Three polymers with very different optical
properties have been shown to be derivable from a
single processable optical quality polymer film (Fig
1b).   So far a range of simple optical devices have

been demonstrated individually, however, there is a
prospect of combining these devices to form
integrated optical circuits. For example a waveguide
could be combined with a HOE for three dimensional
planar and free space optical interconnection
schemes. It may also be possible to define arrays of
non-linear optical or electro-optic elements and
address these via waveguides in the same plane or
from HOEs situated in free space. None of these
scenarios have been investigated yet and their
realisation presents a significant challenge. This
unique set of device properties may ultimately be
attainable from a single polymer film of PFX.

2   CONCLUSION

Polyacetylene is well known for its electronic and
non-linear optical properties. The Durham route to
polyacetylene enables optical quality coherent films
to be produced (via a precursor polymer). However,
the route also offers a further advantage which is
less well known. Ultra violet irradiation of the
precursor polymer induces crosslinking which in turn
prevents the reaction to polyacetylene on heating.
Instead a stable transparent polymer is formed. This
is the basis of a unique form of lithography
involving only light and heat. Simple linear optical
devices such as gratings, holographic optical
elements and waveguides have been fabricated using
this technique. Since these passive devices, as well
as active polyacetylene devices, can all be formed by
treating a single polymer film it has led to the
possibility of using PFX in integrated optic
applications.

3 ACKNOWLEDGEMENTS

I would like to thank Dr P C Allen, Dr D C Bott, Dr M R Drury, Mr N S Walker, Mr S Gray, Dr M M Ahmad, Dr L M Connors and Dr K Walsh of BP, Mr J Andreassen of STC, Professor W J Feast and Dr P I Clemenson of Durham University and Professor Sir G Porter of the Royal Society. I would also like to thank BP for permission to publish this paper.

REFERENCES

1. D.C. Bott, Phys Technol, 1985, 16, 121.
2. M. Bacon, Materials Edge, 16 March 1989.
3. J.H. Burroughes, C.A. Jones, and R. H. Friend, Nature, 1988, 335 (8) 137.
4. P.C. Allen, J.H. Burroughes, R.H. Friend, and A.J. Harrison, Eup Patent, 1988, 304554.4.
5. J.H. Burroughes, R.H. Friend, and P.C. Allen, J. Phys, 1989, D 22, 956.
6. P.C. Allen, J.H. Burroughes, and R.H. Friend, Eup Patent, 1988, 0298628.
7. M.R. Drury, Solid State Communications, 1988, 68 (4), 417.
8. A.J. Heeger, D. Moses, and M. Sinclair, Syn Met, 1986, 15, 95.
9. F. Kajzar, S. Etemad, G.L. Baker, and J. Messier, Sol State Comm, 1987, 63, 1113.
10. L.M. Connors, and M.R. Drury, 1988, US Patent 4,768,848.
11. D.C. Bott, J.H. Edwards, and W.F. Feast, 1985, US Patent 4,496,702.
12. D.C. Bott, J.H. Edwards, and W.F. Feast, Polymer, 1984, 25, 395.
13. S. Gray, Photonics Spectra, September 1989, 125.
14. W.J. Feast, P.I. Clemenson, P.C. Allen, D.C. Bott, C.S. Brown, L.M. Connors, S. Gray, and N.S. Walker, "Springer Series in Solid-State Sciences, Vol 91", Heidelberg, 1989, 456.
15. P.C. Allen, US Patent, 1989, 4,798,782.
16. L.M. Connors, and D.C. Bott, US Patent, 1989, 4,857,426.
17. K. Harper, and P.G. James, Mol Cryst Liq Cryst, 1985, 117, 55.
18. D.C. Bott, and D. White, Eup Patent, 1984, 301773.2.
19. D.C. Bott, C.S. Brown, J.N. Winter, and J. Barker, Polymer, 1987, 28, 601.
20. M.R. Drury, PhD Thesis Queen Mary College, London 1988.
21. M.R. Drury, Private Communication.
22. M.R. Drury, Syn Met, 1989, 32, 33.

Polymer Films for Non-linear Optical Applications

C. A. Jones, J. R. Hill, and P. Pantelis

BRITISH TELECOM RESEARCH LABORATORIES, MARTLESHAM HEATH,
IPSWICH, SUFFOLK, IP5 7RE

1 BACKGROUND

Non-linear optics is the study of the response of a dielectric medium to an intense electromagnetic field, such as that produced by the propagation of laser radiation through the material. This response is a result of the incident electro-magnetic radiation inducing an oscillating electric dipole in the medium, leading to the emission of secondary electro-magnetic radiation, which combines with the incident wave to give rise to new fields altered in frequency, phase or amplitude relative to the incident fields. On a microscopic scale, this is characterised by a dipole moment, p, which is dependent upon the electric field, E, according to a power series:

$$p = \alpha E + \beta E^2 + \gamma E^3 + \ldots \qquad \textbf{1}$$

where α, β and γ are tensors, known respectively as the linear polarisability, and the second and third order hyperpolarisabilities.

The bulk polarisation, P, (dipole moment per unit volume) of a non-linear optical material can be obtained by summing the dipole moments of the individual molecular species. The following equation is obtained:

$$P = \epsilon_0 \ [\chi^{(1)}E + \chi^{(2)}E^2 + \chi^{(3)}E^3 + \ldots] \qquad \textbf{2}$$

where $\chi^{(m)}$ is the mth order electric susceptibility, and ϵ_0 is the permittivity of free space. The odd terms in this expression ($\chi^{(1)}$, $\chi^{(3)}$ etc.) are finite for all materials, but the even terms ($\chi^{(2)}$, $\chi^{(4)}$ etc.) are non-zero only for those materials which lack a centre of symmetry. α, β and γ in Equation 1 are clearly analogous to $\chi^{(1)}$, $\chi^{(2)}$ and $\chi^{(3)}$ in Equation 2; thus, a large value for β will contribute to a high $\chi^{(2)}$, providing the molecular species are arranged non-centrosymmetrically in the bulk solid.

In this paper, we are concerned primarily with materials exhibiting second order non-linear optical effects such as optical frequency mixing, optical rectification, optical frequency doubling and the linear electro-optic effect. The latter two are of particular interest. The linear electro-optic (or Pockels) effect is the phenomenon whereby the refractive index varies linearly with an applied electric field. Optical frequency doubling or second harmonic generation (SHG) is the conversion of light of frequency ω to light of frequency 2ω. A linear electro-optic coefficient, r_{ijk}, and a second harmonic coefficient, d_{ijk}, can be defined thus [1]:

$$2\chi^{(2)}{}_{ijk}(-\omega;\omega,0) = (n_{ii})^2{}_\omega \, (n_{jj})^2{}_\omega \, r_{ijk}(-\omega;\omega,0) \qquad \textbf{3}$$

$$\chi^{(2)}{}_{ijk}(-2\omega;\omega,\omega) = 2d_{ijk}(-2\omega;\omega,\omega) \qquad \textbf{4}$$

where n is the appropriate refractive index at the indicated optical frequency, and the notation $(-2\omega;\omega;\omega)$, for example, refers to two input beams each of frequency ω, and an output beam of frequency 2ω. [2]

The linear electro-optic effect has applications in a fast response modulator, whereby information to be transmitted is applied to the material as an electric field. If light is passing through the material, then the electric field will alter the refractive index such that the emergent light is modulated synchronously with the electric field. Thus, information can be converted between electrical and optical media.

2 POLYMERS FOR NON-LINEAR OPTICS

Large values of β typically occur in organic molecules possessing low lying charge-transfer states, with donor and acceptor groups separated by a conjugated system. However, the large ground state dipole moments of many high β molecules cause them to crystallise centrosymmetrically. This can sometimes be overcome by incorporating a bulky side group, such as CH_3, into the molecule. The simplest such case is 2-methyl-4-nitroaniline (MNA), which is shown in Figure 1. Although high β values can be obtained

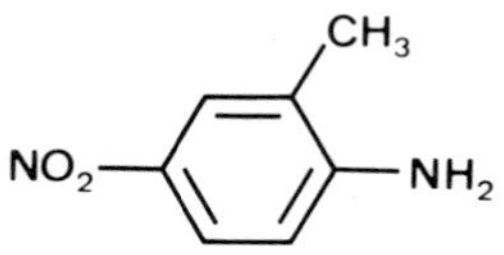

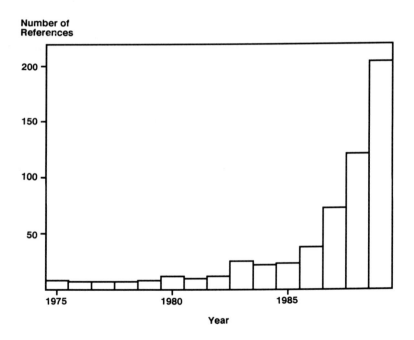

Figure 1   2-methyl-4-nitroaniline (MNA) molecule.

Figure 2   Growth in number of publications on polymeric materials for non-
              linear optics.

in this way, it has proved difficult to produce large optical quality single crystals of many potentially useful substances. An alternative approach has therefore been adopted using polymers with a suitable high β species either dissolved as a guest in an inert polymeric host, or covalently bonded to the polymer as a side-chain. In order to satisfy the requirement for non-centrosymmetry, the species must be induced to point, at least statistically, in the same direction by the application of an external electric field, a process known as poling.

Polymers have several advantages over the inorganic materials, such as lithium niobate, which are currently used for non-linear optical devices. The vast scope of organic chemistry means that a "molecular engineering" approach can be used to design materials with the required properties. Polymers are relatively inexpensive, and can be processed using standard semiconductor processing techniques, such as spin-coating and reactive ion etching. Also, since polymers generally have low DC dielectric constants (ϵ_r ~ 3) compared with inorganic materials (for lithium niobate ϵ_r = 28), the time constants of devices will be reduced [3].

Many research groups worldwide, in both universities and industry, are pursuing projects in non-linear optical polymers [4-14]. Figure 2 shows the increase in interest over the last few years, as exemplified by the number of publications cited in Chemical Abstracts in the field of polymeric materials for non-linear optics.

The choice of polymer for any given non-linear optical device is critical, since many diverse criteria must be fulfilled. It has already been stated that the non-linear species (whether a guest or a side-chain) must have a high value of β, and that the polymer should be capable of being poled to produce a stable, non-linearly active system. Additionally, the linear optical properties, such as refractive index and optical loss, must be suitable at the wavelengths of interest; for optical communications applications these are 850 nm (for local area networks) and 1300 or 1550 nm (for long distance communication). Furthermore, the polymer must be processible into a suitable form for fabricating a modulator.

3 WAVEGUIDES

In order for a modulator to be of practical use, it is desirable to use a structure in which there is a long interaction length between the light and the active polymer, in order to maximise the non-linear effects which occur. Much work has therefore been carried out on producing a modulator in the form of a waveguide, along which light is transmitted for a distance of many millimetres. If polymer waveguides can be deposited onto semiconductor substrates, then there is the possibility of combining both electro-optic and electrical processing functions into a single component.

Apart from optical fibres, there are two generic types of waveguide: planar waveguides (Figure 3a), in which light is confined in one dimension, but is free to diffract within the plane of the film, and stripe waveguides where light is confined in two dimensions. A stripe waveguide may be formed as either a channel in a surrounding medium (Figure 3b), or as a raised rib on a substrate (Figure 3c). In all cases, confinement occurs within a region of higher refractive index than the surroundings.

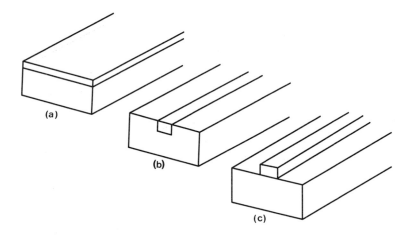

Figure 3   Types of waveguide: (a) planar, (b) channel, (c) rib.

The dimensions of the guiding region are of the order of the wavelength of light, which minimises the effective beam area and hence maximises the input power per unit area in a waveguide. Radiation can propagate only as the eigenmodes of the waveguide structure [15]. The number of eigenmodes and the field distributions within the modes are dependent upon the refractive indices and dimensions of the guiding and surrounding regions. Planar waveguides are useful for assessing the properties of non-linear optical materials, as they can be made rapidly. However, stripe waveguide geometries are required for practical modulator designs, as they are more suited to coupling light to and from optical fibres. This can be achieved simply by firing light directly from a fibre into the end of a waveguide.

4 DEVELOPMENT OF POLYMER DEVICES AT BTRL

In this section we will describe in detail the development of polymer electro-optic waveguide modulators at British Telecom Research Laboratories (BTRL), using one particular approach. This will provide insight into the stringent requirements for non-linear optical polymers and the complexity of the development process, which adds value to these polymers. In the next section we will review alternative methods for forming active polymer waveguides and devices, which have been adopted by other research groups.

The technique we have chosen for depositing polymer films is spin-coating, since this technique is known to produce high quality, large area thin films, and is widely used in semiconductor processing for the deposition of photoresists. The first step in characterising a new non-linear polymer is to study its spin-coating properties. The spinning solvent, solution concentration and spin speed are varied systematically until a uniformly thick, optical quality film is obtained. The film quality is often dependent on ambient conditions such as temperature and relative humidity; spin coating is therefore carried out in an environmentally controlled laminar flow unit. Before spinning, all solutions are filtered to remove any particulate matter which would cause spinning defects and increase the scatter loss of spun films. For optical applications, films of thickness in the region 1 to 5 μm are usually required. Often this can be achieved in a single spin-coating step, but sometimes multiple layers must be built up; in which case, it is important that the polymer dissolves slowly enough in the selected solvent for the previous layer not to re-dissolve during spinning of subsequent layers.

Once uniform films have been obtained, the linear optical properties of the material can be assessed. This is done using slab waveguides of thickness $\sim$1 to 5μm deposited onto silica substrates. To measure refractive index, the technique of prism launching [16] is used, whereby a laser beam is fired into a prism clamped to the film. The angle of incidence is then varied until a guided mode is launched from the prism into the film, and the refractive index and thickness can then be calculated from the incident angles of two or more modes, using standard formulae. Measurements can be made at selected wavelengths in the visible or infrared; we have used 633 nm, 1047 nm and 1321 nm. Optical loss is also measured using light launched via a prism, by using an optical fibre probe to detect light sidescattered from the film. It is desirable that non-linear optical polymers have optical losses as low as possible. For most polymers the minimum value which can be attained is limited, at wavelengths around 1000 nm, by C-H overtone absorptions to values of about 0.1 - 1 dB/cm.

Poling, non-linear activity and stability can also be studied on slab waveguides. This type of study can be carried out in several different ways[17], but one technique is as follows: the polymer is spin-coated onto conducting indium tin oxide (ITO) coated glass onto which a buffer layer of poly(methyl methacrylate) (PMMA) has previously been spun; a second buffer layer of PMMA is spun on top, and the film is then metallised with aluminium. The polymer is then poled by heating to above its glass transition temperature, applying a large DC voltage, and cooling it back to ambient temperature with the voltage applied; a process known as thermopoling. Since optical measurements are to be made in transmission, it is then necessary to remove the aluminium electrode by etching with sodium hydroxide solution. The PMMA layers act as a buffer, protecting the non-linear polymer from the alkaline etch and from electrochemical degradation. A measure of the non-linear optical activity can be obtained by measuring the second harmonic coefficient, d_{33}, using Maker fringe analysis. In the Maker fringe technique, the sample is rotated in the path of a laser beam; frequency doubling occurs through the second order non-linearity of the polymer, and analysis of the second harmonic intensity as a function of incidence angle gives the second harmonic coefficient. Measurements must be made relative to a standard sample, usually a Y-cut quartz plate of known second harmonic coefficient. d_{33} is measured immediately after poling, and also over a period of time after poling. A knowledge of the ageing characteristics of the poled polymers is desirable, since stability is an important attribute for a practical material.

Recently, Eich *et al* [12] have developed a technique for rapidly assessing the poling properties and non-linear coefficients of polymers by measuring d_{33} whilst the polymer is being poled. In this case, the sample is poled by applying a corona discharge whilst the film is heated above its glass transition temperature, and the second harmonic coefficient is again measured using the Maker fringe technique.

Once the fundamental properties of a non-linear optical polymer have been established, a stripe waveguiding structure incorporating the polymer can be designed. Several designs of waveguide have been produced, in which the active polymer stripe is either a rib or a channel on a substrate. In all cases, in order to confine light to the active polymer layer (the "core"), it must be surrounded by media of lower refractive index. At BTRL we have chosen to make rib type structures on glass substrates, where the rib consists of three layers: a low refractive index polymeric cladding layer, the core polymer, and a second low refractive index polymeric cladding layer, which may be the same as or different from the first cladding layer. It is clearly important that the core and cladding polymers must be such that they can be spin-coated from solvents which do not re-dissolve the previous layer. Even a small amount of dissolution can cause problems, as it may produce a polymer-polymer interface with a high scatter loss. The selection of polymers for cladding materials is therefore just as critical as for the core. Mechanical rigidity is provided by bonding a glass cover-slip to the structure, using an epoxy resin, which must also be of lower refractive index than the core material, in order to provide lateral confinement of the light in the waveguide.

The number of optical modes which can propagate along a waveguide at a given wavelength is dependent upon the relative refractive indices of the core and cladding layers, the thicknesses of the layers, the width of the rib waveguide and the refractive index of the overcoating epoxy. For most practical applications, it is essential to fabricate monomode waveguides, *(ie.* waveguides in which only the lowest order optical mode can propagate). We therefore make use of design software, which predicts the number of modes and the optical power distribution for any combination of the above parameters. In practice, the refractive indices are restricted to certain values by the materials available. However, if two polymers can be found which are highly miscible, and can be spun from the same solvent, then a certain degree of tuning of the refractive index can be attained by using mixtures of the polymers.

We will now describe the processing steps for making a polymer waveguide modulator; the process is illustrated in Figure 4. A layer of aluminium is deposited onto the glass substrate as a lower electrode for the application of both poling and modulation voltages. Successive layers of lower cladding, core and upper cladding polymers are then spin-coated from filtered solutions. The sample is baked at an appropriate temperature after deposition of each layer, in order to drive off residual solvent. A top electrode layer of aluminium is then deposited, to give the structure shown in Figure 4a.

At this stage, the entire slide is covered with the laminated structure, and it is then necessary to etch away all but the desired rib of polymer. This is achieved in two steps. In the first step, the rib waveguide pattern is transferred into the upper aluminium layer using conventional UV photolithography; a layer of photoresist is spun onto the sample, and is then exposed to UV light on a mask aligner through a mask which has the desired pattern on it (Figure 4b). The resist is developed and the aluminium etched, leaving the required pattern in the aluminium (Figure 4c). The pattern is then transferred to the polymer by using the top aluminium layer as a mask in a reactive ion etcher. The reactive ion etcher consists of an evacuated chamber with electrodes at the top and bottom. The sample is placed on the bottom electrode, a reactive gas (in this case, oxygen) is admitted to the chamber, and an RF field is applied across the low pressure gas between the electrodes, leading to a plasma being established in the chamber. The combined effect of ion bombardment and chemical oxidation etches away the polymer everywhere except in the region of the rib. The etched structure is shown in Figure 4d.

Wires are attached to the upper and lower aluminium electrodes, using silver loaded epoxy resin, and the sample is thermopoled. The poled sample is then overcoated with low refractive index epoxy resin and bonded to a glass cover-slip. Since light is to be coupled into the ends of the waveguides, it is necessary to prepare high quality end faces. This is achieved by cutting and polishing the ends using conventional mechanical techniques. The waveguide is then ready for launching laser light in and out of the ends using fibres or lenses, and a modulating voltage can be applied between the two electrodes.

Although this technique for fabricating waveguides is very versatile in terms of the range of materials which can be used, it is difficult to produce

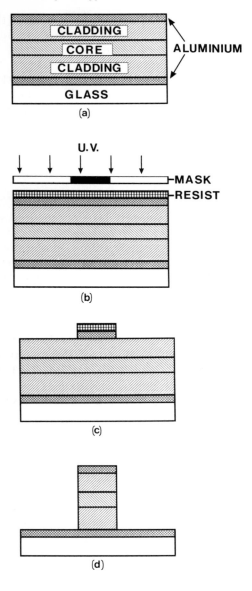

Figure 4   BTRL method for preparing a polymer waveguide by reactive ion etching. (See text for details).

smooth waveguide side-walls by reactive ion etching. If the side-walls are not smooth, then the scatter loss in the lateral direction will be large.

5 OTHER TECHNIQUES FOR MAKING POLYMER DEVICES

We will now review some of the other methods which have been used to define non-linear optical polymer waveguides. While some of these techniques were developed for third order non-linear materials, they may equally be applied to second order polymers.

(a) Selective Poling

This technique was developed by Lockheed Missiles and Space Company, Inc., using materials synthesised by Hoechst Celanese Corporation [18], and allows waveguide definition and poling to be carried out in a single step. The process is illustrated in Figure 5; the arrows indicate the orientation of the dipole moments of the non-linear species. An electrode pattern defining the waveguide is deposited directly onto a substrate photolithographically. The lower cladding layer and the core are then spin-coated, and the top planar electrode is deposited directly onto the core polymer (Figure 5a). The polymer is then thermopoled as described above, but only those regions of the material defined by the lower electrode pattern will be poled (Figure 5b). Because the non-linear polymer is anisotropic after poling, it becomes birefringent, and the refractive index for one polarisation in the poled region will be greater than that for the unpoled polymer; thus light with this polarisation will be confined laterally. The device is completed by etching away the upper poling electrode, spinning an upper cladding layer, and depositing a switching electrode (Figure 5c). This technique has the advantage that the edges of the guiding region will be smoother than those produced by reactive ion etching, reducing lateral scattering losses. However, since the lateral cladding consists of unpoled core polymer and has a fixed refractive index, variation of the refractive index of the core region to produce the required index contrast can only be achieved by altering the poling field and hence the electro-optic coefficient.

(b) Deposition into Etched Channels

Deposition of polymers into channels etched into a substrate has

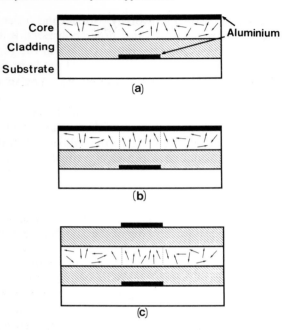

<u>Figure 5</u>   Selective poling method for preparing a polymer waveguide. (See text for details).

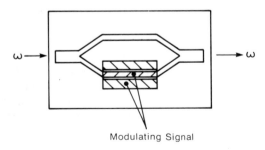

Modulating Signal

<u>Figure 6</u>   Mach-Zehnder interferometer.

been used by researchers at Osaka University [6] for second order non-linear polymers, and by GEC-Marconi [8] for third order polymers. In both cases, the polymer covers the whole substrate, but is thicker in the region of the channel, and this provides confinement of the guided mode to the channel. In reference 6 the channels are chemically etched in a glass substrate, and the non-linear polymer is spun directly into the channels, so that the glass and air respectively form the lower and upper claddings; thus there is little scope for varying the refractive indices of the constituent layers. In reference 8, a layer of SiO_xN_y is deposited onto a silicon substrate, and the channel is reactive ion etched into the SiO_xN_y, which acts as the lower cladding layer. The refractive index of this layer can therefore be varied by adjusting the stoichiometry of the SiO_xN_y. The silicon substrate also provides the advantage that the finished sample can be cleaved, removing the need to cut and polish the end-faces. However, the problem of rough side-walls produced by reactive ion etching of the SiO_xN_y remains.

(c) Other Methods

Researchers working on a project entitled "Polymeric optical switches" within the Research on Advanced Communications in Europe (RACE) program have reported that they have investigated several different methods of forming single mode channel waveguides in electro-optic polymers [7]. They have been particularly successful with a technique in which stripes of a non-linear polymer are embedded in a derivative of the same polymer of slightly lower refractive index, and sandwiched between buffer layers of polyurethane. The details of the fabrication of this type of structure have not yet been published.

6 OBSERVATION OF MODULATION

Applying a voltage to a device containing a second order non-linear polymer will change the refractive index of the polymer, leading to a change in the phase of light propagating through it. One way of observing this phase change is to construct a Mach-Zehnder interferometer, in which the waveguide is defined such that it is not simply geometrically linear, but is of the design shown in Figure 6, where the guide splits into two arms, and re-combines. A voltage is applied to one arm of the interferometer, causing a phase shift of the light in that arm relative to the other, and thus destructive interference occurs when the beams recombine, leading to modulation in the amplitude of the emergent beam. In an ideal modulator, a small voltage will bring about a π radian phase shift, such that there is 100% modulation. Other modulating devices include directional couplers [3], where light is switched between two ports through the electro-optic effect, and cut-off modulators, where the refractive index change is sufficient to cut-off all optical modes when a voltage is applied. Many other types of modulators and switches are possible, and have been reviewed by other authors [19].

7 RESULTS

Early work on electro-optic polymers used active monomeric guests in a polymeric host. One of the first reports of such a system [11] is of the azo dye Disperse Red 1 (Figure 7a) in a matrix of PMMA, in which second harmonic generation was observed. Electro-optic modulators have been made using guest/host systems by several research groups. For example, Haga and Yamamoto [6] made a 10 mm long channel waveguide using MNA in PMMA, and observed 16% modulation depth for an applied voltage of 100V at a wavelength of 633 nm. Two main drawbacks have been found for such systems. Firstly, the amount of guest that can be added is limited, and hence the electro-optic coefficients are usually only of the order of a few pm/V, compared with 30 pm/V for lithium niobate. Secondly, it has been found that the activity drops over a relatively short period of time as the guest molecules return to a random orientation. For example, for PMMA doped with the material shown in Figure 7b, a value for d_{33} of 1 pm/V was measured at 1047 nm immediately after poling; this dropped by 50% after only 2 weeks [17].

Several different approaches have been taken to improve the stability of guest/host systems. Hampsch *et al* [13] have found that guest/host systems become more stable if the guest molecules are physically larger. They have attributed this to the lack of free volume available for re-orientation of the molecules. We have used Foraflon-7030 (Atochem UK Ltd.) as a host polymer; this is a copolymer of vinylidene fluoride and trifluoroethylene in the molar ratio 70:30. This polymer is partially crystalline, and when it is poled with a corona discharge, the carbon-fluorine dipoles in the crystalline regions are brought permanently into alignment with the applied field. If the polymer is doped with non-linear guest molecules, which reside in the amorphous regions, then these also align with the poling field. However, after the poling field has been removed, the permanent internal electric field within the amorphous regions maintains the partial alignment of the guest molecules [20]. Our original study [21] employed the cyanophenylazo-aniline material shown in Figure 7b as the guest. The d_{33} second harmonic coefficient of the mixture was found to increase linearly with guest concentration and to have a value of 2.6 pm/V at a concentration of 10%.

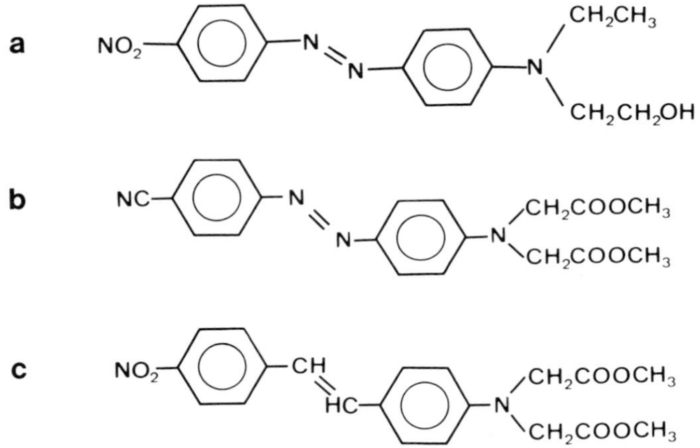

Figure 7   Compounds used as guests in polymer hosts. (See text for details.)

We have shown that the primary ageing mechanism in these films is precipitation of the guest rather than decay of the internal field [17]. Films produced using the more soluble aminonitrostilbene guest shown in Figure 7c showed no precipitation, and their coefficient did not change significantly over a period of more than 300 days. The optical loss of slab waveguides made from such mixtures is about 5 dB/cm at 633 nm, of which some 4 dB/cm can be attributed to scatter from the host. The scatter losses fall rapidly with increasing wavelength, and the bulk loss is about 1.5 dB/cm at 1300 and 1550 nm [22].

In view of the low dopant concentrations and instability of guest/host systems, researchers have tended to progress towards using polymers in which the active species is bonded to the backbone as a side-chain. Singer *et al* [23] have compared a PMMA based guest/host system with the equivalent side-chain polymer, and observed that the second harmonic coefficient of the guest/host system decayed to 25% of its original value over a period of

35 days, whilst the side-chain polymer maintained 90% of its initial value after the same period.

There have been reports of side-chain polymers based on vinyl backbones[10], styrenes [24] and acrylates [25]. Hoechst Celanese Corporation have probably been the most prolific producers of side-chain electro-optic polymers, mainly acrylate copolymers [14]. These materials have been used by Lockheed to make electro-optic devices [18]. One polymer, known as C22, is reported to have a loss of 0.8 dB/cm at 830 nm, and r_{33} = 16 pm/V. Several devices have been made using this polymer, including a travelling wave phase modulator, which operated at frequencies as high as 1 GHz, which is the highest reported frequency for a polymer electro-optic modulator. Celanese have recently produced polymers with an enhanced electro-optic coefficient of 38 pm/V at 1300 nm [26]. We have reported on a demonstrator modulator [25] based on the acrylate homopolymer shown in Figure 8, which is estimated to have an r_{33} of 30 pm/V at 633 nm for a thermopoling field of 20 V/μm. The modulator showed 17% modulation depth, and there was no observable degradation in its performance over a period of 2 years. Researchers working on the RACE project on "Polymeric optical switches" [7, 27] have been successful in developing a polymer with p-dimethylaminonitrostilbene (DANS) side-chains. This polymer has a stable electro-optic coefficient, r_{33}, of 28 pm/V at 1300 nm, and waveguides made from the polymer had a total propagation loss of only 1.2 dB/cm. Monomode phase modulators and Mach-Zehnder interferometers have been made from this material, which required voltages of only 10 V to induce a π phase shift in a 1 cm long device. The devices showed no dispersion in electro-optic coefficient over the frequency range 50 - 500 MHz.

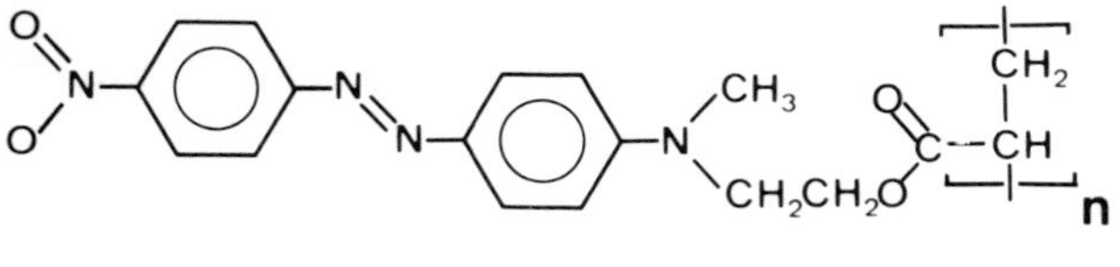

Figure 8   Acrylate side-chain homopolymer used in demonstrator modulator.

Recently, researchers have improved the stability of non-linear polymers further by using cross-linked systems. Akzo Research

Laboratories [28] have developed high thermal stability, high r_{33} cross-linked polymers. Eich *et al* [29] have reported on a polymer which can be cross-linked during poling, which has a final d_{33} of 13.5 pm/V, and does not show relaxation even at 85°C; however, no values for the optical loss of this polymer have been quoted.

8 THE FUTURE

Electro-optic modulators based on lithium niobate are currently commercially available. Although lithium niobate has a loss of only 0.1 dB/cm and an r_{33} of 30 pm/V, optical quality single crystals are expensive to grow, and hence are not well suited to large scale production. There is therefore potential for the development of cheaper non-linear optical materials, such as poled polymers.

Prototype polymeric devices have already been produced, and it has been predicted [30] that inexpensive polymer phase and amplitude modulators operating at frequencies above 10 GHz will be available within 5 years. Lytel and Stegeman have produced a specification of a "target polymer" for electro-optic devices [30], which includes the following properties: $r_{33} > 30$ pm/V, total loss ≤ 0.1 dB/cm, stability over 5-10 years, good performance at 850 nm, 1300 nm and 1550 nm, dimensional tolerance and stability to within 5 nm, refractive index uniformity to within 0.001, low dielectric constant, low electrical conductivity and high thermal conductivity. Excellent progress has been made towards attaining the primary requirements, namely high non-linear optical activity and low loss, and efforts are now being focused on optimising the other properties. Dimensional and refractive index tolerance will probably be the most difficult parameters to control in polymers, since they are dependent upon the precise composition and molecular weight, which are difficult to reproduce from batch to batch.

It seems likely that polymer materials suitable for practical device applications will be developed within the next decade. It is not anticipated [3] that poled polymers will completely supersede lithium niobate devices in optical communications. However, they are likely to compete well in terms of cost, and to find new applications, particularly in the field of optical signal processing, for spatial light modulators and neural networks.

ACKNOWLEDGEMENTS

We would like to thank our colleagues P.L. Dunn and D.G. Smith for performing optical measurements, and G.J. Davies, C.R. Day and R. Heckingbottom for their interest in the work.

REFERENCES

1.   A.F. Garito, K.D. Singer and C.C. Teng, in "Nonlinear Optical Properties of Organic and Polymeric Materials", D.J. Williams (Ed), ACS Symposium Series, Am. Chem. Soc., 1983, Vol. 233, Chapter 1.
2.   D.J. Williams, Angew. Chem. Int. Ed. Engl., 1984, 23, 690.
3.   D.R. Ulrich, Mol. Cryst. Liq. Cryst., 1988, 160, 1.
4.   P. Pantelis, J.R. Hill, S.N. Oliver and G.J. Davies, Br. Telecom Technol. J., 1988, 6, 5.
5.   J.I. Thakara, G.F. Lipscomb, M.A. Stiller, A.J. Ticknor and R. Lytel, Appl. Phys. Lett., 1988, 52, 1031.
6.   H. Haga and S. Yamamoto, Osaka University, Japan, Proc. Integrated Optics and Optical Communications Conference, Kobe, Japan, 1989.
7.   M.B.J. Diemeer, F.M.M. Suyten, E.S. Trommel, G.R. Mohlmann, W.H. Horsthuis, D.P.J.M. van der Vorst, A. McDonach, M. Copeland, C. Duchet, P. Fabre, S. Samso, E. van Tomme, P. van Daele and R. Baets, Proc. European Conference on Optical Communications, Gothenburg, Sweden, 1989.
8.   S. Mann, A.R. Oldroyd, D. Bloor, D.J. Ando and P.J. Wells, Proc. SPIE, 1988, 971 (Nonlinear Optical Properties of Organic Materials), 245.
9.   J.L. Jackel, N.E. Schlotter, P.D. Townsend, G.L. Baker and S. Etemad, Proc. SPIE, 1988, 971 (Nonlinear Optical Properties of Organic Materials), 239.
10.   M.J. McFarland, K.K. Wong, C. Wu, A. Nahata, K.A. Horn and J.T. Yardley, Proc. SPIE, 1988, 993, 26.
11.   K.D. Singer, J.E. Sohn and S.J. Lalama, Appl. Phys. Lett., 1986, 49, 248.
12.   M. Eich, A. Sen, H. Looser, G.C. Bjorklund, J.D. Swalen, R. Twieg and D.Y. Yoon, J. Appl. Phys., 1989, 66, 2559.
13.   H.L. Hampsch, J. Yang, G.K. Wong and J.M. Torkelson, Polymer Comm., 1989, 30, 40.

14.   R.N. DeMartino, European Patent no. 0 316 662, 1989.

15.   G.I. Stegeman and R.H. Stolen, <u>J. Opt. Soc. Am. B</u>, 1989 <u>6</u>, 652.

16.   P.L.Dunn, D.G. Smith, J.R. Hill, C.A. Jones and P. Pantelis, IEE Digest no. 1989/140 (Plastics Materials for Optical Transmission), 1989, p. 7/1.

17.   J.R. Hill, P. Pantelis, P.L. Dunn and G.J. Davies, Proc. SPIE, 1989, <u>1147</u> (Nonlinear Optical Properties of Organic Materials II), 165.

18.   R. Lytel, G.F. Lipscomb, M. Stiller, J.I. Thakara and A.J. Ticknor, in "Nonlinear Optical Effects in Organic Polymers", J. Messier *et al* (Eds), Kluwer Academic Publishers, 1989 p. 277.

19.   R.C. Alferness, <u>IEEE J. Quant. Electron.</u>, 1981, <u>QE-17</u>, 946.

20.   J.R. Hill, P. Pantelis and G.J. Davies, <u>Ferroelectrics</u>, 1987, <u>76</u>, 435.

21.   J.R. Hill, P.L. Dunn, G.J. Davies, S.N. Oliver, P. Pantelis and J.D. Rush, <u>Electron. Lett.</u>, 1987, <u>23</u>, 700.

22.   P. Pantelis, J.R. Hill and G.J. Davies, in "Nonlinear Optical and Electroactive Polymers", P.N. Prasad and D.R. Ulrich (Eds), Plenum Press, 1988, p. 229.

23.   K.D. Singer, M.G. Kuzyk, W.R. Holland, J.E. Sohn, S.J. Lalama, R.B. Comizzoli, H.E. Katz and M.L. Schilling, <u>Appl. Phys. Lett.</u>, 1988, <u>53</u>, 1800.

24.   C. Ye, N. Minami, T.J. Marks, J. Yang and G.K. Wong, in "Nonlinear Optical Effects in Organic Polymers", J. Messier *et al* (Eds), Kluwer Academic Publishers, 1989, p. 173.

25.   J.R. Hill, P. Pantelis, F. Abbasi and P. Hodge, <u>J. Appl. Phys.</u>, 1988, <u>64</u>, 2749.

26.   D. Haas, C.C. Teng, H. Yoon, H-T Man and K. Chiang, to be presented at Topical Meeting on Integrated Photonics Research, Hilton Head, South Carolina, USA, 1990.

27.   G.R. Mohlmann and W.H.G. Horsthuis, to be presented at Topical Meeting on Integrated Photonics Research, Hilton Head, South Carolina, USA, 1990.

28.   W.H.G. Horsthuis, P.M. van der Horst and G.R. Mohlmann, to be presented at Topical Meeting on Integrated Photonics Research, Hilton Head, South Carolina, USA, 1990.

29.   M. Eich, B.Reck, D.Y. Yoon, C.G. Wilson and G.C. Bjorklund, <u>J. Appl. Phys.</u>, 1989, <u>66</u>, 3241.

30.   R. Lytel and G.I. Stegeman, in "Nonlinear Optical Effects in Organic Polymers", J. Messier *et al* (Eds), Kluwer Academic Publishers, 1989, p. 379.

Multilayer Devices

M. Ahlers, H. Bader, R. Blankenburg, A. Laschewsky,
W. Müller, and H. Ringsdorf

INSTITUT ORGANISCHE CHEMIE, UNIVERSITÄT MAINZ, D-6500 MAINZ,
GERMANY

Self-organized assemblies[1] such as monolayers and
Langmuir-Blodgett (LB) multilayers have been investigated in recent years, for their potential use in
various devices[2-5] . These assemblies are appealing
due to their homogeneous thickness, to the anisotropic, well defined arrangement of the amphiphiles
within the layers and to chemically well defined
surfaces. Quality and stability of the layers has
been the focus of research in recent years[6] , and
much progress has been made, in particular by
taking advantage of polymerizable amphiphiles and
of tailored amphiphilic polymers[7-9] .Hence, current
research is aimed towards the preparation of
functional systems, and of functional surfaces[10,11] . In particular selective binding to
surfaces is investigated, with respect to their
potential use in sensor and biosensor devices[5,6] .
As illustrated schematically in Fig.1, binding to
mono- and multilayers can take place via coating or
via insertion, or a combination of both. The latter
process implies the reorganization of the surface
layer.

Fig.1 Binding to mono- and multilayers via
      a) coating or b) insertion

Most often, amphiphiles are functionalized via
their hydrophilic head groups. As LB-multilayers
exhibit hydrophobic surfaces generally, special
preparation techniques are required to expose the
functional groups[1,12] . Conveniently, a protective
layer is placed on top of the desired functional
layer, which is removed prior to use, e.g. by
extensive washing (Fig.2).

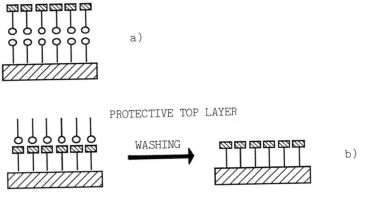

Fig.2 Preparation of functional surfaces
      ▬O = standard amphiphile, ▤ = functional group
   a) functional hydrophobic surface, preparation by
      standard LB-technique
   b) functional hydrophilic surface, preparation
      by protective coating technique

To ensure the selective removal of the protective
coating, and the integrity of the residual functional layer, the use of polymerizable amphiphiles
was shown to be favourable[12-14] . Fig.3 depicts two
such reactive, functionalized amphiphiles.

The binding processes can be distinguished in such
of low specifity, as such due to electrostatic or
to hydrophobic interactions, and in highly specific
ones, such as of receptor-ligand type. Although of
restricted selectivity only, even rather unspecific
binding processes can be exploited in selected
cases. e.g., multilayers made of charged amphiphiles bind efficiently to charged amphiphiles of
opposite sign, enabling the preparation of non-
centrosymmetric structures, as used for non-linear
optical materials or pyro- and piezoelec-
trica[15,16] . Too, the electrostatic interactions of
polyelectrolytes with charged amphiphiles lead to
complexed multilayers of improved deposition
behaviour[8,17] , and of exceptionally good thermal
stabilities[8] . Even binding via the poorly specific
hydrophobic forces may be useful, as exemplified by
the incorporation of pullulanes[18] and dextranes[19]
containing hydrophobic anchor groups into lipid
membranes.

However considering the use in sensor devices[5] ,
highly specific binding processes are needed, as
observed in biological recognition processes. Such
specific binding has been investigated in many
monolayer and LB-multilayer systems, covering
complexation[20-22] ,pairing of complimentary nucleo-
bases[10,23] , host-guest systems[24,25] , receptor-
ligand systems[26-31] and enzyme-substrate binding[32-34] . Examples of such functional amphiphiles
are given in Fig.3. Much insight in the nature of
binding processes has been gained recently, by the

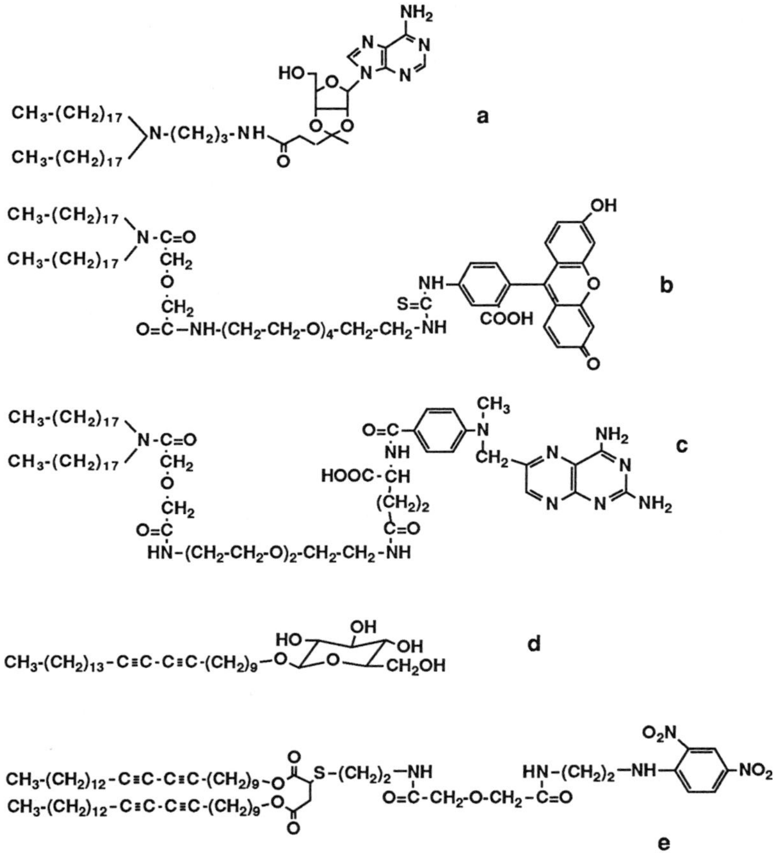

Fig.3
Functional amphiphiles investigated in monolayers or
multilayers for selective binding processes

a) adenosin-Lipid, binding to poly(uridylic acid)[10,23]
b) fluorescein-lipid, binding to anti-fluorescein antibody[11]
c) methotrexat-lipid, binding to dihydrofolate reductase[34]
d) polymerizable glycolipid, binding to lectin concanavalin A[13]
e) polymerizable DNP-lipid, binding to anti-DNP antibody[14]

combination of monolayer studies and simultaneous fluorescence microscopy[35] . e.g., the binding of anti-fluorescein-antibody to its hapten fluorescein has been investigated[11] . On binding, the fluorescence of the hapten is quenched. The use of a second fluorescence label attached to the antibody allows to discriminate between lipid layer and bound protein layer. Thus, extent and kinetics of the binding can be studied and quantified. Furthermore, the lateral organization of the bound protein is visualized, revealing the aggregation into domains, instead of homogeneous binding.

For practical device systems, it is desirable to combine the high binding specificity with a high versatility of the system chosen. i.e., instead of constructing systems which are designed for one special case only, a system which can be adapted to the actual problem by slight modification would be most advantageous.

Such a versatile and specific recognition system is still a challenge. Potential candidates may be based on the receptor-ligand system streptavidin-biotin[29-31] , which is characterized by very high binding constants[36] . Streptavidin was shown to form two-dimensional crystals below the monolayers of selected biotin-amphiphiles[30,31] , thus offering several modes of surface modification and functionalization[30] . First, derivatization of streptavidin e.g. by haptens is feasible. This is exemplified by the derivatization of streptavidin by fluorescein which was shown to enable the selective binding of the anti-fluorescein antibody[11] . Second, as at maximum two of the four binding sites of streptavidin for biotin are occupied in the primary binding to the lipid

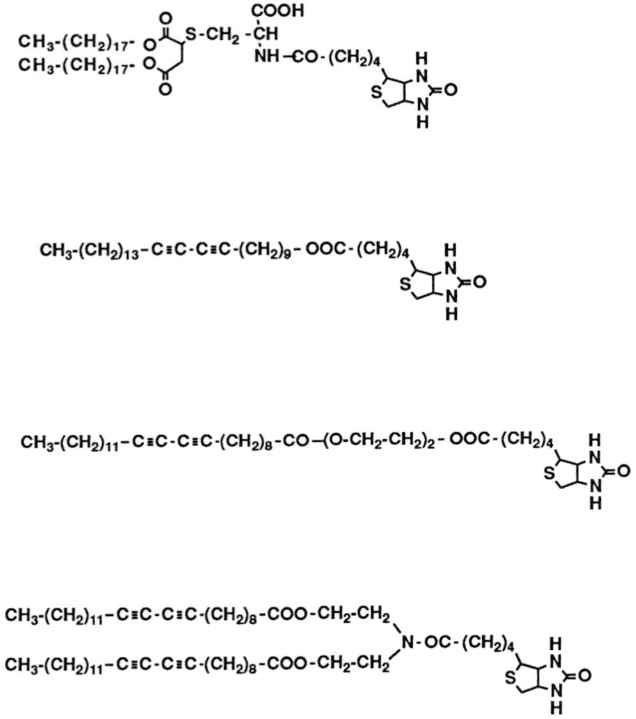

Fig.4
Saturated[29] and polymerizable[11] amphiphiles
functionalized with biotin, for selective
binding of avidin and streptavidin

membrane by the functional lipids, the residual
sites are free for further use. Hence, they are
able to bind selectively biotinylated substrates in
a secondary process. This was shown for biotinylated proteins such as ferritin[31] or C1q[11]. Due
to its crystallinity, the streptavidin layer might
even serve as a template to induce the crystallization of the second protein species. Experiments
to explore the potential as a general adapter
system are currently under investigation including
the use of polymerizable biotin-amphiphiles as
shown in Fig.4.

References

1) H.Kuhn, D.Möbius, Angew.Chem.__83__, 672 (1971)
2) G.G.Roberts, Adv.Physics __34__, 475 (1985)
3) T.M.Ginnai, Ind.Eng.Chem.Prod.Res.Div.__24__, 188
   (1985)
4) M.Sugi, J.Molec.Electronics __1__, 3 (1985)
5) J.D.Swalen et al., Langmuir 3, 932 (1987)
6) Thin Solid Films, Coll.Vol.99, 132-134, 152,
   154, 180
7) R.Elbert, A.Laschewsky, H.Ringsdorf, J.Am.Chem.
   Soc.__107__, 4134 (1985)
8) C.Erdelen, A.Laschewsky, H.Ringsdorf,
   J.Schneider, A.Schuster, Thin Solid Films __180__,
   153 (1989)
9) F.Embs,D.Funhoff, A.Laschewsky, U.Licht, H.Ohst,
   W.Prass, H.Ringsdorf, G.Wegner, R.Wehrmann,
   Angew.Chem. in press
10) H.Ringsdorf, B.Schlarb, J.Venzmer, Angew.Chem.
    Int.Eng. Ed. __27__, 113 (1988)
11) M.Ahlers, W.Müller, H.Ringsdorf, A.Reichert,
    Angew.Chem. in press

12) L.R.McLean, A.A.Durani, M.A.Whittam,
    D.S.Johnston, D.Chapman, Thin Solid Films 99,
    127 (1983)
13) H.Bader, R.van Wagenen, J.D.Andrade,
    H.Ringsdorf, J.Colloid&Interface Sci.101, 246
    (1984)
14) H.Bader, Ph.D thesis, Mainz 1985
15) G.W.Smith, N.Ratcliffe, S.J.Roser, M.F.Daniel,
    Thin Solid Films 151, 9 (1987)
16) M.B.Biddle,S.E.Rickert,J.B.Lando, A.Laschewsky,
    Sensors and Actuators 20, 307 (1989)
17) M.Shimomura, T.Kunitake, Thin Solid Films 132,
    243 (1985)
18) J.Moellerfeld, W.Prass, H.Ringsdorf,
    H.Hamazaki, J.Sunamoto, Biochim.Biophys.Acta
    857, 265 (1986)
19) G.Decher, E.Kuchinka, H.Ringsdorf, J.Venzmer,
    D.Bitter-Suermann, C.Weisgerber,
    Angew.Makromol.Chem.166/167, 71 (1989)
20) M.Sugawara, M.Kataoka, K.Odashima, Y.Umezawa,
    Thin Solid Films 180, 129 (1989)
21) H.J.Winter, G.Manecke, Makromol.Chem.186, 1979
    (1985)
22) Y.Ishikawa, T.Kunitake, T.Matsuda, T.Okutsuda,
    S.Shinkai, J.Chem.Soc.Commun.(1989) 736
23) M.Ahlers, H.Ringsdorf, H.Rosemeyer, F.Seela,
    Colloid&Polymer Sci.268, 132 (1990)
24) M.Tanaka et al, Chem.Lett.(1987) 1307
25) A.Yabe et al, Thin Solid Films 160, 33 (1988)
26) E.E.Uzgiris,R.D.Kornberg, Nature 301,125 (1983)
27) R.A.Reed, J.Mattai, G.G.Shipley, Biochemistry
    26, 824 (1987)
28) H.Haas, M.Möhwald, Thin Solid Films 180, 101
    (1989)
29) R.Blankenburg,P.Meller, H.Ringsdorf, C.Salesse,
    Biochemistry 28, 8214 (1989)

30) M.Ahlers, R.Blankenburg, D.W.Grainger,
    P.Meller, H.Ringsdorf, C.Salesse, Thin Solid
    Films **180**, 93 (1989)

31) S.A.Darst, M.Ahlers, P.H.Meller, E.W.Kubalek,
    R.Blankenburg, H.O.Ribi, H.Ringsdorf,
    R.D.Kornberg, J.Biophys. submitted

32) Y.Okahata, T.Tsuruta, K.Ijiro, K.Ariga,
    Langmuir **4**, 1373 (1988)

33) D.W.Grainger, A.Reichert, H.Ringsdorf,
    C.Salesse, FEBS Lett.**252**, 73 (1989)

34) M.Ahlers, T.Knepper, M.Przybylski, H.Ringsdorf
    in "Pteridin and Folic Acid Compounds",
    M.C.Kurtius Ed., W.de Gruyter, Berlin in press

35) P.Meller, J.Microscopy **156**, 241 (1989) and
    references therein

36) M.Wilchek,E.A.Bayer, Anal.Biochem.**171**, 1 (1988)

Electrocatalytic Polymers for Modified Electrodes in Analytical and Preparative Organic Systems

J. Grimshaw

SCHOOL OF CHEMISTRY, THE QUEEN'S UNIVERSITY OF BELFAST, BELFAST BT9 5A

1. INTRODUCTION

The properties of an electrode can be altered by attaching a useful reactive group to the surface using a polymeric coat. The polymer chosen, besides acting as a glue, also conveys desirable properties to the whole system. The coating must be permeable to the solution so hydrophilic and hydrophobic properties are important. It must adhere to the electrode and be stable for long periods under reaction conditions. The overall system must permit variation of the reactive group so as to make it easy to tailor the redox properties. A bare electrode effects oxidation or reduction of a substrate in solution by an outer-sphere mechanism. The reactive group attached to a modified electrode interacts with the substrate in solution to form an intermediate through which electrons are transferred. Thus, more specific redox properties are achieved while the reactive group is continuously restored to its original state by interaction with electrons from the electrode.

In the following discussion attention is paid to the properties of the chosen polymer, the reactive group and the means for connecting these two together as well as attachment of the system to the electrode. The coating need only be a few molecules thick. Value is added to the system through the choice of reactive groups as catalysts for a specific reaction. This choice may lead to either a desirable synthesis or to an analytical procedure for the detection of one substrate in a complex mixture of compounds.

2. THE DESIGN OF AN ELECTROCATALYTIC SYSTEM

An electrocatalytic system has a stable procatalyst that can be either reduced or oxidised to the active catalyst in a reversible redox transfer reaction. The active catalyst then takes part in a reaction with the substrate to yield the product, together with reformation of the procatalyst. Such catalytic systems are well known and are used in homogeneous solution to promote reaction between an expendable redox reagent and the substrate. The aim of an electrocatalytic system is to replace the expendable reagent by one electrode of an electrochemical cell. This will achieve the desired reaction on the substrate without the accumulation of spent redox reagents in the solution.

Particularly if the reaction between catalyst and substrate is fast, then when the procatalyst is in homogeneous solution only a thin layer of the solution close to the electrode will undergo the turnover between procatalyst and catalyst. Clearly the most effective use of the procatalyst will be to attach it by some method to the electrode surface. The application of this principle generates much of the thrust of the work on modified electrodes.

Molecules designed for use as electrocatalysts must therefore have two types of functional group; first, the procatalyst moiety and, second, either a polymer or a group that is easily polymerised, the two functions being separated by a spacer unit. Historically, polystyrenes were used and the electroactive centre, the procatalyst, was chosen in order to demonstrate the properties of electroactive films.[1] Most recent work has used monomers that can be polymerised onto the electrode surface. Examples are monomers containing a pyrrole group,[2] a thiophene group[3], or an aromatic amine[4], all of which can be oxidatively polymerised. The monomers used here contain a procatalyst centre that is subsequently activated by reduction. We have pioneered[5,6] the use of poly(amino acids) to support a procatalyst centre attached to the side chain functional group of an amino acid such as lysine or glutamic acid. The polymers can be either dip coated or spin coated onto the electrode and then used in a solvent where the polymer is insoluble.

3. POTENTIALLY USEFUL ELECTROCATALYTIC SYSTEMS

Living cells use redox procatalyst – catalyst systems. Examples incorporate iron(III)porphyrin, cobalt(III)corrin, flavin and quinone based systems. One aim of electrocatalysis is the design analogues of these natural systems.

Redox enzymes have been regarded for a long time as potential electrocatalysts. They might act as sensors for specific molecules and also as stereoselective catalysts in reactions. However redox enzymes do not undergo fast electron exchange with an electrode because the active site is within the protein support remote from the electrode surface. Thus there is a need to develop feeder coatings for enzyme activation. Low molecular weight compounds have been used for this purpose[7] as have coatings of a charge transfer salt.[8] Another group has modified glucose oxidase by incorporating ferrocenecarboxylic acid groups as the amide with some of the lysine side chains in the native enzyme.[9] This does not inhibit enzyme activity and provides a channel for electron transport through the enzyme to the active side.

The aim of our work is to develop catalytic systems that will operate in aqueous solution and mimic some of the natural systems mentioned. The use of poly(amino acids) as support for a procatalyst has some advantages. They are not so strongly hydrophobic as, for example, poly(pyrrole) or poly(thiophene). In our experience the latter two polymers tend to make catalyst centres less accessible.

4. CHARACTERIZATION OF POLYMER MODIFIED ELECTRODES

The general methods for characterisation of modified electrodes can be illustrated by examples from our work with poly(amino acids). It was our original idea to have the polymer in the a-helix form and with one electroactive unit per monomer.[5] In order to test the system, poly(N^{ε}-4-nitro-benzoyl-L-lysine), poly(NBL), was prepared. This forms the α-helix in dimethyl-acetamide and from this solvent it is strongly adsorbed onto platinum. The behaviour of the modified electrode can be examined in acetonitrile where the polymer is insoluble.

Cyclic voltammetry was used to characterise this coated electrode. When first scanned the electrode generally gave a very poor response, but that became satisfactory (Figure 1) as the film was swollen by solvent and ions. The mid-point of the anodic and cathodic peaks gave the redox potential of the active species and the shape of the CV curve gave information on the reversibility of the redox process.

Chronoamperometry is useful for saturating the film so as to determine the surface coverage of active sites by integration of the total current passed. From the data current vs time it is possible to extract a combination of two quantities: the diffusion coefficient for charge through the film, (D), and the volume concentration of electroactive species (c). The experimental data is plotted as a graph of current vs $t^{-0.5}$. For short times this graph should be linear and extrapolate through the origin (Figure 2). The graph deviates from linear behaviour at longer times when the flow of charge through the film becomes bounded by the outer boundary of the film. Characterisation of a film by thickness or by the density of electroactive sites per unit volume presents difficulties, particularly with soft and swollen films like those for which we aim.

5. EXAMPLES OF MODIFIED ELECTRODES

The electrochemical response of a charge film of poly(NBL), prepared as above, slowly deteriorates.[5] This deterioration is caused by electrostatic repulsion between the charged side chains overcoming the force of adhesion, so that the polymer film adopts a random coil and buckles away from the metal support. A solution to this problem of stability was found by coating the metal electrode with a thin layer of poly(pyrrole), either before or after dip-coating the poly(NBL).[6] Stronger Van der Waals attraction betwen the amino acid side chains and the poly(pyrrole) stabilised the film. Under the conditions we used the poly(pyrrole) was in its non-conducting reduced form, conduction was due to redox activity of the poly(NBL).

In order to develop a series of potentially useful polypyrroles, we needed a pyrrole plus spacer

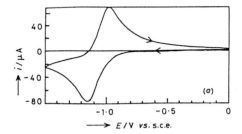

Figure 1. Normal behaviour on cyclic
voltammetry for a thick film. Here poly(NBL) on
Pt; film area 0.087 cm^2, surface coverage of
electroactive groups 21.4 nmol cm^{-2}, scan rate
0.1 V s^{-1}, solvent acetonitrile containing
0.1 M-Pr$_4$NBF$_4$.

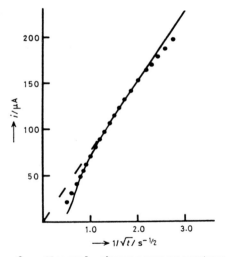

Figure 2. Normal chronoamperometry behaviour
illustrated for a film of poly(NBL) on Pt; film
area 0.087 cm^2, potential step 0 to −1.5 V vs.
s.c.e., solvent acetonitrile containing
0.1 M-Pr$_4$NBF$_4$. Charge passed to saturate the
film was 242.1 µCoulomb. From the linear
portion of the graph D c=16.0 nmol cm^{-2} s$^{-0.5}$.

unit to which various electroactive units could be attached. N-(2-aminoethyl)pyrrole served the purpose.

Attachment of a cyanoanthracene unit gave a monomer that was readily polymerised by oxidation of the pyrrole unit to form a dark coloured film of poly(CAP) on platinum.[10] At negative potentials the film became colourless and the polypyrrole was in its non-conducting state. At a potential corresponding to reduction of the cyanoanthracene group, the film accepts electrons to a limit which is followed by a very sudden cessation of current flow (Figure 3). An explanation for this behaviour can be found if the cyano-anthracene groups are organised in zones and are close to each other in the matrix. Rapid transport of electrons can then be expected, together with ions and solvent molecules. When the limit is reached to which the polymer lattice can expand, transport of electrons into the film must suddenly cease even though unreduced cyanoanthracene units are available.

CAP - monomer   BEP - monomer

$$R = \text{pyrrole}-N-CH_2CH_2-$$

Attachment of pyromellitic anhydride to the N-(2-aminoethyl)pyrrole gives a monomer with two pyrrole units, so that cross-linked oxidative polymerization can occur. This polymer, poly(BEP), shows a different behaviour for reduction of the pyromellitimide group to that observed when the monomer is examined (Figure 4). Addition of two electrons per unit to give the dianion now occurs at less negative potentials so that the waves for the first electron and for the second electron addition overlap. The films were stable and active in water as electrolyte.

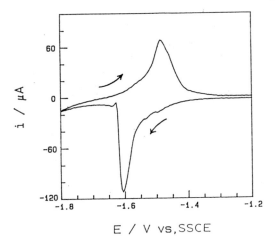

Figure 3. Cyclic voltammetry of a film of poly(CAP) on Pt; film area 0.131 cm^2, surface coverage of electroactive groups 8.7 nmol cm^{-2}, scan rate 0.07 Vs^{-1}, solvent acetonitrile containing 0.1 m–Pr$_4$NBF$_4$.

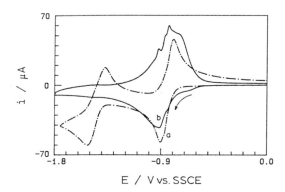

Figure 4. (a) Cyclic voltammetry of the monomer BEP in solution. (b) Cyclic voltammetry of a film of poly(BEP) on Pt, scan rate 0.1 Vs^{-1}, solvent acetonitrile containing 0.1 m–Pr$_4$NBF$_4$.

TBQ - mononer

As thiophene can also be polymerised at an anode, we have used 3-thienylpropanol as a convenient thiophene to which other groups can be attached by esterification. The film of poly(TBQ) prepared with a trichlorobenzoquinone covalently attached shows a reductive redox chemistry quite different from that of the trichlorobenzoquinone monomer in solution.[12] The film shows only the first electron transfer to quinone at a slightly more negative potential to that found for the monomer in solution. Addition of a second electron to the quinone units in the film is not observed. These points indicate that the polythiophene matrix is hydrophobic so that more energy is required to form the radical anion and particularly the di-anion within the film than for the monomer in free solution.

The most useful electroactive polymers in our hands have been the poly(amino acids). The polymers we have available are not soluble in water and electroactive groups attached to the polymers show their expected redox behaviour. Poly(γ-ethyl glutamate) has been one of the most useful because there is a convenient route for attachment of electroactive groups. It can be reacted with a suitable amine to introduce an amide side chain to which redox centres can be attached by coordination. In this way we have attached Ru and Os redox centres to the poly(γ-ethyl glutamate) and also ferrocene groups, all of which show reversible redox behaviour in water.[13] These polymers can be dip-coated onto platinum.

An iron(III)porphyrin has also been attached to the modified polyglutamate through apical coordination to the iron centre.[14] This material is very close in structure to analogues of cytochrome-c. Thus we can expect the coated electrode to undergo a reaction with molecular oxygen when the iron is in the +2 oxidation state. The iron(III)porphyrin doped polymer shows reversible redox chemistry in water when dip-coated onto platinum. There is evidence for reversible replacement of one apical chloride ligand by water, depending on the electrolyte composition.

6. OTHER WORK ON MODIFIED ELECTRODES

Quinones may be involved in natural electron cascade processes so the inclusion of these groups in a polymer film has been the target of several research groups.[3,4,12] There is also a possibility of using the system as a pH sensor. Some of the successful work in this area involves films prepared by the oxidative polymerisation of aminoquinones such as 5-amino-1,4-naphthoquinone.[4]

The oxidation of alcohols to ketones without the use of chromic acid is an important goal of preparative chemistry. Nitroxides will act as a procatalyst for this purpose. They are oxidised to the nitrone which is the catalyst. Thus, considerable interest is shown in the coordinative attachment of such procatalysts to polypyrrole.[16]

There is interest also in attaching low valency transition metal ions to polypyrrole films. In particular Ni,[17] Co,[18] and Re[19] show potential as procatalysts.

7. CONCLUSION

Examples of the use of a glucose oxidase modified electrode were given in section 3. Modified electrodes of this type have been developed for the monitoring of glucose concentration and are commercially available.

In the course of our work with poly(amino acids), we have developed polymer coatings for electrodes that will function in aqueous solution as well as in aprotic solvents. These polymers incorporate transition metal redox centres and the redox potential of the active group can be varied. It is hoped to use some of these materials as electron feeders to redox enzymes. One of the poly(amino acids) has an iron(III)porphyrin group attached which may be of value for specific oxidation processes. Examples of other polymer and procatalyst systems were given in section 6. Work in this general area is presently only in the development stage and such processes have not yet been realised on a large scale.

REFERENCES

1. R.W. Murray, Acc.Chem.Res., 1980, 13, 135;
   G.K.Chandler and D. Pletcher,Chem. Soc.
   Specialist Periodical Reports. Electrochemistry,
   10, 117.
2. A Haimerl and A. Merz Angew.Chem.Int.Ed.Engl.,
   1986, 25, 180; L. Coche, A. Deronzier and J.C.
   Moutet, J.Electroanal.Chem., 1986, 198, 187.
3. J. Grimshaw and S.D. Perera, J.Electroanal.
   Chem., 1990, 278, 287.
4. T. Degani and A. Heller, J.Phys.Chem., 1987,
   91, 1285.
5. A.M. Abeysekera, J. Grimshaw, S.D. Perera and
   D. Vipond, J.Chem.Soc.Perkin Trans.2,1989,43.
6. J. Grimshaw and S.D. Perera, J.Chem.Soc.
   Perkin Trans.2, 1989, 1711.
7. H.A.O. Hill, D.J. Page and N.J. Walton,
   J.Electroanal.Chem., 1987, 217, 141; K.D.
   Gleira, H.A.O. Hill, C.J. McNeil and M.J. Green,
   Analyt.Chem., 1986, 58, 1203.
8. W.J. Albery, P.N. Bartlett, A.E.G. Cass and K.W.
   Sim, J.Electroanal.Chem., 1987, 218, 127.
9. V.K. Gater, M.D.Liu, M.D.Love and C.R.Leidner,
   J. Electroanal. Chem., 1988, 257, 133.
10. J. Grimshaw and S.D.Perera, J.Electroanal.Chem.,
    1989, 265, 335.
11. J. Grimshaw and S.D Perera,J.Electroanal.Chem.,
    1990, 278, 279.
12. J. Grimshaw and S.D.Perera, J.Electroanal.Chem.,
    1990, 281, 125.
13. M. Devenney, J. Grimshaw and J. Trocha-Grimshaw,
    Unpublished work.
14. J. Grimshaw and J.Trocha-Grimshaw, Chem.Commun.,
    1990, 157.
15. P. Audebert, G. Bidan and M. Lapkowski,
    J. Electroanal. Chem., 1987, 219, 165.
16. A. Deronzier, D. Limosin and J.C. Moutet,
    Electrochim. Acta, 1987, 32, 1643.
17. J.C. Collin and J.P. Sauvage, Chem.Commun.,
    1987, 1075.

18. F. Daire, F. Bedioui, J. Devynck and C.B. Charreton, J.Electroanal.Chem.,1987, 224,95.
19. S. Cosnier, A. Deronzier and J.C. Moutet, J.Electroanal.Chem., 1986, 207, 315.

Microlithography – for the Year 2000

R. A. Pethrick

DEPARTMENT OF PURE AND APPLIED CHEMISTRY, UNIVERSITY OF
STRATHCLYDE, CATHEDRAL STREET, GLASGOW, G1 1XL

1 ABSTRACT

Microlithography has played a key role in the development of modern semiconductor technology. This review paper attempts to analyse the factors which are likely to dictate the changes which will occur in lithography up to the year 2000. A number of factors are likely to influence the type of technology which is used and these are not always technical but can reflect the commercial costs of change. Fabrication in future will rely more extensively on dry processing, plasma etch resistance and the contrast available from the resist material. It appears that a variety of problems are likely to emerge and that the search for smaller feature sizes may bottom out at around 0.1 – 0.2 μm. This review attempts an objective assessment of the problems and identification of the challenges to chemists.

2 INTRODUCTION

Over the last thirty years dramatic changes have occurred in the size, sophistication and computational power of semiconductor circuits. In the mid nineteen sixties the typical micro "chip" would contain about ten separate components, whereas today they contain hundreds of thousands of components. A number of factors have helped these changes to be achieved. The design of the structure of each of the individual components has been changed, leading to improvements in reliability and performance of the circuits. Organisation and optimisation of the layout of the components on the chip has reduced the transit time between operations and this also has reduced the power consumption and the amount of heat which needs to be dissipated from the chip. One of the main factors which has helped with the increase in the capacity of each individual chip is the ability to reduce the size of the structure of each element. In the early sixties, the average feature size was of the order of ten microns, this has been reduced until now, in the late eighties, it is of the order of one micron. It has been found that there is an approximately linear variation of the logarithm of the feature size with time, Table (1).

Table 1   DRAM Technology Trends Over the Next Ten Years[2-4]

YEAR	1970	1975	1980	1985	1990		1995	2000
PRODUCTS	1K	4K	16K	64K	256K	1M 4M	16M	64M 256M
FEATURE SIZE μm	10-12	8	5	3	2	1.3 0.8	0.5	0.35
TECHNOLOGY	pMOS	n MOS			CMOS			
ACCESS TIME nS	300	150	125	100	80	70	50	30

* – A glossary of terms is presented at the end of the paper.

Over the period to 1984 there was a yearly growth in the number of components per chip corresponding to a doubling per annum. This growth may be attributed to a factor of 32 to improvement in photo-lithography, a factor of 20 to the use of larger chips and the remaining factor of 100 to improved circuit design and layout[1]. During this period of time, the cost per chip has remained constant leading to a dramatic drop in the cost per component. Projecting this trend to the year 2000 would imply that we would anticipate having to achieve a lithographic capability of the order of 0.2 μm. Clearly there must be a limit to the extent to which change in feature size can continue to follow the trends illustrated in Table (1). In the first section in this review an attempt will be made to analyse the factors influencing the change in the size of component in order that a realistic estimate can be made of the real requirements in terms of lithographic feature size for the year 2000. In order to achieve these targets it will be necessary to change the lithographic methods, the materials used as the resists, their pattern development and also the methods and materials likely to be used in chip manufacture. The aim of this review is to identify the challenges in the area of lithography which need to be faced in the next few years. The resist used in semiconductor fabrication are speciality polymers; 10 mls costing as much as $20, but more importantly they facilitate the current growth in semiconductor technology.

3 ASSESSMENT OF THE LITHOGRAPHIC CHALLENGE

It is clear from Table (1) that if the trends in lithographic feature size continue, then in the next few years production quantities of devices will be commercially available which contain structures less than one micron. However, it is a large assumption that these trends will continue up until the year 2000. The question of what the VLSI manufacturing industry will look like over the next decade has been considered in several reports[2,3,4]. There appears to be general agreement that CMOS will be the dominant type of device structure through to the mid–90s. During this period new technologies are going to emerge and grow, particularly GaAs and optical inter–connects, and these will themselves require sophisticated techniques to be used in the construction of such devices. The commercial effort is therefore likely to continue to be directed to high volume memory products (DRAMs, EPROMs and SRAMs) on which the world has standardised, Table (1). In the mid 1990s, the leading–edge volume production is likely to involve 0.35 μm structures, multilevel–metal (>3), CMOS technology (64 Megabit DRAM) with narrower geometries, shallower junctions, 3D structures and 1.5 times the number of processing steps. The development stage beyond 1995 is likely to involve 256 Megabit DRAM devices which will require the production of 0.25 μm features. Submicron lithography down to 0.5 μm with X–ray and 0.35 μm using optical (deep UV) and electron beam lithography and Masked Ion Beam technology beginning to play a strong role at and below 0.5 μm by the mid–1990s. Electron and ion beam lithography will be used for some circuits requiring fine–line geometries down to 0.1 μm.

The reduction in feature size, development of 3D structures and low temperature processing will necessitate an increased use of three areas of technology:
1) Ion Implantation due to its precision and because its annealing requirements are much less severe than for other doping methods and will become more important in the future.
2) Chemical Vapour Deposition (CVD) will increase in importance with the move towards 3D structures and there will be a requirement to lower the deposition temperature and also develop good isolation layers.
3) Pattern Etching will have become almost entirely dry and will be centred around ion and reactive etch, the latter being favoured to reduce radiation damage[5].

The size of the chip will probably be increased to be between 250 to 300 mm by the mid 1990s; this will help to continue the growth of the component density per chip. One of the major hurdles which will have to be faced is the requirement to increase quality control; increase in chip size leads to a reduction in throughput and to maintain the quality of the product it is necessary to reduce the number of defects to < 0.15 defects/sq.cm. in the range 0.1 to 0.03 μm size. The above predictions are based on the best guesses which can be currently made of the way in which VLSI technology is likely to progress over the next ten years.

Are these objectives realistic? It has been suggested by Brewer[6] that the practical limits to device fabrication are 0.1 μm on the basis of analysis of the use of electron beam technology. The various factors influencing this decision are as follows:

i) <u>Resist related limits</u>. If it is assumed that ultimately electron beam lithography is used for pattern generation then the resist will limit the process. Each resist is characterised by a sensitivity factor S, measured in coulombs per square centimetre. The time (τ_S) required to expose a spot of diameter d_S with a beam having a current density of J_S is $\tau_S = S/J_S$. A typical resist material currently used would have a value of S of 8 x 10^{-5} C/cm^2, which is approximately 5 x 10^{14} electrons/cm^2 to break or crosslink the chemical bonds in the polymer to produce the required degree of contrast. The smaller the value of S, the more sensitive the process and the fewer the electrons required to effect the desired change in the resist.

ii) <u>Machine related limits</u>. The resolution of an electron beam is seriously influenced by the effects of elastic collisions with the atomic nuclei in the resist and substrate[7,8]. The range of the electron is several times the thickness of the resist layer and hence scattering leads to exposure over a region several times the width of the beam diameter. The minimum line–width can be related to the sensitivity factor by; (L min)^2 = N_m e/S, with N_m the number of electrons striking area A, per unit time with S given by eN_m/A. This leads to the prediction that the limit for electron beam lithography is of the order of 0.1 μm for acceptable writing times[9]. This limit is related to the maximum current density which can be concentrated into a spot of dimensions d, and is limited by the transverse thermal velocity effect and acceptable writing speeds[10]. The use of ion beams can in theory reduce this limit to less than 0.1 μm, but the problems of metrology and alignment accuracy would indicate that this is still a realistic limit for all lithographic methods. As the feature size is reduced contamination is likely to become a critical factor through its effect on the number of defects within the structure.

iii) <u>Contamination effects and influences on chip yields</u>. Whilst the ability to produce a particular feature size is critical for the generation of a particular structure, the number of defects will control the practical use of that technology. It is possible that features of the order of 0.1 μm may be possible using electron beam or X–ray lithography, however, the cleanliness of the environment in which this process is performed is critical. Contamination can enter into the fabrication facility in a number of ways;

a) <u>in the air</u>. It is clear that in the future the trend will be towards enclosure of the fabrication facility and the exclusion of personal contact; this will virtually eliminate contamination from workers.

b) <u>from gas supplies</u>. With the elimination of contamination from the air, one of the possible sources of contamination comes from the gas supplies used in the processes. It is possible to limit particles to no more than 20 ($\geqslant$ 20 μm) per cubic foot but significant efforts are now underway to reduce this further in the next few years. It is likely that low single digit concentrations at the 0.1 μm level will be achieved and this will be extended to the 0.02 μm level within the next five years. Practically it is found that the contamination levels are highest with full cylinders of gas and decay exponentially with decreasing pressure[11].

c) from liquids used in processing. The liquids used in processing are themselves acceptably low in levels of contamination, however once they enter the equipment used in their delivery and handling contamination may increase. An experiment in which the purge time for various polymeric diaphragms was investigated indicated that to reach a level of less than 25 particles of 0.4 μm per 100 mls purge times of between 13 and 245 minutes[12], table (2). This indicates that there is a significant problem with achieving the required level of cleanliness with wet processing for sub micron processing. The contamination in most systems arises as a result of fragments of material left over from fabrication but more significantly from wear in the pumping and control diaphragms in the delivery system.

Table 2   Purge Durations Required to Reach a Background Level of 25 particles ≳ 0.4 μm/100 ml.[*]

Material	Purge Duration (minutes)	Material	Purge Duration (minutes)
PVC/EPDM	35	PVDF	34
PFA	13 – 64	PTFE	245

* Precise values depend on type of function as well as material used[12]

iv) Alignment accuracy. In a fabrication process it is necessary to align the new and previous patterns with a precision which is significantly better than the feature size. If the alignment and edge processing errors are equal, the alignment accuracy must be better than 625Å for a 0.5 μm wide line or the lines could miss contact completely[6]. The problem of the ultimate line width limit due to fundamental physical laws lies in the Heisenberg uncertainty principle[13], as the number of features in a circuit increases, the minimum linewidth must be increased to avoid a yield reduction due to the statistical probability of misalignment. Wallmark[13] suggests that a minimum feature size is 0.07 μm.

Ultimately the limit for a feature size may not be determined by the ability to create a particular pattern but may be rather controlled by consideration of the yield of chips. Increasing the chip area rapidly leads to reductions in yield and would indicate that in the race for larger circuits there may also be a limit in the size of chip which can be realistically produced and this is probably of the order of 300 mm. Whilst electron beam lithography is capable of providing 0.1 μm lithography, improvements on this figure could in theory be achieved by use of ion beams; here the back scatter is much less than with electrons and the line broadening limit due to this effect would be greatly reduced. However there still remains the fundamental problem of alignment accurary which limits the feature size to about 0.1 μm. These conclusions have been supported in a series of studies on the problem of metrology using a variety of lithographic approaches[14]. As integrated circuit minimum feature size decreases to 0.5 μm for a 16 Mbit DRAM in the 1990 time frame, the use of X-ray technology is forecast as a lithographic option. X-rays have a virtually infinite depth of focus, compared to conventional optics, so thick single layer resist can be imaged. The minimum feature size is expected to decrease to 0.35 μm in 1993 and 0.25 μm in 1995 in a 3–4 μm thick resist layer. Critical dimension (CD) measurement tools will be required to handle these high aspect ratio structures with near vertical side walls, table (3) and defects are predicted to become limiting at these dimensions[15].

Table 3   DRAM   Design Rules for the Future[15]

YEAR	1985	1988	1990	1993	1995
DEVICE	1 M bit	4 M bit	16 M bit	64 M bit	256 M bit
MINIMUM FEATURE SIZE	1	0.7	0.5	0.35	0.25
REGISTRATION ±	0.2	0.14	0.1	0.07	0.05
CD CONTROL ±	0.1	0.07	0.05	0.03	0.02

It has been suggested that in the future the problem of cleanliness will be addressed by building each element into a self-contained unit which excludes the operator and eliminates handling as far as is possible[16] and achieves the necessary requirements for sub micron operation.

v) Limits of Mask Making. Whilst the lithographic processing on the semi-conductor surface is ultimately the controlling factor for fabrication, mask making is a critical enabling step in the production process. The ability to align the mask with the substrate is limiting and current instrumentation is aiming at an alignment of better than ± 0.05 μm. Scanning electron micrographs of the chrome overlay used in mask manufacture indicate that the grain size is of the order of 0.1 μm, allowing features down to these dimensions before structure in the walls becomes evident. Practical considerations indicate that the limit for mask manufacture is likely to be 0.1 μm.

It is clear from the above data that irrespective of the lithographic method, once the target of 0.35 μm is approached, the problems with further reduction in feature size in a fabrication environment will increase rapidly and probably exclude significant further reduction in feature dimensions, however a real barrier appears to exist at about 0.1 μm beyond which production devices are unlikely to progress. At this point it may be anticipated that GaAs and other III V technology will allow continuation of growth in computing capacity for a further decade, but this discussion is outside of the scope of this article.   In order that the targets indicated in table (1) can be achieved it will be necessary to develop new resist materials and processing schemes.

4 LITHOGRAPHIC PROCESSES AND MATERIALS

The basis of lithography is the ability to selectively expose a sensitive polymer film into which a pattern can be developed by a subsequent solvent exposure or post treatment, Figure (1).

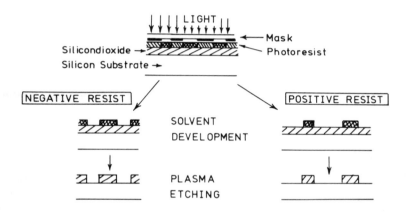

Figure 1   Lithographic process for a resist

The developed image will then be subjected to a processing step in which the mask may be used to selectively implant ions in the substrate, deposit metal contacts, allow selective etching of a substrate or deposition of semiconductor material.   In a real situation the resist materials should have the following properties:
- the material must have sufficient sensitivity to allow exposure using the appropriate methods to allow realistic production rates.
- the material should form pin hole free conformal coatings which exhibit a high degree of adhesion to the substrate and very little swelling on development[17].
- the resist material must be able to stand up to the subsequent treatment which may involve plasma etching or metal deposition.
- the final step in many fabrication stages is the removal of the resist, prior to the deposition of resist for preparation of the pattern for the next stage.

A good resist material allows an optimisation of each of these processes. Depending on the type of process to be used it may be desirable to etch into a resist or to build up resist features. Resists which produce negative images generally undergo crosslinking upon irradiation, rendering them insoluble in the developer solvent. Positive resists experience molecular changes which enhance their solubility in developer solutions such that the exposed regions are removed at a faster rate. Prebaking of the resist will reduce stress, promote adhesion as well as driving off the solvent used in the spinning process. The type of resist material used depends upon the lithographic method used. A variety of different methods are available; optical (300–350 nm), deep UV (200–250 nm), electron beam, ion beam and X–ray lithography.

5 OPTICAL LITHOGRAPHY

The optical lithographic materials used in semiconductor processing have been derived from systems originally used in the printing industry. Despite the major difference in the feature size required, these materials have been surprisingly successful as optical resists.

6 NEGATIVE RESISTS

Certain systems on exposure to light become insoluble, Figure (1), and this is usually associated with the formation of a crosslinked matrix. The image is developed by removal of the unirradiated, uncrosslinked polymer chains which are swollen and dissolved on treatment by the solvent. The most widely used negative resist system is generated from the radiation induced crosslinking of cyclised polyisoprene by bis aryl azide[18–21]. The photo–irradiation of the bis aryl azide generates a nitrene intermediate via elimination of nitrogen, Figure (2). This nitrene is capable of undergoing a variety of reactions that include insertion into carbon–hydrogen bonds, hydrogen abstraction from the matrix backbone to form radicals that may further react or add to carbon–carbon double bonds to form heterocyclic aziridine linkages. It is possible by varying the structure of the bis aryl azide to alter the frequency at which it absorbs light and hence its sensitivity, Figure (2).

The development of the resist must be tightly controlled in order to minimise undercutting of features and maintain line width control. The high optical density of the resist in the region of 250 – 200 nm results in most of the photochemistry occurring in the surface of the resist. Light penetrating deeper into the films will be attenuated due to the absorbance characteristics of the resist and the level of crosslinking therefore decreases with distance from the surface and it is therefore very difficult to use this system below 300 nm.

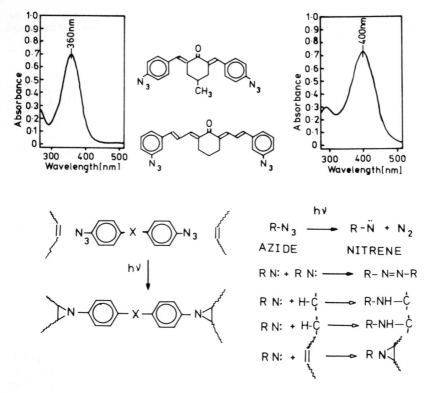

Figure 2   Bis aryl azide

This latter wavelength corresponds to the limit for 1 μm lithography; extension of the operational window down to 200 nm is essential for the problems of diffraction to be overcome and 0.35 μm structures to be generated.

7 POSITIVE RESISTS

The major component is a base soluble resin that is rendered insoluble by the addition of a hydrophobic radiation sensitive material. Upon irradiation, the hydrophobic entity may be either converted to an alkali – soluble species or entirely removed, allowing removal of the irradiated portions of the resist by an alkaline developer solution. The workhorse of the semiconductor industry is the novolac – diazonaphthoquinone, the resin and sensitizer being soluble in organic solvents and forms a hard glassy resin film[22-27]. The solvent inhibitor is present in a concentration of 10–20 wt % and can be modified to absorb at different wavelengths allowing an optimisation of the sensitivity, Figure (3).

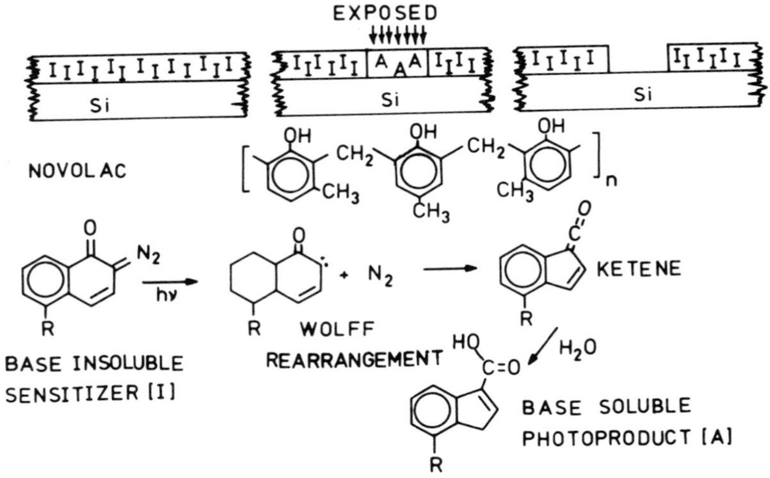

Figure 3   Novolac

As in the case of the negative resists, absorption of the novolac precludes use of the resist below 310 nm and hence makes it difficult to use for submicron lithography.   Considerable progress has been made with improvement of the sensitivity of the diazonaphthoquinone used and also by baking the resist at a temperature of 130°C[28].   This effect is a little surprising since at this temperature the diazonaphthoquinone will have decomposed to about 10%. One of the disadvantages of the novolac resins is their low glass transition temperature (Tg), usually in the range 70 to 120°C.   It would be desirable to use a higher Tg material[29-31], however these all contain aromatic groups which will absorb radiation below 300 nm and hence do not allow the possibility of developing resists operating below the cut off and being usable as UV resists.

The novolac resin has the advantage of a higher plasma etch resistance than the cyclic polyisoprene resin.   Addition of monoazoline, imidazole or triethanolamine into diazonaphthoquinone – novolac resists allows reverse of the toning of a positive resist[32-34]. In this process, the doped resist is exposed through a mask, baked after exposure, flood exposed and finally developed in aqueous base to generate high quality negative tone patterns, Figure (4).   Thermally induced, base–catalysed decarboxylation of the indene carboxylic acid destroys aqueous base solubility of the irradiated material. This is an attractive process since it stays close to existing technology.

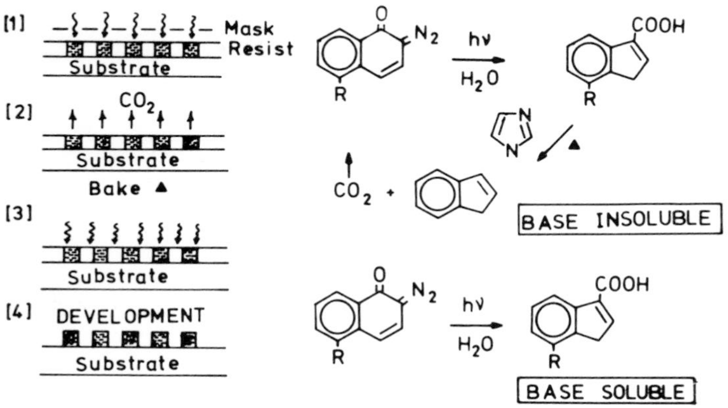

Figure 4   Reverse novolac technology

The problem also exists of producing suitable photosensitive agents for use below 300 nm. Meldrum's acid derivative affords a bleachable chromophore which absorbs at 250 nm, producing a Wolff rearrangement generating a ketene intermediate that further degrades to acetone and carbon monoxide[36−38]. The characteristics of these resists have been improved by use of an amide or lactam functionality. An alternative approach[36−38] is to use an o−nitrobenzyl carboxylate group, which is photo degraded to yield nitrobenzaldehyde, although these systems are not particularly sensitive they are capable of meeting the requirements for submicron resolution, provided a suitable matrix material can be found, Figure (5).

A possible approach to the development of a deep UV resist has been proposed[39,40] based on the incorporation of 3−oximino−2−butanone methacrylate into a methacrylate polymer. The oxime ester chromophore absorbs light leading to effective chain cleavage and increased radical formation enhancing the probability of chain cleavage. Improvements in the sensitivity can be achieved by incorporation of indenone[40], which readily undergoes α−cleavage generating a radical that subsequently undergoes β scission. Both these materials have the capability of being used as UV resists but lack the advantages of the novolacs of dry etch resistance. The problem of enhancing the dry etch resistance will be discussed later in this article.

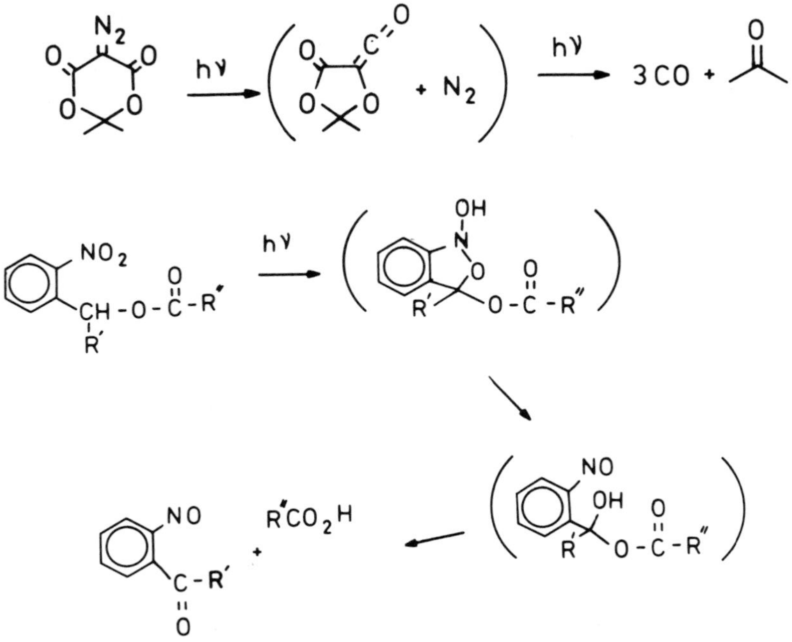

Figure 5 Possible alternative photosensitive agents

In the early 1970's it was recognised that resist sensitivity was not the primary limiting feature size in optical lithography, diffraction of light becoming the controlling factor. Diffraction of light will naturally cause a broadening of the image and also can lead to standing waves, observed as layered regions in the developed features. The source used widely for lithography is the mercury–xenon lamp and it possesses a number of strong lines in the region 450 – 313 nm. The lines typically used are: 436 (G), 405 (H) and 365 (I) nm. Several problems have been encountered in shifting the source frequency to shorter wavelengths. The depth of focus is approaching the thickness of the resist and this leads to a loss in the accuracy of the features produced. New optical systems are required to be used because of the decrease in transparency of standard lens materials in this frequency range, necessitating the use of construction material such as quartz. Conventional optical systems upgraded will allow submicron features to be generated, however truly submicron production is likely to require the use of new high intensity sources.

In the mid eighties considerable interest and effort has been committed to the development of excimer laser sources. Power outputs of the order of 1 kilowatt can be achieved from a Hg–Xe lamp, but the power in the 254nm line is a mere 10–20 milliwatts. Excimer lasers (KrCl, KrCl) which deliver several watts of power over the required wavelength provide a practical alternative to Hg–Xe lamps and together with upgraded optics will allow lithography to be extended down to feature sizes of 0.2 μm using wavelengths of the order of 198 nm[41–43]. It is therefore likely that optical lithography will retain its place as the primary method for semi conductor fabrication, at least in the mid 1990's if not until the year 2000.

8 ELECTRON BEAM LITHOGRAPHY

The problems of producing smaller and smaller feature sizes were considered in the mid 1970's and it was proposed that electron beam lithography was the preferred method for reduction in dimensions and as a consequence the method was commercialised[6]. The role of electron beam lithography over the next decade is likely to be one of primary mask generation rather than fabrication of devices. The use of EB lithography in this role is consistent with its low throughput and high capital costs, the requirement of sophisticated data storage and computer control which allows the possibility of changing the pattern design relatively easily. Scattering processes both in the resist and the substrate make it impossible to achieve features of less than 0.1 μm[44]. A wide range of resists have been produced for EB lithography and as with optical systems allow both positive and negative patterning.

The "classical" positive resist is polymethylmethacrylate PMMA which on exposure to electrons undergoes degradation[45], reducing the molecular weight and enhancing the solubility of the exposed regions. Use of an appropriate developer allows selective removal of the irradiated areas. PMMA is capable of producing the highest possible resolution, nanometer features having been achieved in special conditions, however it is very insensitive ∼ 100 μC/cm² at 20 kV[45]. A wide variety of copolymers have been studied in an attempt to improve the sensitivity of the resist[46–48], a number of these are available commercially, Appendix I. Incorporation of fluorine into the methacrylate chain improves the sensitivity to radiation induced degradation. Poly (2,2,3,4,4,4 – hexafluorobutyl methacrylate) (FBM) has a sensitivity of 0.4 μC/cm², however the adhesion characteristics are poor due to the presence of the large fluorinated side groups[46], and is unsuitable for the generation of submicron features. A compromise is obtained with poly (2,2,2 trifluoro ethyl–α–chloro acrylate) (EBR 9) which inhibits good adhesion and a sensitivity of 5–12 μC/cm²[47,48]. A resolution of 0.25 μm has been acheived, however the plasma etch resistance is marginally worse than PMMA. An alternative method of enhancing the sensitivity is to incorporate methyl methacrylic acid and methacryloyl chloride into the backbone generating a resist which contains both inter and intra molecular anhydride formation with a sensitivity of 5–10 μC/cm^2 at 25 kV[49-51].

Another important class of positive resists is based on the poly (olefin sulfones). These materials are alternating copolymers of an olefin and sulphur dioxide, prepared by free radical solution polymerisation. The relatively weak C–S bond, $\sim$ 60 kcal/mol compared with $\sim$ 80 kcal/mol for a carbon–carbon bond, is readily cleaved with radiation. Poly (but–1–ene sulfone) (PBS) is used commercially, has a sensitivity of 1.6 μC/cm^2 and has a resolution of 0.25 μm[52]. The aliphatic olefin–sulphone resists exhibit poor dry etch resistance and are thus unsuitable for most device process applications. Incorporation of aromatic moieties into the polymer decreases its sensitivity but does increase its plasma etch resistance[53,54].

In a recent article an electron beam system capable of producing 0.1 μm has been described, together with a swell–free resist processing technology[55-57]. The resolution of the conventional direct write machine can be increased by use of a modified electrode system, together with modified electron optics. The resist used with this system consists of poly(p–vinyl phenol) with 3,3' diazido diphenyl sulphone as a sensitizer, developed using a 1.0% tetramethyl ammonium hydroxide solution. The performance of this resist was improved by incorporation of poly(propylene oxide) and alkyl ammonium halide into the resist to act as a monoionic surfactant and a solution inhibitor. This system provides 6 μC/cm^2 sensitivity and 0.25 μm lines and spaces.

9 DEEP UV LITHOGRAPHY

The subject of selection of resists for deep ultraviolet lithography has been reviewed recently[58,59] indicating that whilst equipment exists capable of achieving the desired feature sizes, resists form a current limitation on the development in this area. The main problem with conventional resists is the absorption of the benzene chromophore, conjugated to a strongly donating group, such as the hydroxyl in novolac resins[60-63]. For a successful resist it is desirable either to reduce the absorption, thin the resist or develop a system which on exposure produces a system with a lower optical absorption. A PMMA would meet this definition, however it also would have very poor plasma etch resistance.

The challenge of deep UV resists can be subdivided into two groups; 230 – 300 nm and 193 – 230 nm. As we shall see it is possible to see solutions to the former, but the latter still presents a significant challenge.
i)   Optical Absorbance and Plasma Etch Resistance. Aromatic structures will absorb strongly in the region 193 – 230 nm, however

unsubstituted ring systems such as benzene will give greater than 50% transmission in the region 230 – 300 nm. The absorption problems are associated with substituted ring systems and molecules in which strong excimer interactions are observed. It is therefore possible to find windows of transmission in aromatic materials that are lightly substituted, one such occurring at 240 nm[59]. Lower absorptions can be obtained in molecules containing saturated structures, however these materials demonstrate poor plasma etch resistance. Plasma etch resistance depends on the nature of the plasma but in general can be associated with[64–66]:

a)   the presence of aromatic groups provided chlorine or iodine are absent,

b)   the presence of silicon can substantially reduce etch rates,

c)   the reduction in etch rates for both silicon and aromatic groups is non–additive. Small amounts have a large effect, as little as 20% can produce rates equivalent to those observed in the pure polymers.

d)   reactivity of a material is related to the bond polarity, however in some cases the introduction of a hetero atom can produce increased plasma etch resistance.

It is therefore desirable in the design or selection of a good deep UV resist to have either silicon or aromatic groups, thus increasing the plasma etch resistance and hence allowing the thinner films required to reduce the absorbance[67–70]. Possible changes to conventional resists have been attempted and novolac resins have been developed which have lower absorbances than others[71], however these materials are far from being adequate. Re–examination of the mode of action of the conventional system has revealed that it involves some very subtle photochemistry[72] and it is not easily extended to the DUV. An alternative approach is the use of photolabile acid salts which when activated can catalyse decomposition of some functionality which will render the polymer more soluble. Typical of these systems is the hydroxystyrene with an acid–labile ether or ester protecting group, the loss of which results in base solubility[73]. A more novel approach is the decomposition of poly(phthalaldehyde), chain scission in this case leading to depolymerisation, the polymer itself having a ceiling temperature below ambient[74]. Unfortunately this system falls down on its development characteristics and poor resolution capabilities.

Traditionally, only two mechanisms have been found to be useful in producing high resolution resist patterns; dissolution inhibition and molecular weight degradation. Most crosslinking resists suffer severely from solvent swelling effects[75]. Thus it was surprising to find that poly(vinyl phenol) resin developed in aqueous base did not swell[76], this was because the developer is a non–solvent for the exposed polymer. Based on these observations a trifunctional crosslinking agent, a novolac resin and a photo sensitive acid generator have been combined to produce an interesting resist. During post exposure bake, the acid catalysed the reaction of the crosslinker with the resin; development is with alkaline solution. This is a very acceptable resist at 248 nm giving 0.5 μm line widths[77,78] and furthermore gives a 0.2 μm line width with electron beam exposure, swelling being entirely absent[79].

This concept has been shown to be quite general and only requires that the following conditions are satisfied[80]:
a)   a reasonably electron–rich aromatic moiety is used,
b)   a latent electrophile is present,
c)   a photogenerated acid is present.

The reaction between (a) and (b) is catalysed by the acid and either one or both may be polymeric: if one is a monomer it must be multi–functional and both must be base soluble. Highly reactive acceptors do not appear to be needed; the rate determining step was found to be the generation of benzylic carbonium ions (i.e. transfer of the latent electrophile into an active one). It is not essential that the base soluble ionizable group is attached to the aromatic ring and current research is directed towards obtaining an appropriate compromise in the final materials[81].

It is likely that progress in the DUV area will be made by the use of multilevel lithography, patterning into the thin top layer, development into a sub layer with plasma etching to generate the feature which is transferred to the substrate. A detailed discussion is beyond the scope of this article and has been the subject of a recent review[58,69].

10   ION BEAM LITHOGRAPHY

Ion beam lithography potentially is capable of producing higher resolution patterns because the scattering produced by the heavier ions is smaller than in the case of electrons. The same resists as those used for electrons are used for ion beam lithography, the sensitivity however is a little different reflecting the difference in the chemistry taking place. Focused ion beam (FIB) technology has been explored not only for its potential in the area of microlithography[82,83] but also for ion implantation[84-87], metal deposition[88] and etching[89,90]. Equipment for lithography does now exist but at present is not widely used in fabrication processes, direct writing producing 0.2 μm structure has been demonstrated[82].

11   X–RAY RESISTS

Now X–ray sources based on synchrotron and plasma sources have made commercial lithography in the next ten years a possibility, the extent to which it is adopted will depend on commercial as well as technical constraints. The main advantage with X–ray lithography is that it gives access to in principle very narrow line widths. However, there are a number of factors which suggest that a practical limit may be 0.3 μm[91], a dimension which is also within the scope of excimer laser sources. One of the limiting factors in the development of X–ray lithography is the nature of the substrates available for use as masks. In general they have to be thin, typically of the order of 2 μm, and constructed from a low atomic weight material such as Si, SiC, Si_3, N_4, BN or with a high atomic number layer, tungsten, tantalum or gold of thickness 1 μm absorbed on the surface. This structure is very fragile and sensitive to distortion due to absorber stress[92,93], this can be overcome by the correct choice of material[94]. As a result the contrast through the mask tends to be low compared to optical masks; typical values are about 10. It is becoming clear that throughput is the critical factor determining whether X–ray lithography is likely to become commercially important[94], this will depend on development of resists with a high sensitivity to X–rays. Due to the high photon energies used in X–ray lithography, the primary absorption events occur in the inner non–valence shells of atoms[95,96] are influenced very little by the nature of the chemical bond between these atoms. It has been found that a linear correlation exists between the sensitivity to electron and X–ray radiation[97,98], Appendix I. Chemicals which exhibit chemical amplification[99,100] do not however give high resolution and hence it is clear that a compromise between speed and resolution has to be achieved.

12 PLASMA ETCH RESISTANCE

Perhaps the main challenge in resist design is improvement of the plasma etch resistance of both positive and negative resists. Recent work on a copolymer of vinyl toluene and chloromethyl styrene[101], has shown that considerable increase in sensitivity can be obtained for a negative electron beam resist by the addition of as little as 5% of the chloromethyl groups into the copolymer. The sensitivity of the resist is also found to be a function of molecular weight. Resolution of 0.2 μm has been demonstrated with a sensitivity of 10 μC/cm[2] and a high contrast. The polymer has a high plasma etch resistance[102], comparable to that of the novolac resins. Various attempts have been made to improve the etch resistance of negative electron beam resists based on poly(methacrolyl chloride) which is highly sensitive, but its etch resistance can be increased by treatment with analine or silane. It has also been proposed that a siloxane based polymer with pendant toluene and chloromethyl phenyl groups which can be readily crosslinked with electron beams. The incorporation of either phenyl or silicon groups into the resist very significantly reduces the plasma etch rate and makes it possible by redesign of the resists to achieve materials with significant improvement in the etch characteristics. In a search for plasma etch resistance studies have been carried on a series of polysilanes[103]. One of the most interesting features of the substituted silane polymers is their extraordinary spectral properties[104,105]. Even though the backbone is composed entirely of saturated silicon sigma bonded links, all high molecular weight polysilanes absorb strongly in the U.V. As the chain length is decreased so the absorption shifts to shorter wavelength and also reduces the absorption coefficient. This makes this polymer an ideal contender for a deep UV optical resist in that it undergoes spectral bleaching and allows exposure down to the substrate. Its high sensitivity can be reduced by dispersion of it in a rigid polymer matrix and makes it suitable as an electron beam resist[106,107]. A resolution of 0.2 μm can be obtained and the resist is highly resistant to reactive ion etching. This is an interesting material for future study and may well be part of the answer for the extension of optical lithography to very small feature sizes.

13 CONCLUSIONS

The weight of the evidence presented in this paper suggests that the size of semi conductor features by the year 2000 is likely to approach 0.2 – 0.1 μm. A recent analysis of the design limitations of NMOS and CMOS suggest that the channel lengths are design limited to 0.45 and 0.15 μm respectively supporting the predictions that feature sizes of 0.1 μm represent a realistic lower limit[108,109]. Continuation of the trend of reduction in size appears likely because it will not only require a significant change in the lithographic technology but will also need new levels in cleanliness and quality assurance monitoring. The challenges which remain are primarily concerned with fine tuning electron beam resists, but more importantly extension of the range of excimer resists available for use in the deep UV region. Microlithography has played an important role in the last thirty years and is likely to retain its key role in determining the pace of semi conductor development for the next fifteen years.

14 ACKNOWLEDGEMENTS

The author wishes to thank the SERC/Alvey Directorate for support of research into the development of new electron beam resists. The support of British Technology Group, British Telecom, Plessey, Philips and GEC are gratefully acknowledged. The views expressed here are purely those of the author and do not reflect the attitude of the supporting organisations.

15 REFERENCES

1.   G.E. Moore. <u>Technical Digest</u>. Inst. Electron Devices Meeting, December 103, (1975) pp. 11–13.

2.   J. Robertson. "Semicon Co-op Eyes Limited Production". Electronic News, (1987) Vol. 33 , p.1, Mar 9.

3.   R.C. Dehmol and G.H. Parker. "Future VLSI Manufacturing Environment". Solid State Technology, (1987) May, 115 .

4.   W. Bauer. "Advanced Processing of Electronic Materials in the United States and Japan". National Academy Press, Washington, D.C. (1986).

5.   J.L. Bartett. "Masked Ion Beam Lithography: An Emerging Technology". Solid State Technology (1986), Vol. 29 (5), 215, May.

6.   G.R. Brewer. "Electron Beam Technology in Microelectronic Fabrication". Academic Press, New York, (1980), 24.

7.   O.C. Wells. "Scanning Electron Microscopy" Chapter 3. McGraw–Hill, New York (1974).

8.   T.E. Everhart. J. Appl. Phys. (1960) 31, 1483–1490.

9.   R.L. Seliger, R.L. Kubena, R.D. Olney, J.W. Ward and R. Wang. Symp. Electron. Ion Photon Beam Technol. Boston, Massachusetts, May (1979).

10.   R.W. Keyes. IEEE Spectrum (1969) 6, 36–45, (1969), Proc. IEEE (1972), 60, 1055–1062, ibid Proc. IEEE (1975), 63, 740–767, ibid Science (1977), 195, 1230–1235.

11.   T.K. Hardy, D.D. Christmon and R.H. Shay. Solid State Technology, (1988) October, p.83.

12.   H.J. Warnecke and R. Herz. Solid State Technology, (1988) October, p.1.

13.   J.T. Wallmark. IEEE Trans. Electron Devices (1979) E2–26, 135–142.

14.   S.P. Billat. Solid State Technology, (1989), January, 99. K. Harris, Solid State Technology, (1989) January, 100.

15.   I. Platnik. Solid State Technology, (1989) January, 102.

16.   G. Burns. Solid State Technology, (1988) December, 27.

17.   A.E. Novembre, L.M. Masakowski and M.A. Hartney. Polymer Engineering Science, (1986) 26 (16), 1158.

18.   L.F. Thompson and R.E. Kerwin. Ann. Rev. Mat. Sci. (1976) (Huggins R.A., Bule R.H. and Roberts R.W. Eds.) 6, 267.

19.   E.D. Feit, U.V. Curing. Science and Technology (Pappas S.P. Ed.), Technology Marketing Corp., Stamford, Conn. (1978), p.230.

20.   H. Nakane, A. Yokota, S. Yammamoto and W. Kanai. Proc. Reg. Tech. Conf. on Photopolymers Princ. Processes and Mat. Mid Hudson Sect. SPE, Ellenville, N.Y. (1985), p.11.

21.   M. Hashimoto, T. Iwayanagi, H. Shiraishi and S. Nonogaki. Proc. Reg. Tech. Conf. on Photopolymers Princ. Processes and Mat. Mid Hudson Sect. SPE, Ellenville, N.Y., (1985) p.11.

22.   J. Koser. Light Sensitive Systems. John Wiley and Sons, New York, (1965) p.164.

23.   J. Pacansky and J.R. Lyerla. IBM J. Res. Develop. (1979), 23 (1), 42,.

24.   M. Hanabata, A. Furuta and Y. Uemura. Proc. SPIE Advances in Resist Technology (1986), 681, 76.

25.   P. Trefonas and B.K. Daniels.   Proc. SPIE Advances in Resist Technology (1987), 771.
26.   R.D. Miller, C.G. Willson, D.R. McKean, T. Tompkins, N. Clecak, J. Michl and J. Downing.   Proc. Reg. Tech. Conf. on Photopolymers. Prime Processes and Mat.   Mid Hudson Sect. SPE, Ellenville, N.Y., (1982), Nov. 8–10, p.111.
27.   R.D. Miller, C.G. Wilson, D.R. McKean, T. Tompkins, N. Clecak, J. Michl and J. Downing.   Org. Coat. and Plast. Chem. Preprints (1983), 48, 54.
28.   S.F. Yoon, P.L. Villa, M. Calzavara and G. Degiorgis.   Solid State Technology, (1989), February, 89.
29.   G.J. Carnigliaro and C.R. Shipley.   U.S. Patent 4,439,516 March 27 (1984).
30.   H. Ito, C.G. Wilson, J.M.J. Frechet, M.J. Farrall and E. Eichler.   Macromolecules (1983), 16, 510.
31.   S.R. Turner, R.A. Arcus, L.G. Houle and W. Schleigh.   Proc. Reg. Tech. Conf. on Photophysics.   Princ. Processes and Mat. Mid Hudson Sect. SPE, Ellenville, N.Y., (1985) Oct. 28–30, p.35.
32.   C.G. Wilson, N.J. Clecak, B.D. Grant and R.J. Twieg.   Electrochem. Soc. Preprints, St. Louis, Mo. (1980), p.696.
33.   H. Maritz and G. Paal.   U.S. Patent 4,104,070.
34.   Y. Takahashi, F. Shinozaki and T. Iheda.   Japan Kokai, Tokyo, Koho (1980), 8, 8032.
35.   E. Alling and C. Stanffer.   Proc. SPIE Advances in Resist Technol. and Processing II, (1985), 539, 194.
36.   C.G. Wilson, R.D. Miller and D.R. McKean.   Proc. SPIE Advances in Resist Technol. and Processing IV, (1987), 771.
37.   E. Reichmaris, C.W. Wilkins and E.A. Chandross.   J. Vac. Sci. Technol. (1981), 19 (4), 1338.
38.   C.W. Wilkins, E. Reichmaris and E.A. Chandross.   J. Electrochem. Soc. (1980), 127, 2514.
39.   E. Reichmaris, C.W. Wilkins and E.A. Chandross.   J. Electrochem. Soc. (1980), 127, 2510.
40.   R.L. Hartless and E.A. Chandross.   J. Vac. Sci. Technol. (1981), 19, 1333.
41.   C.G. Wilson, R.D. Miller, D. McKean, N. Clecak, T. Tompkins, D. Hofer.   Proc. SPE Reg. Tech. Conference on Photophysics, Ellenville (1982), 111, Nov.
42.   K. Jain, C.G. Wilson, B.J. Lin.   IEEE Elec. Dev. Lett. EDL 3:3, (1982), 53.
43.   V. Pal, J.H. Bennewitz, G.C. Escher, M. Feldman, V.A. Firtion, T.E. Jewell, B.E. Wilcomb, J.T. Clemens.   Proc. SPIE Conf. "Ceptical Microlithography V", (1986), 633:6.
44.   D. Kyser and N.S. Viswanathan.   J. Vac. Sci. Technol. (1975), 12, 1305.
45.   M. Hatzakis.   J. Electrochem. Soc. (1969), 116, 1033.
46.   M. Kakuchi. S. Sugawara, K. Murase and K. Matsuyama.   J. Electrochem. Soc., (1977), 124, 1648.
47.   T. Tada.   J. Electrochem. Soc. (1979), 126, 1829.
48.   T. Tada.   J. Electrochem. Soc. (1983), 130, 912.

49. E.D. Roberts. ACS Dir. Org. Coat. and Platics Chem. Preprints (1973), 33 (1), 359.
50. E.D. Roberts. ACS Dir. Org. Coat. and Plastics Chem. Preprints (1977), 37 (2), 36.
51. W. Moreau, D. Merritt, M. Moyer, W. Hatzakis, D. Johnson and L. Pedersen. J. Vac. Sci. Techn. (1979), 16 (6), 1989.
52. M.J. Bowden and L.F. Thompson. Solid State Technol. (1979), 22, 72.
53. M.J. Bowden, L.F. Thompson and J.P. Ballantyne. J. Vac. Sci. Technol. (1975), 12, 1294.
54. M.J. Bowden and L.F. Thompson. J. Electrochem. Soc. (1974), 121, 1620.
55. M. Saitou, S. Okozaki and K. Nakamura. Microcircuit Engineering (1986), 5, p.123.
56. S. Okozaki, F. Murai, O. Suya, H. Shiraishi and S. Koibuchi. J. Vac. Sci. Technol. (1987), B5, p.402.
57. N. Saitou, S. Okozaki and K. Nakamura. Solid State Technology (1987), 65.
58. E.A. Chandross, E. Reichmaris, C.W. Wilkins, R.L. Hartless. Solid State Technology (1981), 24, 81.
59. J.R. Sheats. Solid State Technology (1989), (6), 79.
60. T.R. Pampalore. Solid State Technology (1984), 27 (6), 115.
61. N.J. Turro. "Modern Molecular Photochemistry". Banjamin/ Cummings Menlo Park C.A. (1978).
62. K.J. Oruck and M.L. Dennis. Proc. SPIE (1987), 771, 281.
63. J. Pacansky. Polym. Eng. Sci. (1980), 20, 1049.
64. M.P. de Grandpre, D.A. Vidusch, M.W. Legewz. Proc. SPIE 539, (1985), p.103.
65. G.N. Taylor and T.M. Wolf. Polym. Eng. Sci. (1980), 20, 1087.
66. H. Goken, S. Esho, Y. Ohnishi. J. Electrochem. Soc. (1983), 130, 143.
67. L.A. Pederson. J. Electrochem. Soc. (1982), 129, 205.
68. H. Goken, K. Taniguski, Y. Ohnishi. Solid State Techn. (1985), 28 (5), 163.
69. S.A. Moss, A.M. Jolly and B.J. Tighe. Plasma Chem. Plasma Process (1986), 6, 41.
70. K. Karada, O. Kogure, K. Murase. IEEE Trans. Electron. Dev. (1982), ED29, 518.
71. E. Reichmaris, L.F. Thompson. Ann. Rev. Mat. Sci., (1987), 17, 235.
72. D.R. McKean, S.A. McDonald, N.J. Clecak, C.G. Wilson. Proc. SPIE (1988), 922, 60.
73. J.M.J. Frechet, H. Ito and C.G. Wilson. Microcircuit Eng., (1982), 2, 260.
74. H. Ito. Proc. SPIE (1988), 922, 33.
75. L.F. Thompson and R.E. Kerwin. Ann. Rev. Mat. Sci. (1976), 6, 267.
76. T. Iwayanagi. IEEE Trans. Electron. Dev. (1981), ED28, 1306.
77. J.R. Sheats. SPE Preprints (1988), 319.

78.   J.S. Hargreaves, M.M. O'Toole. <u>J. Electrochem. Soc.</u> (1989), <u>136</u>, 225.
79.   H.Y. Liu, M.P. de Grandpre, W.E. Feely. <u>J. Vac. Sci. Technology</u> (1988), B6, 379.
80.   B. Rede. <u>SPE Preprints</u> (1988), 63.
81.   C.W. Wilkins, E. Reichmaris, E.A. Chandross. <u>J. Electrochem. Soc.</u> (1982), <u>129</u>, 2552.
82.   Y. Ochiai, S. Matsui and K. Mari. <u>Solid State Technology</u> (1987), (11), 75.
83.   R.L. Siliger, R.L. Kubana, R.D. Olney, J.M. Ward, V. Wang. <u>J. Vac. Sci. Technol.</u> (1979), <u>16</u> (6), 1610.
84.   E. Miyauchi, H. Arimoto, H. Hashimoto, T. Furuta and T. Utsami. <u>Jap. J. Appl. Phys.</u> (1983), <u>22</u>, L287.
85.   A. Takamori, E. Miyauchi, H. Arimoto, Y. Bamba and H. Hashimoto. <u>Jap. J. Appl. Phys.</u> (1984), <u>23</u>, L599.
86.   S. Shukuri, Y. Wade, H. Masuda, T. Ishitani and M. Tamura. <u>Jap. J. Appl. Phys.</u> (1984), <u>23</u>, L543.
87.   H. Hamadch, J.C. Covilli, A.J. Stechl and I.L. Barry. <u>J. Vac. Sci. Technol.</u> (1985), B3 (1), 91.
88.   K. Gamo. <u>Nucl. Inst. Methods in Phys. Res.</u> (1985), B7/8, 864.
89.   Y. Ochiai, K. Gamo and S. Namba. <u>Jap. J. Appl. Phys.</u> (1984), <u>23</u>, L400.
90.   J. Kato, H. Morimato, K. Saitch and H. Nahatu. <u>J. Vac. Sci. Technol.</u> (1985), B3 (1), 50.
91.   J. Lingnau, R. Dammol, J. Theis. <u>Solid State Technology</u> (1989), (9), 105.
92.   H. Betz. <u>Microelectronic Eng.</u> (1986), <u>5</u>, 41.
93.   A.O. Wilson. "X-ray Lithography: Can it be Justified". <u>Proc. SPIE</u> (1985), <u>537</u>, 85.
94.   G.N. Taylor. <u>Solid State Technol.</u> (1984), <u>27</u> (6), 124.
95.   M. Farhataziz and J. Roberts (eds.). "Radiation Chemistry". VCH Publ. New York (1987).
96.   D. Seligson, L. Pan, P. King, P. Piannetta. <u>Nucl. Industr. Meth.</u> (1988), A266, 612.
97.   J. Pacansky, R.J. Waltman. <u>J. Radiation Curing</u> (1988), 12 October.
98.   M. Tsuda. <u>Materials Sci. Rep.</u> (1987), <u>2</u>, 185.
99.   M.J. Bowden, J.H. O'Donnell. Developments in Polymer Degradation, ed. N. Grassie 617 (1987).
100. K. Nate, T. Inoue, H. Yokono, K. Hatuda. <u>J. Appl. Polym. Sci.</u> (1988), <u>35</u>, 913 .
101. R.A. Pethrick and S. Affrossman. <u>SERC Bulletin</u> (1987), <u>3</u> (9), 16.
102. S. Affrossman, R.A. Pethrick, (unpublished data).
103. R.D. Miller. <u>Polymer News</u> (1987), <u>12</u> (11), 326.
104. R. West. <u>J. Organo. Met. Chem.</u> (1986) <u>300</u>, 327 .
105. C.G. Pitt. "Homo atomic Rings, Chains and Macromolecules of the Main Group Elements". Rheingold A.L. Elsevair, New York (1979).

106. R.D. Miller, D.C. Hoper, G.N. Fickes, C.G. Willson, E. Marincro, P. Trefonas, R. West. Polym. Eng. Sci. (1986), 26, 1129.

107. D.C. Hoper, R.D. Miller, C.G. Willson, A.R. Neureuther. Proc. SPIE, Advances in Resist Technology (1984), 469, 108.

108. J.D. Meindl. Solid State Technology (1987), (12), 85.

109. J.E. Pfiester and J.D. Meindl. IEEE ISSCC Digest (1984), 158.

110. J.E. Meindl. IEEE Tr. Electr. Dev. (1984), ED-31 (11), 1555.

111. J.B. Lin. J. Vac. Sci. Technol. (1975), 12, No. 6, p.1317.

112. Y. Mimura, T. Ohkubo, T. Takeuchi and K. Sekikawa. Japan J. Appl. Phys. (1978), 17, No. 3, p.541.

113. E.A. Chandross, E. Reichmaris, C.W. Wilkins and R.L. Hartless. Solid-State Technol. (1981), 24, No. 8, p.81.

114. Y. Nakane, T.T. Sumori and T. Mifune. Semi-Conductor Intl. (1979), 2, No. 1, p.45.

115. C.W. Wilkins, E. Reichmaris and E.A. Chandross. J. Electrochem. Soc. (1980), 127, No. 11, 1980, p.2510.

116. E. Reichmaris and C.W. Wilkins. ACS Div. Organic Coatings and Plast. Chem. Preprints (1980), 43, p.243.

117. E. Reichmaris, C.W. Wilkins and E.A. Chandross. J. Electrochem. Soc. (1980), 127, No. 11, 2514.

118. R.L. Hartless and E.A. Chandross. J. Vac. Sci. Technol. (1981), 19, No. 4, p.1333.

119. K. Nate and T. Kobayashi. J. Electrochem. Soc. (1981), 128, p.1394.

120. J. Appebaum, M.J. Bowden, E.A. Chandross, M. Feldman and D.L. White. Proc. Kodak Microelectronics Seminar, Interface, (1975), Oct. 19-21, 40.

121. M.J. Bowden and E.A. Chandross. J. Electrochem. Soc. (1975), 122, No.10, p.1371.

122. M. Hatzakis. J. Electrochem. Soc., (1969), 116, 1033.

123. T. Kitakohji, Y. Yoneda, K. Kitamura. H. Okuyama and K. Murakawa. J. Electrochem. Soc., (1979), 126, 1, 1881.

124. T. Tada. J. Electrochem. Soc., (1981), 128, 1791.

125. M. Krikuchi, S. Sugawara, K. Murase and K. Matsuyama. J. Electrochem. Soc., (1977), 124, 1648.

126. S. Sugaware, O. Kogure, K. Harada, M. Kakuchi, K. Sukegawa, S. Imanura, K. Miyoshi. Electrochem. Soc., Spring Meeting, St. Louis, Missouri, May 11-16 (1980).

127. T. Tada. J. Electrochem. Soc., (1979), 126, 9, 1635.

128. C.U. Pittman, M. Ueda, C.Y. Chen, J.H. Kwiatkowski, C.F. Cook, J.N. Helbert. J. Electrochem. Soc., (1981) 128, 8, 1758.

129. J.H. Lai, J.N. Helbert, C.F. Cook, C.U. Pittman. J. Vac. Sci. Technol., (1979), 16, 1992.

130. S. Matsuda, S.T. Suchiya, M. Honma, G. Nagamatsu. U.S. Patent 4,270,984, 1981.

131. T. Tada. J. Electrochem. Soc., (1979) 126, (11), 1829.

132. J.N. Helbert, C.F. Cook, C.Y. Chen, C.V. Pittman. J. Electrochem. Soc., (1979), 126, 4, 695.

133. I. Heller, R. Feder, M. Hatzakis, E. Spiller. J. Electrochem. Soc., (1979) 126, 154.
134. W. Moreau. D. Merritt, W. Mayer, M. Hatzakis. J. Vac. Sci. Technol., (1979), 16, (6), 1989.
135. Y. Hatano, H. Shiraishi, Y. Taniguchi, S. Horigome, S. Nongaki, K. Naraoka. Proc. 8th Intl. Conf. on 'Electron and Ion Beam Sci. and Technol.', Bakish R., Ed., Electron Soc., Princeton, (1978), 332.
136. J.H. Lai, L.T. Shephard, R. Ulmer, C. Griep. Proc. Reg. Tech. Conf. on 'Photopolymers Principles Processes and Materials', Mid Hudson Section SPE, Ellenville, New York, (1976), Oct. 13–15, 259.
137. L.E. Stillwagon, E.M. Doerries, L.F. Thompson, M.J. Bowden. ACS Dir. Org. Coatings and Plast. Chem., Preprints, (1977), 37, (2), 38.
138. E. Gipstein, W. Moreau, O. Needl. J. Electrochem. Soc., (1976), 126, (7), 1105.
139. S. Nonogaki. Proc. Reg. Tech. Conf. on 'Photopolymers, Principles Processes and Materials', Mid Hudson Section SPE, Ellenville, New York, (1982) Oct. 13–15, 1.
140. H. Sacki, M. Kohda. Proc. 17th Symp. on 'Semiconductor and Integrated Circuit Technology', Tokyo, 1979, 48.
141. Y. Taniguchi, Y. Hatano, H. Shiraishi, S. Horigome, S. Nonogaki, K. Naraoka. Japan J. Appl. Phys., (1979), 28, 1143.
142. L.F. Thompson, J.P. Ballantyne, E.D. Feit. J. Vac. Sci. Technol., (1975), 12, 6, 1280.
143. L.F. Thompson, L.D. Yau, E.M. Doerries. J. Electrochem. Soc., (1979), 126, 6, 1703.
144. S. Imamura, R. Tamaura, K. Harada, S. Sugawara. J. Appl. Polym. Sci., (1982), 27, 937.
145. H. Shiraishi, Y. Taniguchi, S. Horigome, S. Nanogaki. Proc. Reg. Tech. Conf. on 'Photopolymers, Principles Processes and Materials'. Mid Hudson Section, SPE, Ellenville, New York, Oct. 10–12, 1979, 56.
146. Z.C. Tau, C.C. Petropoulos, F.J. Rauner. J. Vac. Sci. Technol., (1981), 19, (4), 1348.
147. Y. Yoneda, K. Kitamura, J. Naito, T. Kitakohji, H. Okuyama, K. Murakawa. Proc. Reg. Tech. Conf. on 'Photophysics: Principles Processes and Materials', Mid Hudson Section, SPE, Ellenville, New York, Oct. 10–12, (1979), 44.
148. H.S. Choong, F.J. Kahn. J. Vac. Sci. Techn., (1981), 19, 4, 1121.
149. K. Sugegawa, S. Sugawara. Japan J. Appl. Phys., (1981), 20, L583.
150. K. Kugegawa, T. Tamamura, S. Sugawara. Proc. 10th Intl. Symp. on Electron and Ion Beam Sci. and Technol., (1983), 83–2, 193.
151. W.M. Moreau. "Semiconductor Lithography – Principles, Practices and Materials". Plenum Press, New York, (1988).
152. F. Asmussen, W. Gaenzler, W. Wunderlich. DE 3446074 (1986).

153. R.A. Pethrick. <u>Prog. Rubber Plast. Technol.</u>, (1987), vol. <u>3</u>, p.11.

154. Data sheet Daikin Kogyo.

155. R. Redaelli, et al. <u>Microelectronic Eng.</u>, (1987), vol. <u>6</u>, p.519.

156. A. Eranian, et al. <u>Br. Polym. J.</u>, (1987), vol. <u>19</u>, p.353.

157. M. Tsuda, et al. <u>J. Vac. Sci. Technol.</u>, (1986), vol. <u>B4</u>, p.256.

158. E. Tai, B. Fay, C.M. Stein, W.E. Feely. <u>Proc. SPIE</u>, (1987), vol. <u>773</u>, p.132 .

159. H. Liu, M.P. deGrandpre, W.E. Feely. <u>J. Vac. Sci. Technol.</u>, (1988), vol. <u>B6</u>, p.379.

160. N. Yoshioka, Y. Suzuki, T. Yamazaki. <u>Proc. SPIE</u>, (1985), vol. <u>537</u>, p.51.

161. W.M. Moreau. "Semiconductor Lithography – Principles, Practices and Materials", Plenum Press, New York, (1988).

162. F. Asmussen, W. Gaenzler, W. Wunderlich. DE 3446074 (1986).

163. R.A. Pethrick. <u>Prog. Rubber Plast. Technol.</u>, (1987), vol. <u>3</u>, p.11.

164. Data sheet Daikin Kogyo.

165. R. Redaelli et al. <u>Microelectronic Eng.</u>, (1987), vol. <u>6</u>, p.519.

166. A. Eranian et al. <u>Br. Polym. J.</u>, (1987), vol. <u>19</u>, p.353.

167. M. Tsuda et al. <u>J. Vac. Sci. Technol.</u>, (1986), vol. <u>B4</u>, p.256.

168. E. Tai. B. Fay. C.M. Stein, W.E. Feely. <u>Proc. SPIE</u>, (1987), vol. <u>773</u>, p.132.

169. H. Liu, M.P. deGrandpre, W.E. Feely. <u>J. Vac. Sci. Technol.</u>, (1988), vol. <u>B6</u>, p.379.

170. N. Yoshioka, Y. Suzuki, T. Yamazaki. <u>Proc. SPIE</u>, (1985), vol. <u>537</u>, p.51.

171. K. Nate, T. Inoue, H. Yokono, K. Hatada. <u>J. Appl. Polym. Sci.</u>, (1988), vol. <u>35</u>, p.913.

172. R. Dammel. <u>Microelectr. Eng.</u>, (1987), vol. <u>6</u>, p.503.

ACRONYMS USED IN THIS ARTICLE

DRAM	Direct Random Access Memory
pMOS	positive Metal Oxide Semiconductor
nMOS	negative Metal Oxide Semiconductor
EPROM	Erasable Programmable Read Only Memory
CVD	Chemical vapour Deposition
VLSI	Very Large Scale Integration
PVC	Poly(vinyl chloride)
PTFE	Poly(tetrafluoroethylene)
PVDF	Poly(vinylidene fluoride)
PFA	Perfluoroalkoxy resins

APPENDIX 1

SUMMARY OF DATA OF VARIOUS TYPES OF RESIST

Selected positive deep UV photo resists

Polymer resist	Sensitivity range (nm)	Relative selectivity	Ref.
Poly(methyl methacrylate)	200-240	1	110,111
Poly(methyl isopropenyl)	230-320	5	112,113
Poly(methyl methacrylate-co-3-aximino-2-butanone methacrylate)	240-270	30	114,115
Poly(methyl methacrylate-co-indenone)	230-300	35	116,117
Poly(p-methoxyphenyl isopropenyl iketone-co-methyl methacrylate)	220-360	166	118
Poly(styrene-co-acenaphthalene sulphones)	250-330	500	119,120
Poly(styrene sulphones)	240-280	1000	120

Summary of some acrylate based resist materials — Electron Beam Resist

Resist	Sensitivity at 20 kV ($\mu C/cm^2$)	M_w	Tg	Ref.
Poly(methyl methacrylate) (PMMA)	8	5×10^5	$104\,^0C$	121
Poly(hexafluorobutyl methacrylate) (FBM - 110 (DAIKIN KOGYO))	0.4	5×10^5	$50\,^0C$	124
Poly(t-butyl methacrylate)	0.5	5×10^5	$19\,^0C$	122,123
Poly(dimethyl tetrafluoropropyl methacrylate) (FPM (DAIKIN KOGYO))	3-12	$10^6 - 10^7$	$93\,^0C$	125
Poly(trichloroethyl methacrylate) (EBR 1 (TORAY INDUSTRIES))	1.25	5.7×10^5	$138\,^0C$	126

Poly(trifluoroethyl methacrylate)	4.5	9.2×10^5		123,127
Poly(methyl-α-chloro-acrylate) (PMCA)	4.6	1.6×10^6	130°C	128
Poly(ethyl-α-cyanoacrylate)	1.5	2×10^5		129
Poly(ethyl-α-carbox-amido acrylate)	1.4	2.5×10^5		130
Poly(trifluoro-α-chloro-acrylate) (EBR-9 (TORAY INDUSTRIES))	0.8 -6.4	2.5×10^6	133°C	131
Poly(methacrylonitrile)	30	1.3×10^6	120°C	132
Poly(methyl methacrylate - or methacrylic acid) P(MMA-MAA) (4:1)	20	1.10^5		132
Poly(methyl methacrylate - co-methacrylic acid - or methacrylic anhydride) P(MMA - MAA - MAH) (70:15:15)	6.5	3×10^5		133
Poly(methyl methacrylate - co-acrylonitrile) (P(MMA-AN) (89.11)COEBR-1013 (TOKYO OKHA)	6	6×10^5		134
Poly(methyl methacrylate - co-methyl-α-chloroacrylate) P(MMA - MCA) (1:C:1)	7.5	1.2×10^5		135
Poly(methacrylate - co-methyl acrylonitrile) P(MMA–MAN) (1:1)	8	1×10^6		136
Poly(methyl α-chloroacrylate - co-methacrylonitrile) P(MCA-MAN)	24	1.10^5		125
Poly(ethyl α-cyanoacrylate - co-ethyl carboxamido acrylate) (9:1) (FMR – E101 (FUJI CHEMICALS))	1.5	2×10^5		129
Poly(methyl methacrylate - co-isobutylene) P(MMA – IB) (75:25)	6	10^5		137
Poly(hexafluorobutyl methacrylate - co-glycidyl methacrylate)P(FBM-GMA (99:1))	0.4	10^5		138
Poly(methyl methacrylate - co-1-butyl methacrylate) (CP-3)	1.6	10^5		139

Styrene and methacrylate based negative electron resists

Resist	Sensitivity ($\mu C/cm^2$)	M_w	Ref.
Poly(glycidyl methacrylate) PGMA (OEBR-100 (TOKYO OHKA))	0.5	1.25×10^5	140
Poly(glycidyl methacrylate-co-ethyl acrylate) (COP (MEAD CHEMICALS))	0.6	1.8×10^5	141
Poly(chlorinated methylated styrene (40% chlorinated) (CMS-EX (TOYO SODA))	39-0.4	6.8×10^3 - 5.6×10^5	142,143
Poly(2-hydroxy-3 (methyl fumarate) propyl methacrylate-co-3 chloro-2-hydroxyl propyl methacrylate)(SEL-N (SOMAR CORP))	0.8	-	138
Polyiodo styrene (IPS) (-70% IODINATED) (RE-4000N (HITACHI CHEMICAL))	2	3.8×10^5	144
Poly(allyl methacrylate-co-2-hydroxyethyl methacrylate) (3:1) (EK-88 (KODAK))	0.4	3.5×10^4	145
Poly(diallyl ortho phthalate) (PDOP)	56-0.9	1.1 - 11.1×10^4	146
Poly(chloromethyl styrene) (PCMS)	7-0.35	2.10^4 - 4.5×10^5	147
Poly(4-chlorostyrene) (PCS)	2.5-1.5	7×10^5	148
Chloromethylated Poly (α methyl styrene) (-95% chloromethylated)	42-3	8.10^3 - 1.9×10^5	148,149
Chlorinated poly(vinyl toluene) (CPMS)	0.4-1.6	1.2 5.8×10^5	149

X-ray Resists – Sensitivity Data

No	Resist	Type	Company	Source	X-ray sens. (mJ/cm²)	Ref.
1	poly(methylmethacrylate)	950k	KTI	1GeV	1800	165,163
2	poly(2-fluoroethyl-methacrylate)				75	161
3	poly(2,2,2-trifluoroethyl-methacrylate)				50	161,163
4	poly(tetrafluoroisopropylmethacrylate)	FPM-210	Daikin	MoL	160	164
5	poly(tetrafluoroisopropylmethacrylate)	FPM			100	161,163
6	poly(hexafluorobutyl-methacrylate)	FBM-110	Daikin	MoL	50	164
7	poly(hexafluorobutyl-methacrylate)	FBM-120	Daikin	Mol	35	164
8	poly(fluorinated methacrylate)	FBP	Thomson	1GeV	280	165
9	poly(2-chloroethylmethacrylate)				75	161
10	poly(2-bromoethylmethacrylate)				75	161
11	poly(methacrylonitrile)				200	161
12	poly(butylcyanoacrylate)			754MeV	120	166,163
13	poly(methacrylic acid)-co (methacrylonitrile)			13.34A	100	162
14	poly(methylmethacrylate)-co (methacrylic anhydride)			754MeV	100	161,163
15	poly(butenesulfone)	PBS	Mead		100	161,163
16	poly(methylpentenesulfone)/novolak	RE5000P	Hitachi		14	163
17	poly(trimethylsilylsilylpropional)-co (3-phenylpropanal)				360	171
18	naphthoquinonediazide/novolak	WX242	Hitachi	MoL	200	165
19	naphthoquinonediazide/novolak	AZ1450J	Olin-Hunt		2100	161
20	3CS/novolak	RAY-PF	Hoechst		65	172
21	poly(methyl-isopropenylketon)/bisazides		Hoechst	754MeV	1380	167
22	poly(glycidylmethacrylate)	OEBR-100	TokyoOhka		50	161,163

23	poly(glycidylmethacrylate)-co (ethylacrylate)	COP	Mead	5	161,163
24	poly(glycidylmethacrylate)-co (ethylacrylate)	COP	Mead	116	161,163
25	poly(glycidylmethacrylate)-co (ethylacrylate)	COP	Mead	185	161,163
26	poly(glycidylmethacrylate)-co (methylmaleate)	SEL-N	SOMAR	10	161
27	epoxy poly(butadiene)	EPB		2	161
28	chlorinated poly(methylstyrene)	CPMS		8	161,163
29	chlorinated poly(p-methylstyrene)	CPMS		19	161,163
30	chlorinated poly(p-methylstyrene)	CPMS-X		17	170
31	poly(iodostyrene)	RE4000N	Hitachi	10	161
32	poly(tetrathiafulvenylstyrene)			50	161
33	negative tone novolak	SAL601	Shipley	Pd	40	168,169
34	negative tone novolak	RAY-PN	Hoechst	754MeV	100	172

New Chemistry and Technology in the Development of Isocyanate-free Paint Systems

A. R. Marrion and A. B. Port

COURTAULDS COATINGS LIMITED, STONEYGATE LANE, FELLING,
GATESHEAD NE10 0JY, TYNE AND WEAR

1. INTRODUCTION

Amongst polymer based industries, the Surface Coatings Industry is unusual for its long history. One legacy of that history is a public expectation of technically unsophisticated, low cost materials.

In reality, surface coatings make a major contribution to the value of a great many manufactured and other items in films varying from less than a micron (electronics photoresists) or a few microns (can linings) to many millimetres (deck coatings).

Another unusual feature of the Coatings Industry is that the final and crucial stage in coatings' manufacture is conducted by their suppliers' customer, who may not have a high level of skill or equipment related to polymer processes. Consistency and ease of application, and safety in use, therefore often rank with performance as key requirements of a system.

A surface coating is required to undergo a transition after application from a liquid or solution of specified viscosity (or a fusable powder) to an intractable film with the required performance properties. The necessary film properties are generally achieved with polymers of appreciable molecular weight $(10^4 - 10^5)$. Such polymers impart high viscosity to their solutions. For example, nitrocellulose which was formerly used for car finishes, can only be satisfactorily sprayed when diluted to about 12% non-volatiles. As a result it yields thin films and many applications are required.

One approach to the application of high molecular weight polymers is to use a dispersion in water or organic non-solvents, when viscosity is related to the viscosity of the continuous phase and the volume fraction of the disperse phase. Such systems have found wide acceptance in, for example, latex building paints, but present formidable technical problems if the performance and appearance of a high quality enamel is required.

The more usual technique is to use "convertible" coatings which undergo chemistry leading to chain extension, polymer-isation, or crosslinking of low molecular weight prepolymers (M_n 2000-20000) when required.

The crosslinks must not compromise the performance of the polymer, yet are formed outside the direct control of the manu-facturer. As legislative pressure for lower solvent contents continues in various parts of the world, molecular weights of the prepolymers are decreased and the relative importance of crosslinking increases.

Suitable items coated under production line conditions are stoved at temperatures between 120° and 250° and reliable form-ation of crosslinks is relatively easy. However, when substrate size, sensitivity, or other factors preclude a stoving oper-ation, network formation is more difficult, since:

-   Ambient temperature and humidity are notoriously variable.
-   Highly reactive functional groups must be used.
-   Premature reaction must be taken into account.
-   As crosslinking and solvent loss proceed, the glass transition temperature of the coating approaches ambient and mobility constraints greatly impede further reaction.

Against that background, it is not surprising that relativ-ely few ambient cure chemistries have been successful.

2. EMERGENCE OF AMBIENT TEMPERATURE CURING COATINGS

Self-curing natural media such as blood and egg-white have been used for painting caves and people since pre-historic times. Vegetable "drying" oils, such as linseed oil, have been in use for 1500 to 2000 years. They contain multiple allylic unsaturation and "dry" by an auto-oxidative mechanism. Such materials were often pre-reacted or "bodied" by prolonged heating in contact with air and possibly catalysts.

The incorporation of fatty acid residues into synthetic polymer backbones led to "alkyds", "urethane alkyds" and "acrylic alkyds". Alkyds based on glyceryl phthalate polyesters with pendant drying fatty acid residues were introduced in the late 1920's[1] and, with increasing competition from latex paints, have dominated the decorative paint market ever since.

Unfortunately, the auto-oxidation process continues long after acceptable cure is achieved and leads eventually to embrittlement and network degradation, particularly when transition metal driers have been used. Despite their convenience, auto-oxidatively cured systems appear to suffer from inherent performance limitations. (The remarkably weatherable Chinese and Japanese laquers based on the catechol derivative "urushiol", may be an exception but do not cure exclusively by auto-oxidation[2]).

A more satisfactory approach to ambient cure is the "two package" system, where two components with reactive functional groups are mixed and used within a short period, known as the "pot life".

Two useful chemistries appeared in the 1950's and were used quite widely by the 1960's. Isocyanate curing, pioneered by Bayer, involves the reaction of an isocyanate functional curing agent with a hydroxyl functional polymer, usually polyester or acrylic, to form a urethane linkage:

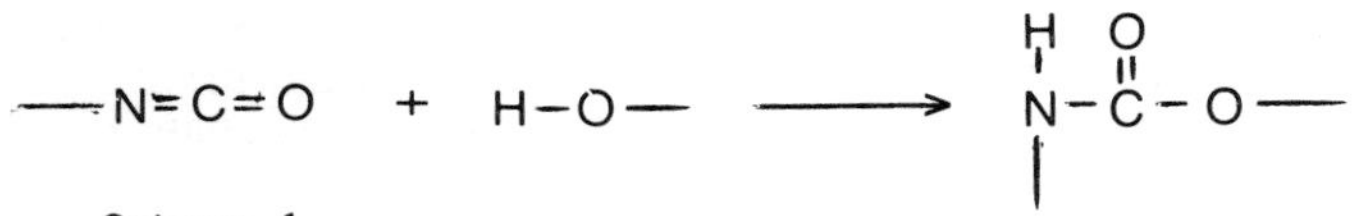

$$\text{---N=C=O} \quad + \quad \text{H--O---} \quad \longrightarrow \quad \overset{\overset{\displaystyle H}{|}}{N}-\overset{\overset{\displaystyle O}{||}}{C}-O\text{---}$$

Scheme 1

Isocyanate cure has gained great popularity since it occurs at a very convenient rate, which is adjustable with tin or amine catalysts, it forms very stable linkages, and it imparts particular mechanical properties to its networks, apparently because secondary crosslinks are formed by hydrogen bonding.

The even faster isocyanate-amine reaction, which gives even more stable urea linkages, is gaining in popularity. It can be adapted to moisture cure by using protected amines such as ketimines.

The great drawback of isocyanate systems is toxicological. They can cause respiratory sensitisation in some individuals, who then respond with asthma-like symptoms to extremely low airborne concentrations.[3] Although vapourisation is minimised by use of oligomeric isocyanate prepolymers, an aerosol hazard still exists in spray painting and isocyanates must be used under conditions where operators and others are completely isolated. Even then, a hazard to by-standers or other workers cannot be discounted in some operations.

Epoxy-functional systems were introduced by Ciba Geigy inter alia. Epoxy resins derived from epichlorhydrin and bisphenol A react readily with primary or secondary amines at room temperature:

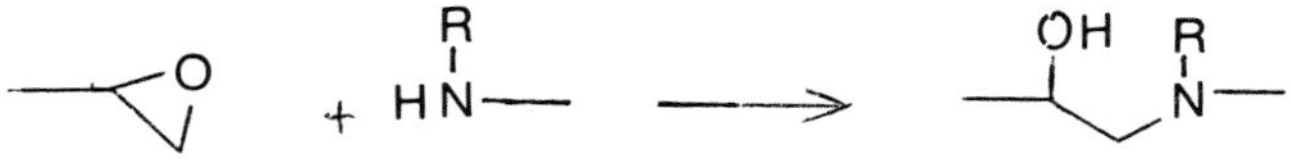

$$\text{—}\triangleleft\!\!\!O \;\; + \;\; HN\!\!-\!\!\overset{R}{\underset{}{|}} \;\; \longrightarrow \;\; \text{—}\overset{OH}{\underset{}{|}}\text{—}\overset{R}{\underset{N}{|}}\text{—}$$

Scheme 2

They are very useful in primers and interior coatings, but unfortunately are very susceptible to photolytic degradation and cannot be used in exterior applications. Low molecular weight epoxy compounds moreover are often skin irritants and some are suspect mutagens[4].

3.   THE COMMERCIAL POSITION

Ambient cured isocyanate surface coatings in particular have established a very firm and expanding place in world markets. It has been recognised since the late 1970's that a substitute of acceptable performance and improved toxicology would make a deep penetration into many such markets.

Some estimates of the scale of a number of industries using isocyanate cured sysetms are given in Table 1. Despite the necessarily approximate nature of the estimates, it is clear that very large markets exist.

Different markets make different demands on technical performance, technical support, and the variety of qualities and colours required. They also differ widely in scale, making it particularly difficult to relate the cost of the coatings to the value of the polymers.

The exact extent to which an isocyanate free material could take market share is controversial, particularly if it proves to fall short of the urethane in some technical respect, or be significantly more costly.

The picture is also confused by different national and industrial perceptions of the danger to workers from exposure to toxic materials and the possibility of ensuring adequate protetction by physical means such as extraction and compressed air masks.

Nevertheless, it is clear that considerable enthusiasm exists in many parts of the globe for an isocyanate-free coating system. Some industries would readily pay a premium and accept a reduction in certain aspects of performance.

Table 1 Volume and Value of Two Pack Polyurethane Coatings in the US Predicted for 1990.

MARKET SECTOR	PREDICTED 1990 (TONNES SOLID)	INCREASE OVER 1985 (%)	VALUE OF COATING (TOTAL) ($M US)	(PER TONNE) ($K US)
Automotive Clearcoat	360		4	11
Marine	2200	19	43	19.5
Aviation	1050	-16	28	27
Other Land Transport	450	11	12	27
Metal Furniture	130	0	5	38
Machinery	4000	28	72	18
Conformable Coatings	1200	61	15	12
Pipe coatings	490	72	7	14
Other Metal Finishing	1600	4	39	24
Plastics	2800	70	44	16
Roof,Tank,Deck	9200	21	188	20
Industrial Maintenance	6100	69	114	19
Automotive Refinish	3600	14	192	53
Fire Retardant	100	10	5	50
Miscellaneous	250	14	7	28
TOTAL	33530	33%	777	23 (av.)

With increasing attention being paid worldwide to environmental issues, the pressure for more "user friendly"

coatings systems can only increase, as witnessed by the ever
increasing patent activity in the area.

4.   APPROACHES TO ISOCYANATE REPLACEMENT

Ring Opening Reactions

Some of the earliest approaches attempted to exploit epoxy
chemistry whilst eliminating the poor weathering properties of
bisphenol A derivatives. Cycloaliphatic epoxies, epoxidised
oils and glycidyl methacrylate copolymers have all been used.

Epoxy-amine systems have enjoyed considerable success[5] and
by a suitable choice of amine, their tendency to form carbonates
or carbamates at the surface by reaction with atmospheric carbon
dioxide, or to exude from the network with consequent stickiness
can be minimised.

Epoxy-thiol systems which can avoid carbonation and give
excellent technical properties have often been discounted on
grounds of smell, however low odour thiols and their use in
curing epoxy resins have been claimed.[6]

The epoxy-acid reaction would normally be expected to be
far too sluggish for ambient temperature cure, but satisfactory
commercial sytsems have been devised and make use of enhanced
reactivity acids, internal catalysis, and proprietary epoxides.[7]

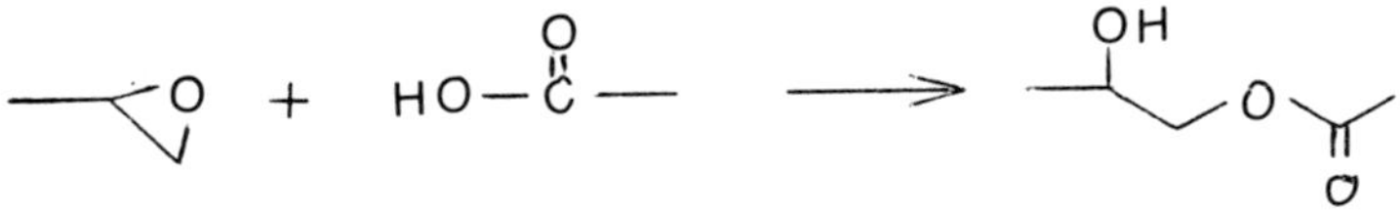

Scheme 3

In a related system,[8] carboxylic acid and tertiary amine are
present in one molecule and apparently both react with an
epoxy.

The reaction of cyclic anhydrides with epoxies is a step
growth process, which proceeds readily in the presence of tertiary amine. With the optional addition of hydroxy and carboxyl
groups, the chemistry has been claimed in several publications.[9]

Recently reaction of anhydrides with amines has been
disclosed.[10] Since the reaction is extremely rapid, amine is
presented in protected form only available for reaction on
hydrolysis by amient moisture. The protecting groups were
ketimine, aldimine, oxazolidine or 1,3-diazacyclopentane or
1,3-diazacyclohexane. Epoxy groups were also optionally
included.

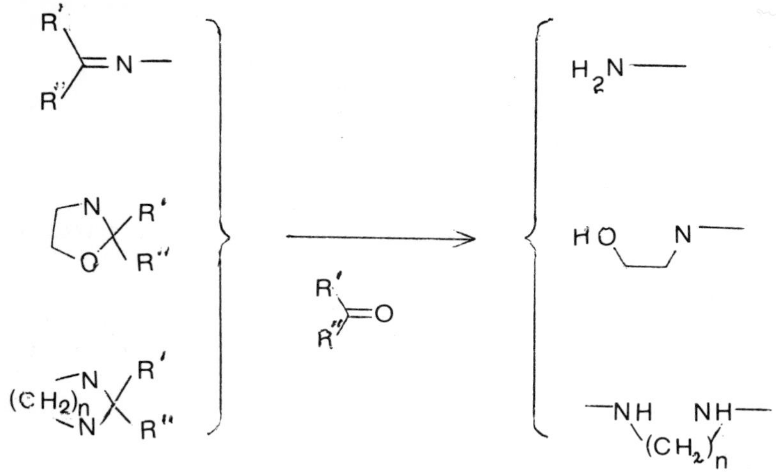

$$n = 2 \text{ or } 3$$
$$R' = \text{Alkyl}$$
$$R'' = \text{Alkyl or H}$$

<u>Scheme 4</u>

The Michael reaction and its heteroatom analogue have been embodied in a remarkable variety of inventions. In both, an activated multiple bond is saturated by reaction with a species containing an activated hydrogen atom.[11]

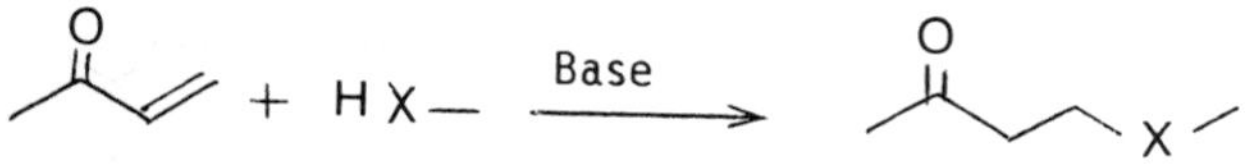

X = C(CO)$_2$ for the Michael reaction proper.
N or S- for the Michael-type reaction.

<u>Scheme 5</u>

The reactions with carbon-acids generally require strongly basic catalysts, KOEt, Bu$_4$NOH, amidine, etc. In one ingenious scheme, catalyst is generated <u>in situ</u>, from tertiary amine and epoxide.[12] Reactions using thiols are catalysed by tertiary amines, whilst primary and secondary amines can be regarded as self-catalysing.[13]

Unsaturated materials have ranged from low molecular weight molecules such as trimethylolpropane triacrylate to acryloyl functional acrylic polymers and unsaturated polyesters based on maleic or itaconic residues, and maleimides.

Reactants have ranged from acetoacetyl functional acrylics and malonate functional poyesters to low molecular weight polyamines, polythiols, acrylic primary and secondary amines, polyamidoamines, and polyamines protected as their ketimines.[14]

Some other ring opening reactions recently exploited are:

The addition of acids to carbodiimides[15]

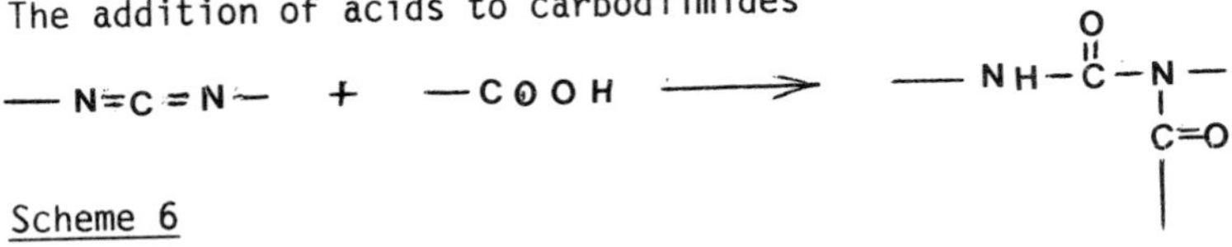

$$— N=C=N— \quad + \quad —COOH \quad \longrightarrow \quad — NH—\overset{\overset{\displaystyle O}{\|}}{C}—\underset{\underset{\displaystyle C=O}{|}}{N}—$$

Scheme 6

(which has the advantage of being applicable to water based coatings).

The reaction of pendant cyclic carbonates with amines (as two pack systems) or ketimines (as one pack ambient cured system);[16]

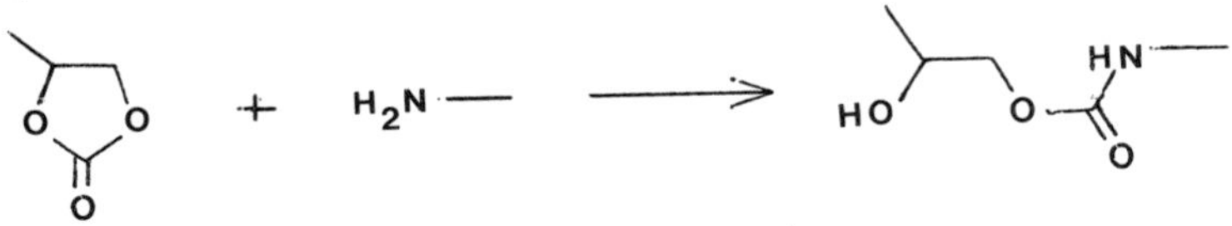

Scheme 7   and

The familiar dihydropyran hydroxyl reaction,[17]

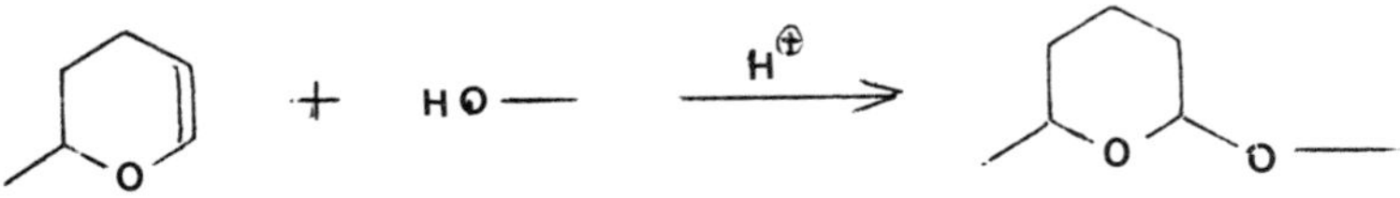

Scheme 8

It is also worthy of note that hindered isocyanates which are said not to act as pulmonary sensitisers have been proposed.[18]

Condensation Reactions

Condensation reactions are generally best suited to elevated temperature curing, where removal of volatile species is relatively easy, but some ambient curing systems have been proposed. The most successful are those which depend on the condensation of silanol groups,

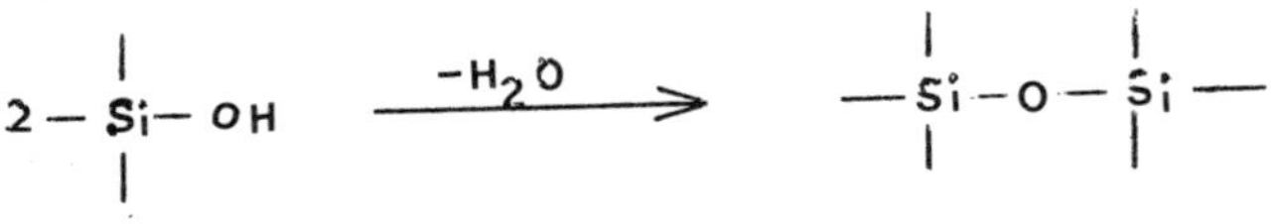

$$2 — \underset{|}{\overset{|}{Si}}— OH \quad \xrightarrow{-H_2O} \quad — \underset{|}{\overset{|}{Si}}—O—\underset{|}{\overset{|}{Si}}—$$

Scheme 9

either generated <u>in situ</u> by atmospheric hydrolysis of alkoxy-
silanes or acyloxysilanes[19] or preformed[20] and condensing <u>inter
alia</u> under the influence of a catalyst.

Attempts have also been made to condense melamine formaldehyde resins with hydroxyl functional polymers[21] at ambient temperature[22] Careful selection of acid catalysts[21] or the amino
resin[22] have been claimed to achieve the required reactivity.
The use of activated esters susceptible to ammonolysis has
received particular attention. The monomer MAGME (methyl acryl-
amidoglycolate, methyl ether) (1)

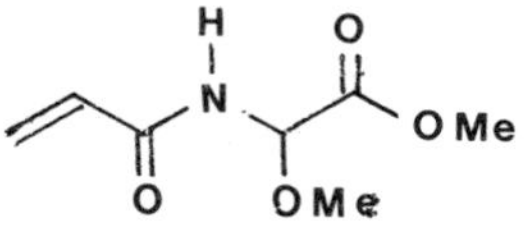

(1)

has been claimed to form copolymers with other acrylic monomers
which could then be cured with polyamines.[23]

A condensation between a polymeric ketone such as an acetoacetyl functional polymer and a polyamine as its ketimine to
yield a polyketimine which rearranges to an enamine has also
been claimed.[24]

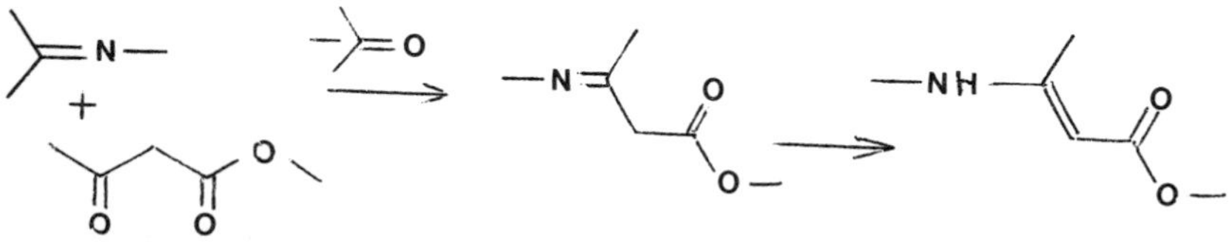

<u>Scheme 10</u>

Amines can also be used to displace a blocking agent such
as phenol from an isocyanate to form a urea linkage at ambient
temperature.

Free Radical Reactions

Apart from the auto-oxidation systems noted above, free
radical curing systems have occupied their own particular <u>niche</u>
and have not been widely applied.

Solutions of unsaturated polyesters dissolved in styrene,
cured by peroxides, have been the backbone of the glass fibre
composite industry and have in the past been adapted to high
gloss wood finishes. Their more recent counterparts, mono- and
diesters of acrylic monomers, cured by peroxides, ultraviolet,

or electron beam radiation, are severely limited by toxicity, and by the surface stickiness which results from oxygen inhibition of the curing process.

An alternative, which was thought to be unique, was the addition of thiols to non-activated double bonds under the influence of free radicals.[25] The odour of thiols has played its part in preventing the thiol-ene reaction from becoming popular in coatings, but a recent disclosure[26] proposes use of an analogous reaction between tertiary amines and olefins. Activated hydrogen atoms from positions adjacent to nitrogen can be abstracted under free radical conditions.

5.   COURTAULDS COATINGS CROSSLINKING CHEMISTRY

A reaction which promised to fulfil the necessary criteria for ambient curing was that between an alcohol and a cylic anhydride. It is known to take place readily at ambient temperature, in the presence of a suitable catalyst and to stop at the half-ester stage.

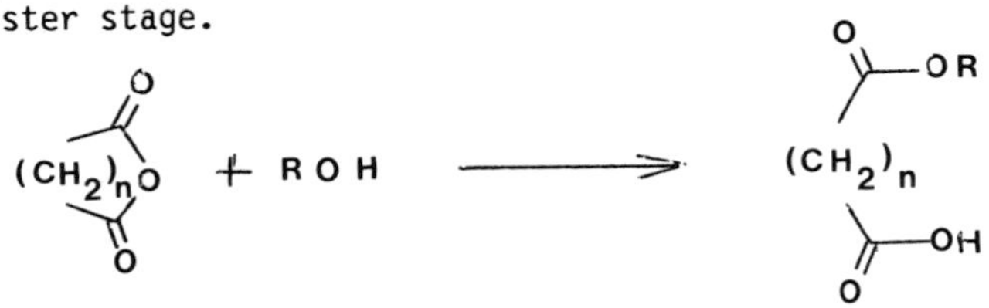

$$(CH_2)_n O \quad + \quad R\,O\,H \quad \longrightarrow \quad (CH_2)_n$$

Scheme 11

Anhydride-hydroxyl reactions have been little used for curing in the past, perhaps because the liberated acid group was expected to be troublesome, and the half-ester linkage was known to be somewhat labile to hydrolysis.

Prior Art

One early example of such a reaction,[27] involved anhydride functional polyesters derived from trimellitic anhydride (2) and titanium and aluminium chelates as catalysts. The chemistry seems best suited to curing at elevated temperatures.

Hydroxyamines have been proposed as thermally reversible curing agents for ethylene maleic anhydride copolymers used in moulding compositions.[28]

Examples of epoxy-anhydride or epoxy-hydroxy-anhydride cure chemistry are more numerous[9] but a pure hydroxy-anhydride chemistry avoids the toxicological associations of epoxies.

Selection of Poly(cyclic anhydrides)

The only readily available cyclic anhydrides are five and six membered rings. Succinic (3) and glutaric (4) anhydrides were reacted with a hydroxy-amine,

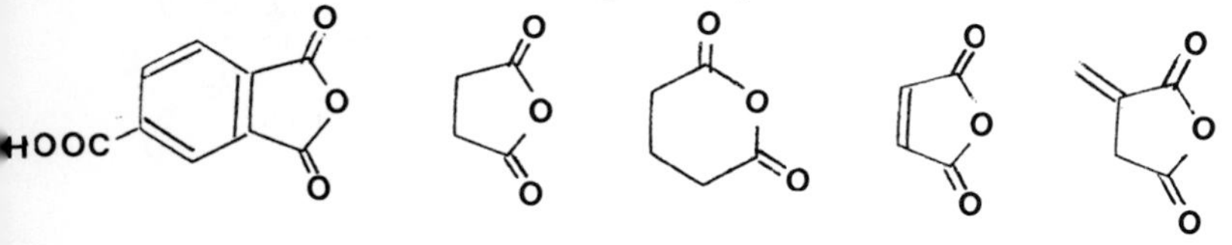

(2)   (3)   (4)   (5)   (6)

and the reaction was followed by infrared spectroscopy. As predicted,[29] the six membered structure was the more reactive of the two. Nevertheless, since they were much more readily available in polymeric form, five membered rings were selected for further study.

The formation of 4-esters of trimellitic anhydride has been reviewed.[30] The acidolysis of a polymer acetate with (2), or direct thermal esterification of a polyol involved high processing temperatures and consequent discoloration, whilst the reaction of acid chloride of (2) with polyols lacked the required specificity, and led to premature molecular weight advancement. Thermal reaction of a 1,2,3-carboxy-anhydride with re-ring closure was easier than the thermal reaction of (2), but still showed a tendency to molecular weight advancement.[31]

Diels Alder reactions of maleic anhydride were discounted for want of suitable conjugated precursors.

Ene-adducts of maleic anhydride and molecules containing allylic unsaturation such as polybutadiene could be readily made by established techniques[32] and proved useful as reactive plasticisers after hydrogenation of the residual unsaturation:

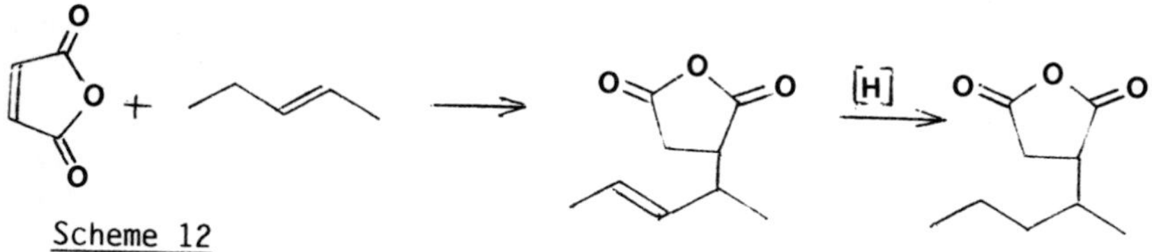

Scheme 12

Another approach, useful for forming flexible poly(cyclic anhydrides) was to carry out a Michael type addition between a thiol tipped polyester and maleic anhydride:

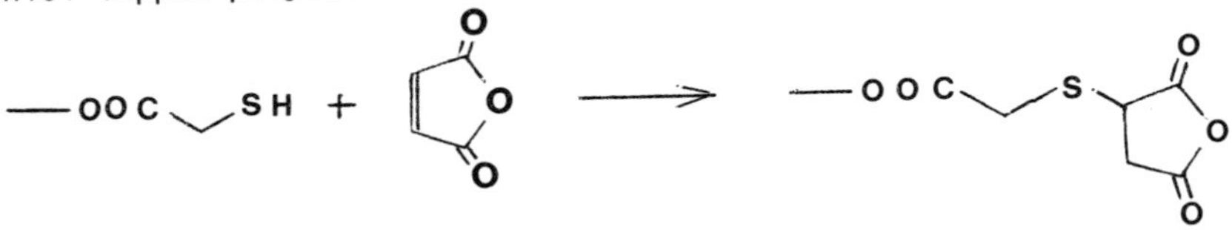

Scheme 13

For most purposes though, free radical polymers of maleic[33] anhydride (5) or itaconic anhydride (6) proved most suitable.

Both (5) and (6) have a marked tendency to alternate with styrene though ostensibly homogenous co- and terpolymers could be made by drip feeding monomers.

Styrene has generally been included since it displays a marked moderating effect on the crosslinking reaction as compared with methyl methacrylate. However, a certain level of acrylate or methacrylate was required to confer compatibility with the curing polymer.

The choice between (5) and (6) is made on the basis of cost and performance.

Maleic anhydride is a very low cost monomer. Its polymers show initially higher reactivity, but cured systems are much more susceptible to weathering. Photolytic susceptibility can be rationalised since the maleic residue contains two activated methine groups compared with the methylene and quaternary carbon of itaconic residues.

High initial reactivity but poor ultimate conversion has[34] been recognised in maleic/styrene copolymers for some time, and attributed to the existence of residues of varying reactivity. The existence of threo and erythro structures in such a copolymer was recently demonstrated.[35] The two forms are hindered towards nucleophilic attack to quite different extents. No such differentiation is possible for an itaconic residue, perhaps accounting for its "smoother" reaction.

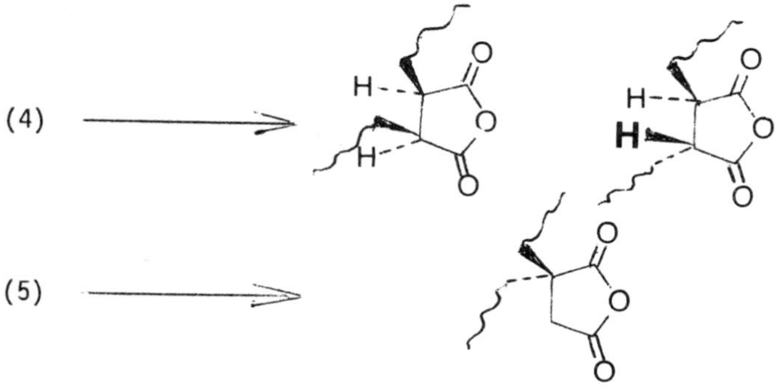

(4) $\longrightarrow$

(5) $\longrightarrow$

Scheme 14

Unfortunately, not being commercially available in bulk, (5) is costly. Its high cost can be reduced by copolymerising itaconic acid which dehydrates readily under azeotropic conditions, once polymerised. It shows a marked reluctance to copolymerise with methacrylates, however.

Selection of Polyols

Primary hydroxyl groups have proved the most satisfactory and can be present as acrylic, polyester or polyether polyols. Secondary hydroxyls, as expected, react more slowly.

Nitrogeneous bases such as tertiary amines have proved to be effective catalysts, though they need to be present in molar quantities comparable to those of the reactants, rather than strictly catalytic quantities. They can, in principle, be introduced as separate species or as part of the polyol molecule. Amines randomly incorporated into acrylic polyols are generally less effective than when free, presumably because of mobility constraints.

An acrylic system containing hydroxyethyl acrylate and diethylaminoethyl methacrylate in one component however has been successfully commercialised and has given excellent service, particularly in regard to weathering properies.

If the amine group is orientated β-to hydroxyl, extremely high reactivity results.

Acrylic systems were made by reacting glycidyl functional copolymers with secondary amines such as dibutylamine or particularly diethanolamine. Epoxy-amine reaction was also exploited to make polyether based materials; amine tipped polyethers were reacted with phenylglycidyl ether.[36]

Polyester systems were prepared by direct condensation with triethanolamine, albeit with some chain extension.

A hydrocarbon based material was prepared by mesylating a hydrogenated hydroxyl tipped polybutadiene and displacing the mesyl ester with triethanolamine.

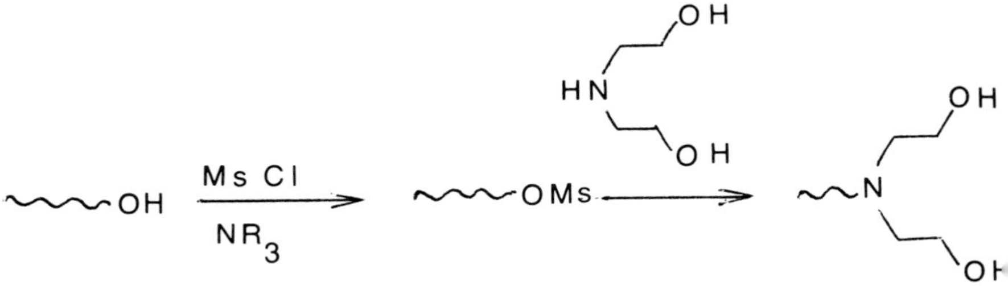

Scheme 15

All such systems showed very high reactivity with poly(cyclic anhydrides). In some cases, to achieve a sufficiently long "pot life", it was necessary to remove a proportion of the β-hydroxy-amine character by reaction with caprolactone.

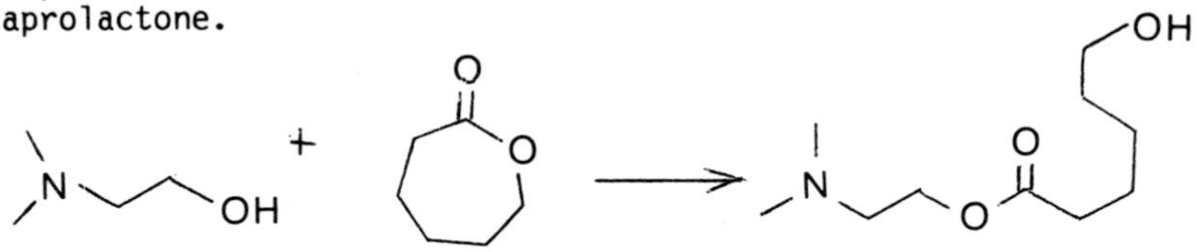

Scheme 16

Reactions were slowed by increasing molar proportions of caprolactone but the completeness of the reaction was little changed until more than half of the hydroxyl groups had been extended.

Mechanism

The mechanism of the reaction of anhydrides (7) with alcohols has been studied by analysts and others to further investigations of hydroxyl determination and acylations.[37,38,39,40] Nucleophilic catalysis by amines has generally been presumed:

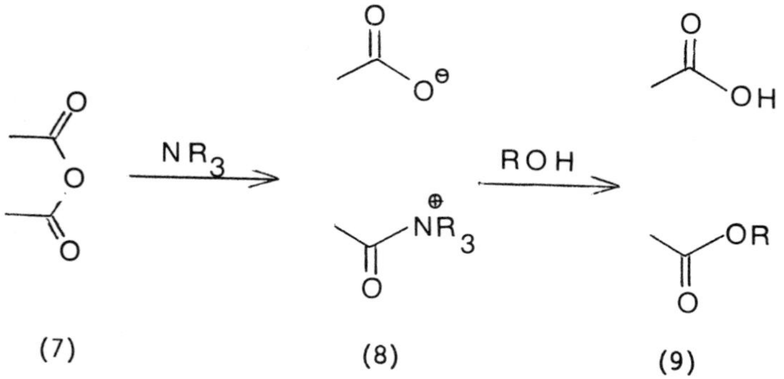

$$
\text{(7)} \qquad\xrightarrow{\ NR_3\ }\qquad \text{(8)} \qquad\xrightarrow{\ ROH\ }\qquad \text{(9)}
$$

Scheme 17

The effectiveness of a particular catalyst has been discussed in terms of ease of formation and stability of the acylammonium intermediate (8). For example, the unsubstituted pyridinium intermediate is highly reactive, but its formation is rate limiting. By contrast, the 4-(dimethylamino)pyridinium intermediate is stabilised by charge delocalisation and DMAP is said to be 10^4 times more catalytically effective than pyridine. Again, a non-N-substituted imidazolium intermediate can be deprotonated to so stable a form that imidazoline actually retards the alcoholysis of linear anhydrides, whereas N-methylimidazole is an effective catalyst.[40] Cyclic anhydrides behave differently, apparently because of the anchimeric effect of the vicinal carboxylate groups, and imidazoles and N-substituted imidazoles show very similar catalytic behaviour. Spectroscopic and kinetic evidence has been adduced to support a nucleophilic mechanism at least for imidzole catalysis in DMF solution.[37]

Alternatively, catalytic amine can be envisaged as a general base, converting the hydroxyl group to an alkoxide ion of enhanced nucleophilicity. Though it would apply equally to randomly positioned catalyst, a general base mechanism has been proposed for β-hydroxy-amine systems[28] (Scheme 18) and bears a striking resemblance to the first stage in well supported schemes for the reaction of epoxies with cyclic anhydrides (scheme 19).[41]

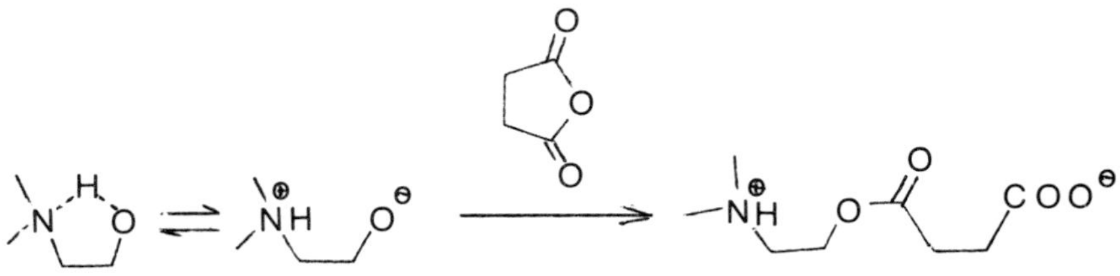

Scheme 18

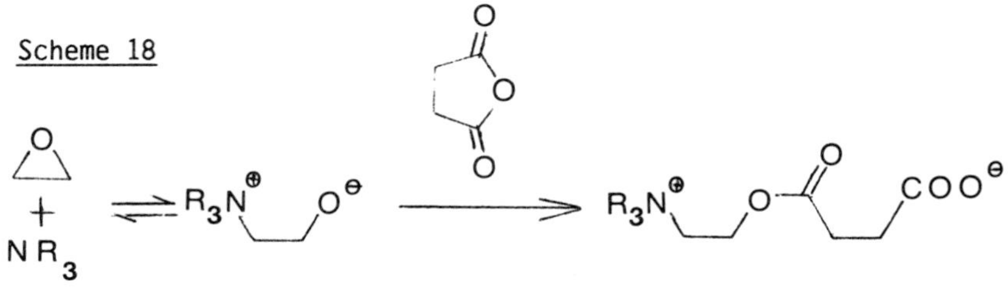

Scheme 19

In an attempt to discriminate between the two mechanisms, the influence of base strength on catalytic effectiveness was explored. The results shown in Table 2 are somewhat inconclusive except inasmuch as they show a certain level of basicity to be necessary for any useful reaction to occur. The efficient catalytic behaviour of DMAP and imidazole were predictable and favour the nucleophilic catalyst mechanism. However, it should be noted that under somewhat different conditions, strong bases like sodium methoxide and tetrabutyl-ammonium hydroxide have provided well cured films. The speed of evaporation of the methanol solvent in which they were used may be an important factor. In fact, the volatility of the amines themselves is a significant variable when considering thin coating films.

Table 2 Influence of catalyst on degree of cure of a
proprietary maleic copolymer/HEA copolymer system after 8 days
at ambient temperature.

CATALYST (14.2 Mole%)	pKa	Gel Time (h)	Solvent* Rubs (MEK)
NaOMe	14	>4	25
Bu$_4$NOH	14	-	15
Dimethyl-Alkyl amine	(9.8)**	2	85
4-(Dimethyl-amino)pyridine	9.70[39]	0.5	>200
Imidazole	6.95	2	100
Pyridine	5.25	-	25
N,N-dimethyl-aniline	4.23	-	20

* Number of to and fro rubs with a butanone soaked cloth
required to penetrate film.
** pK$_a$ value of triethylamine.

Whichever mechanism is operating, the greatly enhanced
reactivity of β-hydroxyamines can be rationalised in terms of
5-membered cyclic transition states or intermediates - (10) for
nucleophilic catalysts, (11) for general base catalysis.

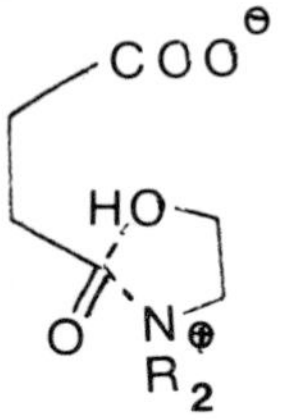

     (10) 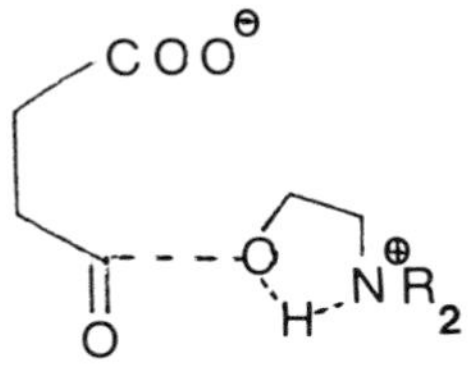   (11)

It also follows that similar rate enhancement can apply to
hydrolysis of the linkage once formed and may account for the
disappointing weathering properties observed with some systems.

Toxicology

The commercial all-acrylic system has been submitted to a
battery of toxicological tests and showed effects little more
severe than would be produced by its solvents (xylole, mixed
hydrocarbons, and butyl acetate) cf. Table 3.

Table 3   Toxicological tests carried out on a commercial
            all-acrylic system based on itaconic anhydride and
            intramolecular amine, ex. ref. 33, example 2,
            (Inveresk Research International, 1984).

TEST	OECD NO.	COMPONENT	RESULT
Acute Oral Tox. (Rat)	401	Polyhydroxyamine	$>5000 mgkg^{-1}$
		Polyanhydride	$>5000 mgkg^{-1}$
Acute Dermal Tox (Rat)	402	Mixed	$>2000 mgkg^{-1}$
Acute Dermal Irritation (Rabbit)	404	Polyhydroxyamine	Slight irrit[n.]
		Polyanhydride	"
		Mixed	"
Acute Eye Irritation (Rabbit)	405	Mixed	Severe irrit[n].
Buehler Sensitisation (Guinea Pig)	406	Mixed	No evidence of sensitisation
Ames Test	471	Polyhydroxyamine	No evidence of mutagenicity

Market Acceptance

Systems based essentially on acrylic poly(cyclic anhydrides) and
acrylic polyols, with catalytic amine as a separate component or
randomly incorporated in the polyol molecule, have been quite
widely commercialised. They have been shown to be of low toxicity, to have excellent cosmetic and weathering properties and
easy application. They are satisfactory replacements for isocyanate cured systems in markets like car refinish, commercial
vehicle finishing, and painting of land and sea based metal
structures. They do however, fall short of the performance of
polyurethanes in certain demanding applications.

6.   HIGH PERFORMANCE THROUGH MOLECULAR ARCHITECTURE

Mechanical Properties

In some applications, such as anti-abrasion or aircraft
finishes requiring tough, hard but flexible coatings, and others
requiring film formation at low ambient temperatures the all-
acrylic systems above compare unfavourably with polyurethanes
and a different polymer architecture has been investigated.

A high glass transition temperature (T_g) anhydride functional acrylic (reference 36, example 1 (itaconic), example 5, (maleic)) was combined with more flexible, lower T_g linear polyesters. This architecture is comparable with the hard urethane and soft polyether or polyester blocks in some polyurethanes. The terminal functionality in the polyesters also tends to improve mechanical properties as compared with the randomly distributed functionality in the acrylic polyols. Table 4 shows that the mechanical properties of these novel polyester-acrylic systems and of a related telechelic hydrocarbon acrylic approach those of some typical commercial polyurethane coatings. By contrast, the all-acrylic films could not be tested for tensile properties because of their low strains to failure.

Table 4   Physical and Mechanical Properties of Selected Systems

Polymer Sample	Tg (days) 7	30	Youngs Modulus (MPa)	Elongation (%)	UTS (MPa)	Time(Days) to Sol/Gel @ 35° 100%RH 0.5:1	t1:1
A	40	70	900	2.5	16	29	37
B	30	55	220	20	12	27	32
C	28	47	140	10	10	35	48
D	–	67	600	2.5	12	150	200
E	90	100	500	10	15	250	–
F	35	47	141	110	16	>>250	–
G	67	78	550	60	13	>>250	–
H	55	65	–	–	–	>>250	–

A = Adipic polyester hydroxyamine
B = Adipic polyester hydroxyamine, 50% extended with caprolactone
C = Azelaic polyester hydroxyamine
D = Cycloaliphatic polyester hydroxyamine
E = Hydrocarbon hydroxyamine
    All used to crosslink a high T_g itaconic copolymer.
F,G = Proprietary polyesters cured with aliphatic tri-isocyanates
H = Proprietary system based on maleic copolymer and acrylic polyol

Reactivity

The greater reactivity of the β-hydroxyamine polyester/ anhydride functional acrylic combined with the greater flexibility of the polyester polyols resulted in both a faster rate and a higher extent of cure than for the all-acrylic polymers. For some applications, higher reactivity resulted in an

unacceptably short pot life and was moderated by partial
extension of β-hydroxy groups with caprolactone as described
above. Figure 1 shows the change in rate of cure with caprolactone capping from Tg measurements determined by DMTA. The
pot life of the mixed anhydride-polyol polymers was
extended from approximately thirty minutes to 3-4 hours by
extending up to 60% of the β-hydroxyamine groups with
caprolactone.

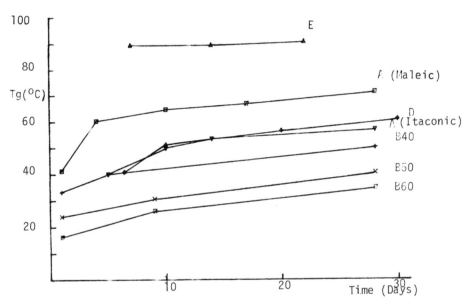

FIGURE 1:Curing of acrylics with telechelic hydroxyamines:
         Tg from Dynamic Mechanical Analysis vs. time @ 25°C.
         (A,B,D,E refer to Table 4, Maleic and Itaconic versions
         of A; 40,50,60 refer to % caprolactone extension).

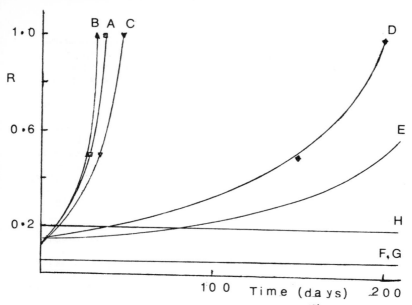

FIGURE 2: Hydrolytic stability. Sol:Gel ratio, R vs Time @ 35°C, 100%RH (Idealised Curves) (A-H refer to Table 4).

Polyesters, both caprolactone modified and otherwise, have shown satisfactory cure with acrylic anhydrides at 0°C as determined by solvent rubbing. Similarly, the polyester-acrylics cure to higher extents than the all acrylic systems as demonstrated by their lower sol fraction in solvent swelling measurements (figure 2) and anhydride concentration measurements by IR spectroscopy.

Comparison of an hydrocarbon crosslinker with a relatively high Tg cycloaliphatic polyester with similar end group chemistry strongly suggests that molecular mobility is a much more significant factor than reactivity in completion of the network (figure 1).

Weathering Properties

        Coatings intended for external use require an early indic-
ation of durability. Durability here means the retention of
those properties important to the function of the coating and
includes changes in appearance (colour and gloss differences),
film integrity (cracking and peeling), and degradation of physical and mechanical properties (embrittlement, loss of adhesion,
reduced barrier properties). The problems of predicting
durability from accelerated tests are widely recognised[42] and
such tests can only be used as guides for screening potential
products before long term natural exposure results are obtained.
For anhydride cured polyols, QUV testing was used to indicate
resistance to the combined effects of sunlight and high humidity.
Simultaneously, painted panels were exposed at a number
of sites worldwide including Florida, Saudi Arabia, Singapore,
Houston (Texas) and Tyneside in the UK.

        Exposure results for the all-acrylic systems showed excel-
lent resistance to sunlight and high humidities whereas early
laboratory results for the anhydride acrylic hydroxyalkyl amine
polyesters indicated poorer durability (later confirmed by
natural exposure data). This poorer performance was manifested
by loss of gloss, cracking, embrittlement and water spotting
(ie. small matt areas on the paint surface occurring in condensing humidity conditions).

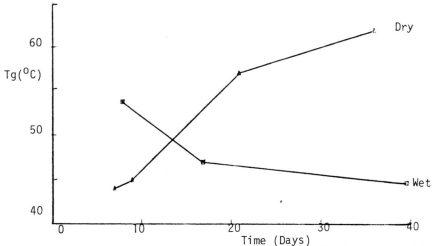

FIGURE 3:   Effect of humid ageing on dynamic mechanical properties.
            Tg (from peak in Tan) vs. Time (35°C, 100%RH)
            System A ex. Table 4).

A series of ageing experiments was carried out to identify the principle causes of degradation and it quickly became clear that high humidity was the most aggressive environment. Polyester-acrylic samples aged at 35°/100% RH showed evidence of cumulative, irreversibe degradation. Clear (unpigmented) films became white and opaque and DMTA showed a progressive lowering of Tg and reduced rubbery modulus compared to similar ageing in dry conditions (figure 3). Measurement of the increasing sol fraction clearly demonstrated that reversion of the networks was taking place and proved to be a powerful tool in assessing hydrolytic instability (figure 2). Table 4 and figure 2, show[43] that for the more hydrolysis resistant polyester based films, the onset of reversion was delayed, ie. poly(ethylene-co-butylene adipate) < poly (neopentyl azeleate) < cycloaliphatic polyester. The acrylic-hydrocarbon system was most resistant of all although reversion occurred eventually. For comparison, evidence of reversion was neither seen for any samples maintained dry at 35° nor for the commercial all-acrylic/anhydride or polyester/urethane samples kept at 35°/100%RH.

Since the data on hydrolytic ageing could not identify the site of degradation, spectroscopic and chromatographic analysis of the sol fractions was undertaken. The sol fraction proved to be a complex mixture of products and analysis was difficult. A complete description of the analytical investigation is beyond the scope of this article and the results are not unequivocal. It seems likely however that hydrolyis in the polyester-acrylic systems occurs at the polyester-triethanolamine link or at the triethanolamine half ester bond (I and II below):

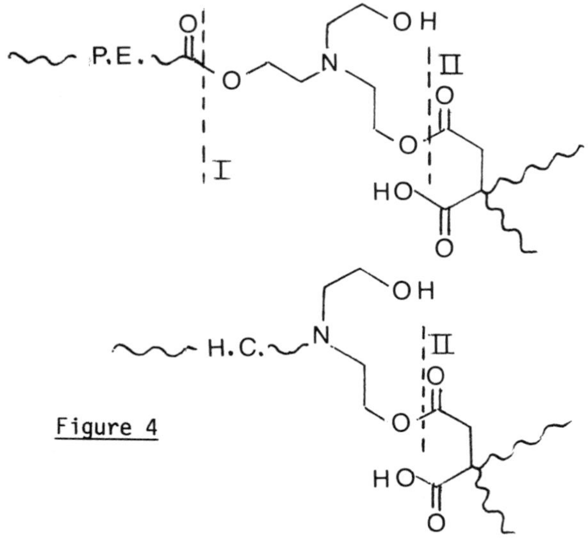

Figure 4

Hydrolysis of the polyester backbone does not appear to take place and linkage I is more susceptible than II as shown by the sol-gel data for the series of polyesters studied. The fact that the hydrocarbon-acrylic system eventually reverts indicates that linkage II will slowly hydrolyse.

7.   CONCLUSIONS

Useful commercial systems with interesting technical properties can be based on anhydride chemistry. Within certain limits, their properties and cost can be tailored to suit a particular market.

An ideal hydroxyl/anhydride system showing both hard-urethane type mechanical properties and acceptable stability to hydrolysis as well as rapid but controlled cure under the chosen conditions, should be accessible by changing the nature, position or environment of the catalyst.

8.   EXPERIMENTAL

Preparation of acid (methyl ester) tipped polyesters

The required quantities of diacid (or their diesters) and diols to give a molecular weight of 1000, were heated with stirring under an N_2 blanket until the theoretical amount of water was collected in a Dean and Stark trap (ca. 5 hours). Molecular weight was confirmed by acid content analysis (where possible) and GPC relative to styrene.

Table 5

	Charge (moles)		
Ethylene glycol	2.21		
Butane-1,4-diol	2.21		
Neopentyl glycol		3.17	
Cyclohexanedimethanol			2.86
Adipic acid	5.42		
Azelaic acid		4.17	
Dimethylcyclohexanedioate			3.86
(Condensate, water or methanol)	(8.84)	(6.34)	(5.72)

Hydroxyamine tipping of Polyesters

The acid or ester tipped polyester (1 mole, 1000g), xylole (1000g) and triethanolamine (2 moles, 298g), were heated under reflux (145°C) with stirring until theoretical water (36g) had been removed (ca. 4 hours). Xylole was then removed under vacuum and heating continued @ 180°C to obtain an acid value below 5 (where appropriate). The product was characterised by its amine equivalent weight, assumed to be 2x the effective hydroxyl equivalent weight.

Caprolactone Extension of Hydroxyamine Tipped Polymers

The hydroxyamine tipped polymer (1 mole, 1262g), caprolactone (2 moles, 228g) and "Fascat 4101" (a proprietary tin catalyst) (1.5g) were held @ 150°C for 4 hours with stirring. The absence of significant amounts of free caprolactone could be confirmed by gas chromatography. The product was the so-called 50% extended version. Other levels of caprolactone could be introduced by the same procedure.

Mesyl Ester Tipped Hydrocarbons

Nisso GI1000 (a hydroxyl tipped hydrogenated 1,2-polybuta-
diene mw ca. 1000, equivalent weight 897, ex. Nippon Soda),
(1244.7g), xylene (1244.7g), triethylamine (140.2g) were stirred
at ambient temperature. Methanesulphonyl chloride (158.9g) was
added over 1 hour whilst maintaining the product at room temperature. The mesyl ester was recovered as a 51% solution in
xylole by filtration. It was used without further purification.

Hydroxyamine Tipped Hydrocarbon

The mesyl ester solution (2201.9g), diethylamine (121.2g)
and potassium carbonate (500g) were refluxed with stirring until
the evolution of water ceased (6 hours). The product was recovered as a pale yellow oil by filtration, and removal of
solvents under vacuum. It was characterised by its amine equivalent weight (1258.9) which was taken as 2x its hydroxyl
equivalent.

Film Preparation

Stoichiometric amounts of the component polymer solutions
were weighed, mixed and degassed under vacuum. Films were
produced by drawing down onto p.t.f.e. covered glass such that
dry films of 150-200μm thickness were produced. Samples were
allowed to cure for 24 hours in a dust free enclosure at ambient
temperature and humidity before removing them from the panels.
The films were then stored at 25°, 30% RH for 6 days prior to
environmental exposure or mechanical testing.

Mechanical Properties

Dynamic mechanical properties were measured using a Polymer
Laboratories DMTA operating at 1Hz, heating rate 4°min^{-1} and
double cantilever test geometry. Complex modulus, storage
modulus and damping (tan δ) were recorded and the temperatures
of maxima in tan δ were used as relative measures of Tg in
figures 1 and 3.

Tensile mechanical properties were measured according to
ASTM D2370-82 using a JJ Lloyd M5K tensile tester at 25°, strain
rate 12.5% per minute.

Solvent Swelling

A more rapid, simplified procedure for classical solvent swelling measurements was developed to give a relative measure of changes in crosslink density on ageing. Specimens of polymers approximately 25 x 10 x 0.02mm were cut from films and weighed. Immersion in a large excess of butan-2-one solvent for 5-6 hours was followed by a change of solvent and the solvent then discarded. The specimens were dried overnight at 35-50° (dependent on the sample) and reweighed. Results expressed as sol:gel ratio, ie. weight of extracted material to weight of insoluble network polymer (pigment excluded).

Environmental Exposure

Free films after 6 days at 25°/30%RH were placed in an environmental cabinet at 35°/100%RH and removed periodically for assessment. Mixed polymer solutions were sprayed onto aluminium panels and mounted in a QUV cabinet operating with QUV-B fluorescent tubes. The cabinet cycled at 4 hours dry (temperature ca. 55°) and 4 hours dark, 100% RH (temperatures ca. 44°).

9.   REFERENCES

1.   W.M.Kraft, E.G.Janusz, D.J.Sughrue in "Treatise on Coatings", Ed. Myers, Vol.1, 71.
2.   R.Oshima, Y.Yamauchi, C.Wanatabe, J.Kumanotani, J.Org. Chem 1985, 50 2613.
3.   "Isocyanates in the Paint Industry", Paintmakers Association of Great Britain Limited.
4.   "Mutagenic Effect of Aromatic Epoxy Resins", A compilation and discussion of toxicological data, by K.I.Darmer, of Shell, 9.11.79.
5.   US 4525521, US 4785054 Dupont; EP 240460 Ciba Geigy.
6.   US 4472569 Philips Petroleum.
7.   EP 179399, BP 2161164 Coates Bros; DE 1494405 Ashland Oil.
8.   JP 1153714, EP 123793 Dainippon Ink; EP 247402 BASF.
9.   EP 134691 ICI; US 4816500 Dupont; EP 316873, EP 316874 Sherwin Williams.
10. EP 302373, EP 284953, EP 307701, EP 319864, EP 346669 Bayer.
11. US 460261 Akzo; WO 88/07555, WO 88/07556 BASF; EP 227454 Cook Paint & Varnish; EP 310011 Hoechst.
12. EP 326723 Rohm & Haas.
13. BP 2166749 ICI; US 4529765 Dupont; EP 267004, US 4698406 Dow Corning; EP 262720 Akzo.
14. EP 203296 Akzo, Nooman, Prog.Org.Coatings 1989 17 27
15. EP 121083, EP 274402, EP 259511 Union Carbide.
16. JP 1146968, JP 1146966 Dainippon Ink.
17. US 4387271 Hoeschst.
18. R.A.Davis, M.A.Friedman, The Pharmacologist, 1986, 28 184
19. BP 2197325 ICI; US 4499150, US 4499151 PPG.
20. BP 2202538, BP 2212164 Kansai Paint;
     O. Isozaki, Proc 9th Intl Conf on High Coatings 23/25 Oct 1989, 123.
21. BP 2097409 ICI.
22. US 4837278 BASF.
23. US 4530960 American Cyanamid
     Lucas, J.Coatings Technol, 1985, 57 49
24. EP 199087, EP 264983, US 4772680 Akzo.
25. US 3661774, US 4008341 W R Grace; US 4119617 Showa High Polymers; US 4308367 Ciba Geigy.
26. EP 262464 De Soto
     G.K.Noren, E.J.Murphy, Proc ACS Divn. of PMSE 1989, 60 228
27. FR 2392092, BP 1561828 Hoescht.
28. J.C.Decroix, J.M.Bouvier, R.Roussel, A.Nicco, C.M.Bruneau J.Polymer Sci. Symposium, 1975, 52 299.

29. H.C.Brown, J.H.Brewster, H.Shechter, J.Amer. Chem. Soc., 1954, 76 467.
30. I.Puskas, E.K.Fields, Ind. Eng. Chem. Prod. Res. Develop. 1970, 9 403.
31. EP 209377 A2 International Paint.
32. B.C.Trivedi, B.M.Calbertson, "Maleic Anhydride", Plenum Press, NY, 1982 Chapter 11, p.459.
33. EP 48128 International Paint.
34. US 3245933 Sinclair Research.
35. M.Ratzsch, S.Zschoche, V.Steinert, J.Macromol Sci-Chem, 1987, A24, 949
36. EP 259172 International Paint.
37. B.H.M.Kingston, J.J.Garey, W.B.Hellwig, Anal.Chem, 1969, 41 86.
38. G.Hoffe, W.Steglich, Synthesis, 1972, 619.
39. G.Hoffe, W.Steglich, H.Vorbruggen, Angew. Chem, Int. Ed. Engl. 1978, 17, 569.
40. K.A.Connors, N.K.Pandit, Anal.Chem, 1978, 50, 1542.
41. L.Matejka, J.Lovy, S.Pokorny, K.Bouchal, K.Dusek J.Polym. Sci-Polym. Chem., 1983, 21 2873.
    M.Fedtke, F.Domaratius, Polymer Bull. 1986, 15 13.
42. G.P.Bierwagen, Prog. Org. Coatings, 1987, 15 179-95.
43. E.T.Turpin, J.Paint Technol. 1975, 47 40.

Polymer Microgels in Paints

S. B. Downing

ICI PAINTS, SLOUGH, BERKSHIRE, SL2 5DS

1 INTRODUCTION

The world market for paints and related coatings was estimated at a staggering 12 billion litres in 1989, worth over £20Bn. This was split between major trading areas, with North America and CWE being responsible for approximately a third each and Japan heading up the 'rest of the world' which accounted for the remainder. Decorative or Architectural paints comprise just over half the total world sales by volume. The other half of the market, termed Industrial Paints, can be usefully sub-divided into a number of key markets, particularly Motors, Auto Refinish, Can, Coil and Powder.

Solvent-based paints comprising solutions of resins or polymers in moderate to strong solvents have dominated the paint industry historically. Increasingly, these are being replaced by polymer dispersions carried in mainly aqueous diluents, or powder coatings.

The Decorative market in particular has moved strongly towards waterborne paints, mainly emulsion polymers, over several years, with the USA leading the way. However, there are still some areas of this market where solvent-based paints are still required, provided certain deficiencies can be overcome. In answer to this need, ICI Paints has developed and commercialised Non-Aqueous Microgel technology offering a range of high performance durable products for decorating and protecting timber and masonry.

The Automotive industry has seen many changes in paint technology, driven particularly by the need to reduce pollution and achieve high performance coatings. The major changes have been Electrocoat, metallic finishes, Non-Aqueous Dispersion polymers, clear-over-basecoat and, most recently, waterborne basecoat. ICI Paints has been at the forefront of all these technologies and world leader in three of them. The most recent innovation for the Automotive market is Aquabase, a waterborne basecoat utilising aqueous microgel technology. Non-Aqueous Dispersion (NAD) and Aquabase have both won Queen's Awards for Technology Innovation testifying to the strength of ICI Paints in polymer dispersion technology. This success in innovating high value polymers for the paint industry derives from the ability to formulate unique concepts, based on a deep understanding of the market and customer needs and underpinning good technology with scientific understanding to produce unique products with highly attractive cost-performance properties.

2 NAD MICROGELS FOR HIGH PERFORMANCE EXTERIOR DECORATIVE COATINGS AS FILM FORMERS.

In the 1960's, ICI Paints pioneered the development of Non-Aqueous Dispersion (NAD) polymers. The work was initially stimulated by Automotive market requirements for pigmented topcoats containing high molecular weight durable polymers which could be applied at higher solids than lacquers and use less strong solvents, such as petrol, to comply with legislation on reducing pollution (USA Rule '66).

The result of this work was a high level of understanding of the principles of stabilisation and the kinetics of polymerisation of polymer dispersions in organic media [1]. Development of the technology led to a series of patents and licences and a range of solventborne thermosetting finishes including basecoats and clearcoats for the Automotive market worldwide.

In recent years, interest in NAD as film formers, as a way of improving the performance of a variety of current coating systems, has stimulated further exploitation of this versatile technology [2].

<u>Architectural Air-Drying Systems</u>. One of the key benefits of solventborne systems for the Decorative market is their robustness to application conditions, particularly damp, cold climates.

Most solventborne systems rely on the autoxidation
of alkyds which form initially tough, protective coatings.
Unfortunately, the catalytic metal driers remain active
in the film, which still retains some unsaturation, so
that on ageing the film becomes further crosslinked. As a
result, alkyd paints, typically, lose toughness and flexibility after a relatively short period of time, eventually
cracking and flaking off the substrate after only a few
years. Many conventional alkyd gloss finishes have only
a few percent extensibility to break after one or two
years exposure to the weather. This is more marked in
undercoats and primers, where the level of pigmentation
is much higher compared to gloss paints, and they are
therefore inherently less flexible. Other factors, such
as adhesion, elasticity, modulus, plasticisation by
moisture and movement of the substrate, which in the case
of timber can be as high as 10%, also have an important
influence on overall performance.

Aqueous latex-based paints are capable of rapid film
formation under ideal conditions and do not embrittle on
ageing. These systems rely on physical coalescence of
high molecular weight polymer dispersions and, in the
case of acrylic polymers, can be particularly good for
durability. However waterborne systems have considerable
application and early performance limitations in cold and
wet climatic conditions, which exist in many parts of the
world, sometimes throughout much of the year. A potential
solution to the market requirements for a durable coating
for timber or masonry which could be applied under a wider
range of climatic conditions was to produce an acrylic
dispersion polymer in white spirit using NAD technology.

NAD Polymer Dispersions. The basis of NAD technology
is the use of steric stabilisation to prevent the flocculation of fine particle polymer dispersions in non-polar
diluents, where ionic stabilisation is inadequate because
of the low dielectric constant of the medium.

A variety of polymers can be used as steric stabilisers with either chemical or physical bonding to the
the particle surface. Effective stabilisers are the
so-called amphipathic graft copolymers. These consist
of a relatively polar anchor component and an oleophilic
component which is soluble in, or solvated by, the
diluent. The most efficient type of graft copolymers are
known as "comb stabilisers". These are prepared by polymerising several vinyl-terminated poly(12-hydroxystearic
acid) (PHS) chains with a molecular weight of about 1700
into a backbone consisting mainly of poly(methyl meth-

acrylate) [1]. The insoluble polymethyl methacrylate backbone is physically adsorbed onto the surface of the polymer particle leaving the PHS chains ("teeth") in solution
to act as a steric barrier to the close approach of
similarly stabilised particles (Figure 1). Levels of 2-4%
comb stabiliser are adequate for stabilising high T_g, submicron sized polymer dispersions.

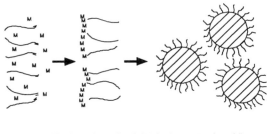

⌒ Vinyl-terminated poly(12-hydroxy stearic acid)

M Monomer Units

Figure 1. Steric Stabilisation of NAD Acrylic Microgels.

Although the soluble chains used in these systems are
usually PHS, it is possible to vary the nature of the
anchor group, the ratio of the anchor/soluble groups and
molecular weights of the overall comb, anchor or soluble
chain components.

Initiation of polymerisation for acrylic systems is
similar to solution polymerisation. The growing polymer
chain becomes insoluble in the diluent which is a good
solvent for the monomers and only a moderate solvent for
the stabiliser. It is usually beneficial to use azo
initiators, such as azo-diisobutyronitrile rather than
peroxides to avoid potential problems due to excessive
grafting of the comb stabiliser.

A seed and feed process is normally employed to
produce a high solids, fine particle, polymer dispersion.
The seed stage is typically produced using 10% of the
monomers with about 25% of the total stabiliser and
initiator to be used. This gives a very fine polymer
dispersion (c. 80nm) which can then be grown to the final
particle size (c. 300nm) by feeding in the rest of the
monomers, stabiliser and initiator over a few hours,
whilst maintaining a temperature of about 80°C under
inert gas or reflux conditions.

NAD for Ambient Application Conditions. The adaptation of NAD technology from automotive uses, involving spraying high T_g polymer compositions and then stoving at moderately high temperatures, to use in architectural paints where application and film formation at ambient temperatures on substrates demanding flexible coatings, required considerable reformulation and experimentation. Attempts to follow the conventional practice of using polyester plasticisers with high T_g acrylic compositions failed to give good durability on exterior exposure due to leaching out of the plasticiser.

The use of acrylic monomers to achieve low T_g polymer compositions initially presented problems in the quality and stability of the polymer dispersions. The range of monomers commercially available at an acceptable cost for producing these softer polymers are by nature oleophilic, e.g. butyl acrylate (T_g -56°C), 2-ethyl hexyl acrylate (T_g -70°C), lauryl methacrylate (T_g -65°C). The use of these oleophilic monomers resulted in poor quality coarse latexes or opaque semi-solutions when high boiling aliphatic hydrocarbons, e.g. white spirit, were used as the diluent. The problem can be alleviated to some extent by choosing diluents which are weaker solvents. Normal white spirit contains 15-20% of aromatic hydrocarbons and is best replaced with a lower aromatic containing variant. A further improvement is to choose a monomer composition which combines low T_g with high polarity. A mixture of methyl methacrylate (T_g 100°C) and ethyl acrylate (T_g -22°C) was found to be particularly effective. Small amounts of polar hydroxyl or carboxyl-functional acrylic monomers are also helpful in achieving the right balance.

NAD Microgels. Improvements in key properties, such as wet adhesion and mechanical performance of films derived from NAD latexes were sometimes observed when the polymer particles were internally crosslinked to form microgels.

Wet Adhesion. An important characteristic of coatings on wooden substrates is whether they exhibit good wet adhesion. This can be conveniently assessed using a standard Blister Box Test [3]. Uncrosslinked NAD latexes and derived paints gave unimpressive results on this test. However, when some crosslinking was introduced into the particles by various means, improvements in wet adhesion could be observed in some formulations. The most direct route of introducing a gel fraction into the particles was to use small amounts of multifunctional monomers during preparation of the NAD latex, resulting in ready-made microgel. Several such monomers were evaluated, e.g. allyl methacrylate, trimethylolpropane triacrylate and triallyl isocyanurate (Table 1). Other commonly used difunctional monomers e.g. ethylene glycol dimethacrylate, fail to achieve an adequate level of crosslinking. The use of larger amounts of these crosslinkers merely results in formation of poor quality, flocculated products.

Table 1. Effect of Crosslinkers on Gel Level and Wet Adhesion.

Crosslinking Monomer	% Level	% Gel	Wet Adhesion
None	—	0	Poor
Allyl Methacrylate	0.5	57	Good
Ethylene Glycol Dimethacrylate	3.0	13	Poor
Divinyl Benzene	0.5	23	Poor
Hexane Diol Dimethacrylate	5.0	23	Poor
Triallyl Isocyanurate	0.2	10	Poor
Trimethylol Propane Triacrylate	0.5	64	Good

The gel fraction (% of polymer insoluble in tetrahydrofuran) achieved in the NAD latex particle, which is related to the proportion of crosslinker employed, correlates well with the performance of the film on the Blister Box Test (Figure 2). Diallyl phthalate (DAP) was particularly effective in improving the wet adhesion and even small amounts gave significant gel contents. Practical experience has led us to associate good wet adhesion, as determined by tests such as the Blister Box, with good long term exterior durability on wood.

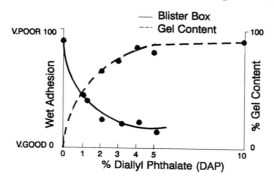

Figure 2. Gel Level and Wet Adhesion with DAP.

Mechanical Performance. The long term mechanical
performance of coatings, such as flexibility, elasticity
and tensile strength, is particularly important for
durability over dimensionally unstable substrates, such
as timber. Stability of the polymer binder to UV, water
and oxygen is also essential, hence the choice of acrylics
for the NAD microgel. Polymer characteristics, such as
T_g and molecular weight, are important for mechanical
performance.

One of the advantages of NAD technology is that
the mechanism and kinetics of polymerisation inherently
favour the formation of much higher molecular weight
polymer than is normally achieved in solution. The
principal reason for this is that the locus of polymerisation is essentially in the polymer particle. The
growing polymer chain is easily fed by mobile monomer
entering the swollen particles. Termination reactions
are significantly reduced by the restriction of the
polymer chain mobility within the high viscosity medium
of the particle.

High molecular weight ensures that the acrylic polymer
films prepared from NAD are durable and that they exhibit
adequate tensile strength and toughness to be good
protective coatings.

A polymer coating must have adequate extensibility
if it is to accommodate movement of a timber substrate.
Based on experience, the polymer film should be capable of
at least 300% extensibility at low strain rates to allow
normal levels of pigmentation for primers and undercoats.
The T_g of the polymer is an important factor and must be

Table 2.   Film Extensibility at 0 deg C.

Tg of NAD	NAD/ALKYD Ratio	
	100/0	70/30
+2°C	44%	16%
-2 °C	58%	18%
-6 °C	176%	38%
-21 °C	217%	107%

low enough to allow for this level of extension even in crosslinked films. If the T_g is too low, the film will be weakened and tend to pick up dirt too easily. Table 2 shows how the extensibility of DAP crosslinked microgels increases as the polymer T_g is reduced. The effect of crosslinker (DAP) level on the extension to break for films of the same main monomer composition or T_g is shown in Figure 3. A further benefit of having some crosslinking present in the particles is to provide a sufficient degree of resistance to the diluent such that short recoat times are possible.

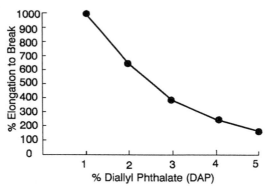

Figure 3.   Crosslinking and Extensibility.

The properties of a typical acrylic NAD microgel are shown in Table 3. Particle size distribution as measured by the Brookhaven Disc Centrifuge initially developed by Glidden[4] appears to be somewhat bimodal.

Table 3.   Typical NAD Microgel Properties.

N/V Content	59-60%
Particle Size	0.2-0.3um
Solution Polymer	<3%
Gel Content	73-74%
Film Extensibility	400-500%
Tg(Theoretical)	2°C
Tg(DSC)	3-4 °C
MFT	-4 °C
Residual Free Monomer	<10 p.p.m.

Application of NAD to Paint Systems

As described earlier, NAD technology is not new to
the Coatings Industry. However, the application of low
T_g Microgels to the architectural coatings market is
novel [2]. In recent years, ICI Paints has introduced a
number of new products based on NAD Microgel technology
into the UK market.

NAD Exterior Undercoat. A few years ago ICI Paints
launched a new improved exterior gloss paint system (Dulux
Weathershield Exterior Gloss [5] following an extensive
re-investigation into durability requirements of coatings
over timber substrates [6]. Although this was initially
based on a reformulated alkyd resin technology particular
attention was given to addressing flexibility of the overall system, fungal protection and permeability to moisture
to and from the substrate. The role of the undercoat,
which requires high levels of pigmentation to provide
opacity, in determining overall flexibility and therefore
durability clearly indicated that further improvements
could be made if the NAD acrylic microgel system were to
be used.

For practical application of the paint, it was found
desirable to introduce minor amounts of long oil alkyd
solution polymer into the undercoat to improve application
rheology, without significantly affecting the flexibility
of the coating (Figure 4).

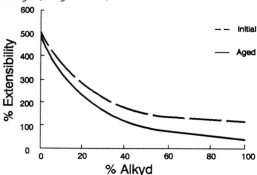

Figure 4. Extensibility of NAD/Alkyd Blends.

Extension to break of coatings based on a 75/25 NAD/
Alkyd by weight composition at 20°C with various levels of
pigmentation show (Figure 5) how rapidly extensibility
falls with increasing pigment volume concentration (PVC).

Coatings at 35 and 45% PVC were compared to a standard alkyd undercoat (45% PVC) using accelerated weathering techniques and ultimately 'natural weathering' tests.

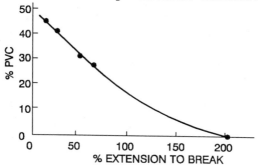

Figure 5.   Effect of PVC (TiO$_2$ + Chalk) on Extensibility of NAD/Alkyd (3/1 by Wt) Paint Film.

Early indications of the significant improvement in performance of the NAD system, particularly at 35% PVC, were obtained using an Acoustic Emission technique. Acoustic Emission is a simple and non-destructive technique traditionally used in the Aerospace Industry for detecting early signs of failure of materials under strain.   The technique depends on the use of sensitive transducers directly attached to the coating. These detect stress waves generated by micro cracking, produced when the applied stress is relieved by film failure.   The technique has been usefully applied to studying the performance of paints and other coatings. Its main advantage is the ability to detect the onset of failure early on in the life of a coating and, in combination with accelerated weathering, speeds up the evaluation of experimental systems in broad terms. It can also be helpful in designing balanced combinations of coatings in multicoat systems.

The technique was applied to panels of complete gloss paint systems after exposure for various times in an Atlas XWR Weatherometer. The results [Table 4] show that the flexibility of the NAD variants were retained over an extended period of exposure.

Results from exterior weathering tests over the past five years have confirmed the excellent durability of paint systems based on NAD microgel technology.

A further advantage of NAD primer undercoats is a significant reduction in drying times with touch-dry times typically reducing from 2 hours to 1 hour and recoat times from 5 hours to 2½ hours, over conventional alkyd systems.

Table 4. Strain Level for Film Breakdown.
(Accelerated Weathering)

Undercoat Used	Initial	300hrs	600hrs
Conventional Alkyd 45% PVC	6-7%	4-5%	—
NAD Variant 45%PVC	>20%	16%	7%
NAD Variant 35% PVC	>20%	24%	22%

NAD Exterior Woodstains. Woodstains are being used increasingly as an alternative to conventionally pigmented gloss paint systems on timber substrates. Being semi-transparent, they enhance the natural beauty of the wood used in the construction of windows and doors. However, the low level of pigmentation used means they are more vulnerable to weathering, due to the lack of protection offered by opacifying titanium dioxide. Normally alkyd binders are used for woodstains but these have limited durability due to degradation and embrittlement. When the alkyd is blended with a major proportion of NAD microgel, a much superior system is obtained (Figure 6).

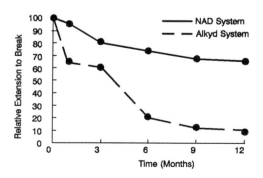

Figure 6. Natural Weathering-Extensibility of Woodstain.

The acrylic NAD provides a durable polymer system which retains flexibility. The alkyd component penetrates the timber substrate helping to stabilise it and improving adhesion. Further evidence for the improved performance of NAD microgel based woodstains has been obtained from a number of natural exposures on timber substrates on several UK test sites[8].

NAD Exterior Masonry Paint. Waterborne paints are increasingly being used for decorating masonry. These systems offer good durability, particularly if formulated to be flexible, smooth coatings and preferably acrylic-based. However, damp cold conditions often prevail in Northern European climates and this constrains the use of such waterborne masonry paints. Solventborne systems are available commercially but are often deficient in flexibility, alkali resistance or UV resistance. NAD acrylic microgel masonry paint introduced recently by ICI Paints provides a superior balance of properties. Laboratory tests show that film formation takes place quickly and water resistance is achieved within a matter of 15-30 minutes after application since only physical coalescence of the polymer dispersion is required, and not curing, to achieve a resistant film. The choice of acrylic polymer provides non-yellowing and alkali resistance as well as good UV stability.

Cost and Performance Benefits.

Although the raw material costs of the 'NAD Exterior Undercoat', 'Added Durability Woodstain' and 'All Seasons Masonry Paint' described above are up to 50% more expensive than conventional alkyd or aqueous emulsion paints, the benefits offered command corresponding premiums on the selling prices, and to date several million litres of the paints have been sold.

For the trade decorator, paint is a minor proportion of the total cost of a job, typically 10-20%. A superior product offering much improved durability and robustness to application conditions makes good economic sense, particularly on maintenance contracts where the time between repainting or preparation time is an important factor. The DIY decorator also prefers to be offered systems which extend paint lifetimes, so long as the additional cost is justified. Overall, there is a growing demand for quality and high performance in the UK decorative paint market, so long as it is cost-effective, and this is a challenge to the inventiveness of all paint chemists involved.

3   AQUEOUS MICROGELS FOR LOW POLLUTING METALLIC AUTOMOTIVE FINISHES.

The final appearance and quality of finish on a car is a major selling feature. This is particularly true of the paintwork where standards of finish have continued to improve and major new styling aspects, such as metallic finishes and clearcoat-over-basecoat (COB), have been introduced from time to time. Over 70% of cars produced in the USA, and a similar percentage in CWE and Japan, are now finished with metallic COB systems. Metallic finishes display a difference in brightness and hue when observed from different angles, highlighting different styling features on the car, emphasising various curvatures or lines of the bodywork. This is known as "flip-tone" effect, "flop" or geometric metamerism.

The Metallic Effect. The highest standard of metallic finish is typically offered by low solids polyester basecoat with a thermosetting acrylic clearcoat applied wet on wet. For maximum effect, i.e. highest flip, the metallic finish should appear much darker (Figure 7) when observed at a low angle to the surface (C) than when viewed at an angle (A) near normal to the surface. The effect is due to the orientation of the aluminium flakes, used in the basecoat, parallel to the substrate. The observer at (A) sees incident light (B) reflected from the surface of the aluminium flake through the clearcoat. When observed at (C), the reflected incident light (D) appears less intense because of travelling a longer path length through the coating.

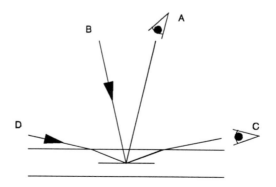

Figure 7.   Reflection of Light in a Metallic Flake Pigmented Film.

Measurement of Metallic Orientation. The metallic
effect can be measured by a goniophotometer (Figure 8).

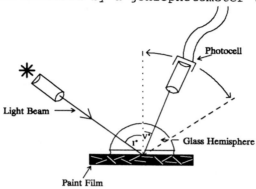

Figure 8. Measurement of Metallic Flake Orientation.

The paint film is covered by a glass hemisphere with an
intermediate oil film to exclude air and therefore eliminating specular reflection. A well-collimated beam of
light is incident at the panel at angle I. The intensity
of the reflected light at angle V is measured by a photo-
cell. The angle V is varied for a given incident angle I.
The reflected light intensity is plotted against the
viewing angle and gives a curve as shown in Figure 9 for
a silver metallic finish with good flip (Paint A). At the
chosen angle I = 45°, the relative height of the peak at
V = 45° correlates well with flip as judged visually.

Figure 9 also shows a curve for a silver metallic of
poor flip (Paint D). At an angle V = 45°, the paint film
A reflects light at an intensity some two times greater
than paint film D, and at V = 25° the curves cross showing
the wider scatter of the incident light emitted by paint
film D, i.e. it looks a little whiter compared to the much
darker film A. Micrographs of cross sections of the two
basecoats show that the aluminium flakes in Paint A are
very nearly parallel to the substrate, whereas those in
Paint film D are more randomly orientated.

Orientation of Metallic Flakes. Although various
factors influencing the degree of orientation of the
aluminium flakes have been examined [9,10], it is well
known that the application solids of the basecoat play
a major role. High levels of shrinkage from low solids
basecoats, as shown by A, B, Figure 9, tend to produce
well-orientated flakes and the maximum flip effect.

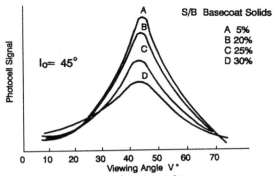

Figure 9.   Goniophotometric Results.

Conversely higher solids systems containing less solvent
(C, D) tend to produce randomly-oriented flakes and poorer
metallic flip.   To maximise these effects current solventborne basecoats are formulated at low solids, typically
12-14%.

   <u>Solvent Emission and Pollution</u>.   Painting operations
in car plants are the single largest users of organic
solvents in the coatings industry.   The average amount of
volatile organic solvent, per car painted, emitted to the
atmosphere is 12-15 litres, or 2-3,000 tonnes per year for
a moderate-sized car plant producing 150-200,000 cars per
annum.   This represents about 400,000 tonnes of solvent
used by the automotive industry worldwide every year, 80%
of which currently ends up being emitted to the atmosphere.   The application of (metallic) basecoats accounts
for about half of this solvent used (Figure 10).

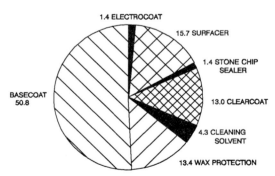

Figure 10.   % Relative Emission of Solvent in Painting
                a Car

Such emissions are under scrutiny and there are mounting pressures and legislation on the automotive industry to reduce or eliminate this source of pollution. Solvent removal and abatement technology can be used to address these pollution reduction requirements in part, but the cost is high, e.g. after-burners are effective on ovens, but do not cope well with exhausts from spray booths which are at much higher dilutions. Removal or reduction of the problem at source could be tackled by three basic routes - (a) higher solids systems, (b) powder coatings, and (c) waterborne basecoats.

Given that it is essential to maintain the high standard of appearance and performance of metallic finishes, ICI Paints pioneered waterborne basecoat with solventborne clearcoat technology in the 1980s to meet these legislative demands without sacrificing the high performance of the basecoat, and with minimum investment to both manufacturer and user.

The use of waterborne spray coatings in the automotive market is not new, General Motors used such coatings in the early 1970s for anti-pollution reasons. However, with these systems, several successive thin coatings were required under very narrow temperature and controlled relative humidity conditions, due to the lack of a robust technology.

Paint Rheology. During application, a liquid coating requires low viscosity at high shear rates ($>10^4 sec^{-1}$) to achieve satisfactory atomisation in the spray gun followed by a rapid and substantial rise in viscosity at low shear rates (c. 1 sec^{-1}) after landing on the substrate. The latter is to limit movement of the aluminium flake under the effect of gravity or convection currents in the coating and to resist sagging of the coating overall. However, in a finish coat, the low shear viscosity must not be so high as to prevent adequate levelling of the paint film. The rapid change in viscosity characteristics is achieved in conventional low solids solventborne basecoats by carefully blending solvents of different evaporation rates and solubility parameters. Although it is partly possible to engineer such viscosity characteristics in waterborne paints by blending with various solvents, these systems tend to be extremely sensitive to relative humidity and temperature during application. According to several investigators, compensation for loss of control of viscosity by solvent evaporation can be achieved by producing systems which exhibit pseudoplastic behaviour[11,12].

ICI Paints has developed a unique waterborne basecoat technology [13], known as "Aquabase", which relies on the use of an aqueous microgel polymer dispersion to produce the pseudoplastic rheology required. The new basecoat minimises the dependency of application on relative humidity such that it can be satisfactorily applied under a wide range of ambient conditions, normally encountered in car plants. It is also resistant to disruption or 'strike back' by the subsequent application of the clearcoat.

Figure 11 shows the characteristic pseudoplastic viscosity profile obtained for Aquabase systems for paint circulation in pipelines and on spray application.

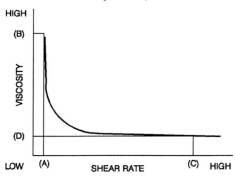

Figure 11.   Characteristic Viscosity of Aquabase.

At low shear rates (A) typically for the paint in a container, or immediately after spray application, the viscosity is very high (B), resulting in (a) excellent anti-settling properties, (b) optimum metallic flake orientation with good aesthetic appearance, and (c) excellent sag resistance. The latter is virtually independent of ambient relative humidity and retained water in the coating (as much as 50% retained water has been determined in coatings sprayed at 70-80% RH). This obviates the need for special control of relative humidity and temperature control in car plants which would be expensive because of the circulation of large volumes of air.

At high shear rates (C) typical of spray atomisation, the viscosity is very low (D) resulting in (a) excellent atomisation of the basecoat, (b) excellent metallic flake orientation, and (c) good circulation characteristics in pipelines.

The high degree of pseudoplasticity characteristic of Aquabase would be unacceptable in a single coat finish because of poor levelling leading to low Distinction of Image (DOI). However, the high degree of shrinkage of the basecoat helps to produce a relatively smooth surface which is then overcoated with a much thicker clearcoat.

The pseudoplastic viscosity obtained can be expressed by Ostwald's power law

$$\tau = K \dot{\gamma}^n$$

where τ = shear stress
$\dot{\gamma}$ = shear rate
K is a constant
and n = the viscosity index

Figures 12 and 13 illustrate the rheological characteristics of two Aquabase paints obtained using a Rheomat 30A-cup (Contraves, A G Zurich). Figure 12 shows the viscosity plotted against shear rate for a metallic silver basecoat.

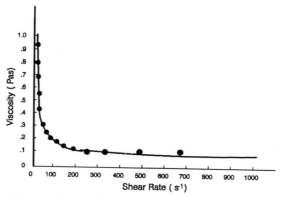

Figure 12.   Viscosity of Silver Basecoat.

Figure 13 shows the same data but as a plot of the log shear stress against log shear rate. It can be seen that the log/log plot is a good fit to a straight line of slope n.

For pseudoplastic materials, the value of n will vary between 1 and 0. When n = 1, the rheology is Newtonian and the lower the value of n, the greater the degree of pseudoplasticity.

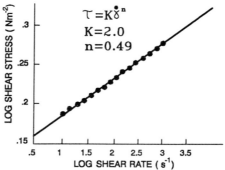

Figure 13. Viscosity of Silver Basecoat.

The value of K is equal to τ when $\dot{\gamma} = 1$. It is therefore equal to the extrapolated value of the viscosity of the paint in **PaS** at a shear rate of 1 sec^{-1}. This is termed the low shear viscosity (LSV).

Characterisation of a white basecoat pigmented with titanium dioxide gave the value of K = 0.57, lower than in the case of the silver where K = 2.0; and the value of n = 0.7, higher than the silver where n = 0.49. The white paint is therefore less viscous at low shear and less pseudoplastic.

In general, high metallics are formulated at higher values of K and lower values of n than is the case for solid (non-metallic) colours. In the case of metallics flow within the film must be controlled to a greater degree to prevent disruption of aluminium flake alignment parallel to the substrate, than is the case in solid colours, where it is only necessary to prevent sagging.

In addition, solid colours often have to be applied at greater film thicknesses to achieve opacity, particularly when bright organic pigments are used. Too high a value of K will prevent air bubbles escaping, leading to "popping" or "pin-holing" which will spoil the final appearance.

In practice, the viscosity of Aquabase paints is usually expressed at two extreme shear rates, the high shear viscosity (HSV) typical of spray atomisation and the low shear viscosity (LSV) typical of the paint immediately after being sprayed on to the car body. Table 5 shows the typical values of HSV and LSV for metallic and solid colour Aquabase paints.

Table 5.   Viscosities of Aquabase Basecoats.

	H.S.V. (10,000 sec⁻¹) (ICI cone & plate)	L.S.V. (1 sec⁻¹)
Metallic basecoats	0.028-0.036 PaS	1.5-3.0 PaS
Solid colour basecoats	0.03-0.04 PaS	0.5-2.0 PaS

Aqueous Microgel Polymer Dispersion

The key to the success of Aquabase as a low
polluting basecoat system is the aqueous acrylic microgel
polymer dispersion which imparts the required pseudoplastic viscosity and other essential mechanical and film
properties.   The average particle size of the microgel is
about 90 nm.   Each particle consists of an acrylic
copolymer which is crosslinked to render it insoluble and
resistant to the solvents contained in the clearcoat to be
applied over it.   This crosslinked core is encapsulated
with a mantle of acrylic polyelectrolyte, containing
functional groups (e.g. hydroxy) for subsequent crosslinking of the basecoat.   Post-neutralisation of the mantle
with organic bases (amines) results in considerable
swelling and the development of pseudoplastic rheology as
shown schematically in Figure 14.   The mantle is normally
chemically bonded onto the core to ensure it is not
removed by solvents or shearing forces.

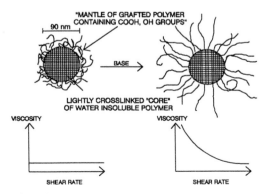

Figure 14.   Rheology of ICI Aquabase Dispersion.

The composite microgel particles are produced by a multi-stage emulsion polymerisation process followed by neutralisation to swell the polyelectrolyte. The polyelectrolyte mantle confers pseudoplastic rheology characteristics which enable paint to be sprayed under a range of application conditions. The gelled polymer core particles pack down in the applied basecoat film which can be coated after only a short drying stage with solventborne clearcoat without an intermediate stoving or curing stage. The gelled particles help prevent disturbance of the coloured basecoat layer by solvents in the clearcoat.

Typically, the basecoat will contain certain polar and non-polar solvents to assist development of the rheology and application characteristics. A crosslinker will also be used to improve mechanical and resistance properties of the basecoat in the final system.

The monomer composition of the core and mantle and their relative proportions can be adjusted to engineer the desired polymer properties.

In summary, the core provides:

Resistance to attack (disturbance from subsequent application of the clearcoat).

Mechanical toughness and flexibility.

Exterior durability associated with the chemical nature and high molecular weight of the polymer.

Chemical resistance associated with the high crosslink density.

The mantle provides:

adhesion to substrate and clearcoat

crosslinking potential

rheological characteristics

stabilisation to static and dynamic forces.

Extensive automotive production line experience confirms that the basecoat can be applied satisfactorily by a variety of methods commonly used in the industry, i.e. air-assisted spray and electrostatic (bells) application.

Appearance of Aquabase Metallic Basecoats. Earlier (Figure 9), it was shown that only low solids (solvent-borne) metallic basecoats were capable of displaying good flip. The appearance of Aquabase basecoats, formulated typically at 16% solids, also shows good flip (Figure 15), comparable to the best solventborne systems.

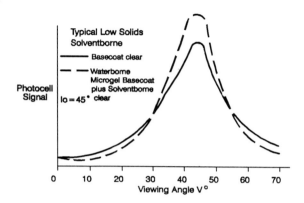

Figure 15. Goniophotometric Characterisation of Basecoats.

Legislation Limiting Emission of Volatile Organic Solvents. In the USA, emission levels are described in terms of Volatile Organic Content (VOC) per US gallon of applied solids, the targets being 14.9 and 12.2 lbs for existing and new facilities respectively.

In West Germany present emission levels are expressed in terms of weight of solvent emitted per square metre of total body painted, the targets being 120g for a metallic finish and 60g for a solid colour. It has recently been stated that these levels could be reduced to 70g and 40g per square metre respectively.

Transfer Efficiency. The one parameter which more than any other dictates the amount of solvent emitted to the atmosphere is the transfer efficiency of the process, i.e. how much of the paint actually transfers onto the substrates, e.g. by spraying, compared to losses to the atmosphere etc. Obviously, if the transfer efficiency is high, both emission levels and paint costs will be kept to a minimum. Some application methods are much better than others for transfer efficiency. Typical values for air-atomised spray are 40% and for electrostatic (bells) application as much as 70%.

Emissions from Metallic Basecoats. Table 6 shows
the effect of transfer efficiency on the solvent emission
(directly emitted to the atmosphere) of a series of
solventborne basecoats at 13, 20 and 35% solids, in
comparison with Aquabase at 16% solids when applied at 15
micron thickness. The results highlight that Aquabase is
much lower in solvent emission over the range of transfer
efficiencies normally found in practice. The solventborne
basecoats, on the other hand, need to be sprayed at much
higher solids, which will adversely affect their appearance, and at the highest transfer efficiencies if they are
to approach the standard of Aquabase.

Table 6. Weight of Solvent Emitted (g/m^2 body painted).

% Transfer efficiency	70	60	50	40
1. 13% N.V. solvent	69.1	80.7	96.8	121
2. 20% N.V. solvent	41.3	48.2	57.8	72.3
3. 35% N.V. solvent	19.2	22.4	26.9	33.6
4. Aquabase	9.0	10.5	12.7	15.8

Development and Commercialisation of Aquabase

The development of Aquabase with its unique aqueous
microgel technology was a major undertaking by ICI Paints
over many years with what was eventually a large international team.

The research was centred in the UK with the development and technical service located close to major potential customers in North America and Germany. The key to
success was overcoming the industry prejudice against
waterborne topcoat finishes and obtaining worldwide
approval from leading international customers. Since
then, commercialisation of the technology has proceeded
in both North America and Europe with licences granted to
other major paint companies around the world. Escalating
activity in this area by other significant paint companies
supplying the automotive market confirms the validity of
the original concept and the superior approach of using
aqueous microgel technology. To date nearly one million
vehicles have been finished in Aquabase.

The basecoat is not only the highest value component of the total car paint system but provides a key opportunity to offer a complete, high performance, low polluting clearcoat-over-basecoat system. Further development of the Aquabase technology is currently underway to reduce solvent levels still further and to maximise mechanical performance and application properties as well as extending the colour range to meet the latest styling requirements.

No doubt pressures will continue on the car manufacturers to reduce solvent emissions continually over the next few years. These will be addressed by a combination of good plant engineering and innovative chemistry, such as Aquabase.

4 SUMMARY

ICI Paints has developed novel polymer microgels with unique high performance characteristics in solvent and waterborne coating systems for the Decorative and Automotive markets.

A range of new Dulux paints and woodstains have been formulated using solventborne microgel and commercialised in the UK Decorative market. The polymer microgel was developed from ICI's Non-Aqueous Dispersion technology, suitably modified for use under ambient conditions. These paints offer significant advantages in application and early coating performance due to their robustness to cold, damp conditions, typical of Northern Europe. In particular, they have much improved exterior durability over conventional systems, remaining flexible and tough with outstanding adhesion, thereby providing much better protection for timber and masonry.

In the Automotive market, a major advance in paint technology has been achieved with the introduction of ICI Aquabase, a waterborne basecoat containing ICI's unique aqueous polymer microgel. This technological breakthrough was the result of a novel concept, researched and developed over several years by a large, international ICI Paints team working closely with major motor manufacturers in a number of leading countries. The aqueous microgel is stabilised with polyelectrolyte and produces a pseudo-plastic rheology which provides anti-sag and metal flake control (for metallic finishes) whilst still being easy to spray. The waterborne basecoat, overcoated with a protective, high gloss clearcoat, provides the final colour and aesthetic appeal required for automobiles.

The new system contains a much smaller amount of solvent compared to conventional systems and therefore offers the best prospects for meeting present and future solvent emission requirements whilst maintaining the high standards of performance and aesthetic appearance required for today's modern car industry.

Acknowledgement

The author wishes to acknowledge the valuable contributions of C W A Bromley and Dr D W Taylor to the article on NAD microgels and Z Vachlas and Drs A J Backhouse, A Frangou and J L Pearson to the article on waterborne basecoats.

REFERENCES

1. K.E.J. Barrett (Ed), "Dispersion Polymerisation in Organic Media", J. Wiley, London, 1975.

2. C.W.A. Bromley and J.A. Graystone, GB2,164,050A.

3. BS 5082; 1974, Appendix C.

4. T. Provder (Ed), Am. Chem. Soc. Symposium Series, 322, Chapter 12, (1987).

5. "Two Years Extra Life", Painting and Decorating, 20 (1985).

6. J.A. Graystone, "The Care and Protection of Wood", ICI Paints, Slough, 1985.

7. T.A. Strivens and R.D. Rawlings, "The Application of Acoustic Emission to the Study of Paint Failure", J. Oil & Colour Chemists' Assoc. 63, 412 (1980).

8. R. Gray, Products, Applications, Industry News, (A J Focus), April 1987.

9. A. Toyo, "The Orientation of Aluminium Pigments in Automotive Finishes", Polymers Paint Colour J. 796-798, Oct 29, 1980.

10. H. Van Oene and S.S. Labana, "Colour of Aluminium Flake Orientation in Metallic Colour Paints", VIIth Int. Conf. in Org. Coatings Sci. and Tech., 13-17th July 1981.

REFERENCES (Continued)

11. S. Wu, "Rheology of High Solids Coatings - Analysis of Sagging and Slumping", J. Appl. Polymer Sci., 22, 2769-2782 (1978). ACS Div. of Organic Coatings and Plastics Chemistry, Preprints, 37, No 2, 314-377, (1977).

12. L.O. Kernum, "Rheological Characterisation of Coatings with regard to Application and Film Formation", Rheol. Acta, 18, 178-192 (1979).

13. A.J. Backhouse, USP 4,403,003.

Commercial and Technical Aspects of Poly(alkyl-2-cyanoacrylate)s

J. Guthrie

LOCTITE (IRELAND) LTD., WHITESTOWN INDUSTRIAL ESTATE, TALLAGHT,
CO.DUBLIN, REPUBLIC OF IRELAND

INTRODUCTION

Poly(alkyl-2-cyanoacrylate)s form the basis of the so called instant adhesives or SUPER GLUE. Their ability to undergo what appears to be an instant polymerisation is responsible for their attractiveness as bonding materials. Alkyl-2-cyanoacrylates are not strictly one part adhesives as the substrates being bonded form the "second" coreactive material required to bring about polymerisation. The substrate borne active material is usually some form of ionic base such as a metal oxide. The activity of most substrates is enhanced in the presence of atmospheric moisture.

They may be regarded as high value polymers in as much as they are probably the most expensive adhesive materials routinely available to both industry and the consumer. Table 1 illustrates the average prices of a range of common adhesives.

It can be seen from Table 1 that Anaerobic and UV curable adhesives are as expensive as cyanoacrylates, but it should be noted that anaerobics and UV curable materials are complex formulations of monomers, resins, initiating systems and property modifiers. In contrast cyanoacrylate adhesives (CA's) consist, essentially, of pure liquid monomer. In addition the final cured material is a homopolymer unlike other adhesives which invariably result in the formation of copolymers

Table 1 Relative prices(Industrial) of common
adhesives [1]

ADHESIVE	COST($£kg^{-1}$)
Epoxy	5-11
Polyurethane	2-2.5
Acrylic	13-15
Phenolic	3-5
Polyester	3-4
Anaerobic	30-200
UV curable	60-250
Cyanoacrylate	40-250

and/or polymer blends. Given the high cost of instant
adhesives in the consumer market it is worth noting that
the cost of the homopolymer of ethyl-2-cyanoacrylate
formed in an adhesive joint is in the region of
£200 kg^{-1}(excluding packaging)!

Cyanoacrylate adhesives were discovered by H.W.Coover
at Eastman Kodak.Coover was in charge of a research group
engaged in examining the properties of new acrylate
polymers. Attempts to measure the refractive index of
ethyl-2-cyanoacrylate(I)

$$CH_2{=}C\begin{array}{c} \diagup CN \\ \diagdown C{=}O \\ \diagup \\ EtO \end{array}$$

(I)

in an Abbe refractometer inadvertently led to the
observation that this monomer was highly susceptible to
polymerisation and more significantly that the polymer
formed was an excellent adhesive!

Not until the 1970's did cyanoacrylate adhesives
become readily available. This was due largely to
difficulties in manufacture, packaging and transportation.
These problems were overcome by several manufacturers
including the Loctite Corporation. The resulting products
were considered particularly useful as they are one part,
solvent free and cure or polymerise rapidly at room

temperature with excellent adhesion to most substrates.
Disadvantages include poor impact, thermal and moisture
resistance.

CHEMISTRY OF ALKYL-2-CYANOACRYLATES

Adhesives based on alkyl-2-cyanoacrylates consist
essentially of 100% cyanoacrylate monomer. The patent
literature abounds with references to many different
monomers but only a restricted group are used in
commercial products. The structures of the most common
monomers are shown in Table 2.

Table 2 Alkyl-2-cyanoacrylate monomers

Commercially Important Esters

	Name	R
	Methyl	CH_3-
	Ethyl	CH_3CH_2-
	n-Butyl	$CH_3(CH_2)_3-$
	Allyl	$CH_2=CH.CH_2-$
	-methoxyethyl	$CH_3OCH_2CH_2-$
	-ethoxyethyl	$CH_3CH_2OCH_2CH_2-$

Structure: $CH_2=C$ with CN and $C=O$ groups, RO attached.

Monomer Synthesis. Alkyl-2-cyanoacrylates can be
prepared by several synthetic procedures. The only method
of importance involves the Knovenagel condensation of an
alkylcyanoacetate with formaldehyde. As this is a base
catalysed reaction, the monomer is rapidly polymerised to
give a low molecular weight polymer. The resulting polymer
is retropolymerised by heating under controlled conditions
to yield monomeric cyanoacrylate (Scheme 1).
Depolymerisation is carried out under vacuum in the
presence of an acid such as sulphur dioxide.

The monomer which distils from the reaction mixture
is collected in a vessel containing radical and ionic
polymerisation inhibitors. On paper this scheme looks
relatively simple but in practice considerable care must
be taken to produce high yields of monomer of the required
purity on an industrial scale. Repeated distillations are
required.

It should be noted that the reaction proceeds by
monomer formation followed by polymerisation and not by a

step growth condensation of the alkyl cyanoacetate with
formaldehyde [2,3].

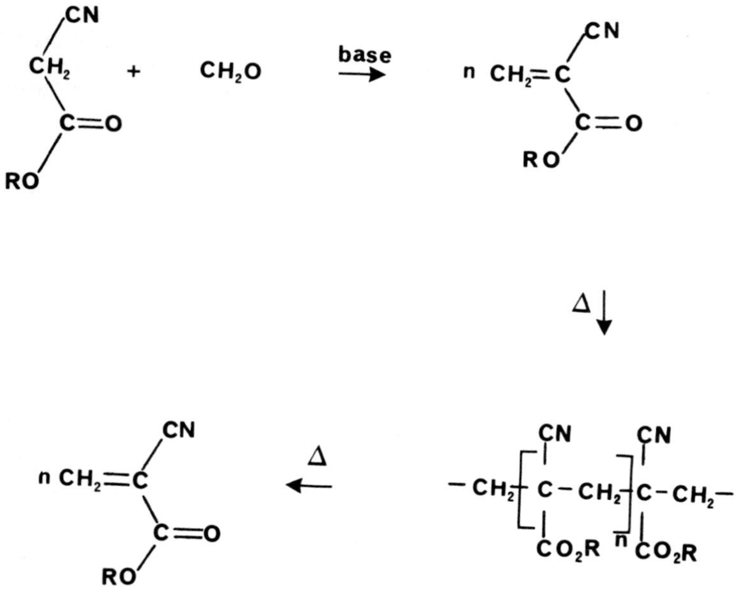

Scheme 1

Polymerisation of Alkyl-2-cyanoacrylates.
Cyanoacrylate monomers can be polymerised by both free
radical and ionic addition reactions. In adhesive
application ionic polymerisation is the most important.
The cyanoacrylate π electron system is under the influence
of two strongly electron attracting groups. This results
in a reduced electron density on the β carbon and an
enhanced susceptibility to nucleophilic attack (Scheme
2).The carbanion formed at the α carbon is stabilised by
delocalisation (II) and (III). The characteristic features
of a highly electrophilic β carbon, a stable carbanion and
an unhindered β carbon confer on alkyl-2-cyanoacrylates
their unique activity.

The mechanism of ionic polymerisation of
cyanoacrylates has been extensively studied by Pepper and
co-workers [4-8]. Ethyl and butyl cyanoacrylates(ECA,BCA)
have been found to be polymerisable with a wide range of
simple anions such as bromide, iodide, cyanide and
acetate. The reactions were extremely rapid with half
lives of the order of 1s at room temperature.

The "living" character of the polymerisation of BCA
initiated with butyl lithium was demonstrated by a
"repeated monomer addition" experiment in an adiabatic
calorimeter (Fig.1)[4]

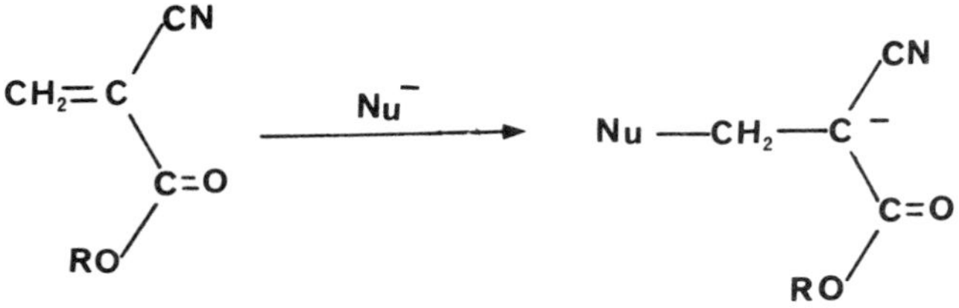

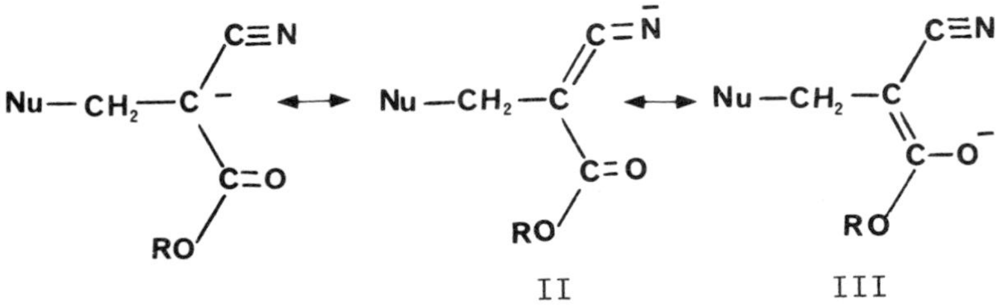

Scheme 2

After polymerisation of a first aliquot of monomer,
five further aliquots of monomer were added at intervals.
Rapid reactions and near identical thermograms were
observed at each addition even in cases where additions of
oxygen, carbon dioxide, water and alcohols were made.

Termination of polymerisation could only be brought about by the addition of gaseous hydrogen chloride.

The inability of these reagents to suppress polymerisation is indicative of the high stability of the growing carbanion. It should be pointed out that both water and alcohols could in fact terminate the growing polymer chain and that the resulting hydroxyl or alkoxy anions would initiate new polymer chains.

The molecular weight distributions of anionically polymerised poly(BCA) are found to be quite broad indicating that although the rate of initiation is fast, it is not fast enough to result in a classical "living polymerisation".

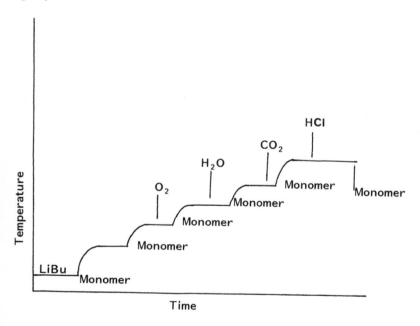

Fig.1. Evidence for the "living" nature of cyanoacrylate polymerisation.

Due to the electrophilic character of the β carbon it is possible to initiate polymerisation of cyanoacrylates with covalent bases. Studies by Pepper[4] have shown that,

with covalent bases such as amines and phosphines,
polymerisation proceeds by a zwitterionic mechanism
(Scheme 3).

Evidence for this type of behaviour has been
found from UV spectroscopy which shows the presence of
both onium and anionic end groups. Fig. 2[5].
The chemistry of initiation by both amines and phosphines
appears to be identical but the kinetics are quite
different. Both involve addition to monomer. The
subsequent kinetics observed are however different.

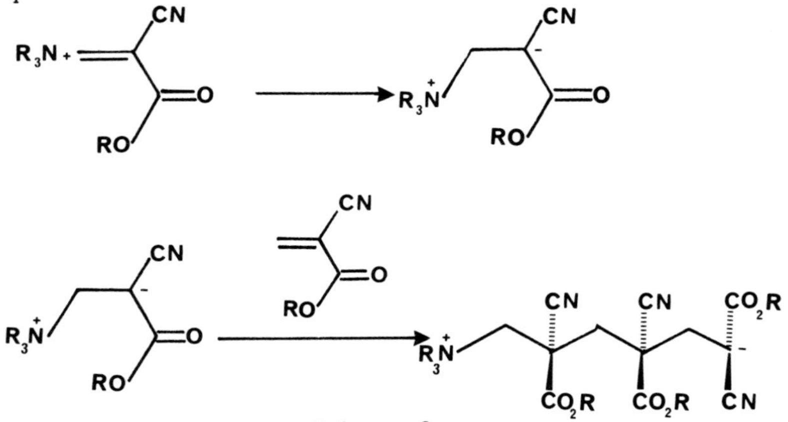

Scheme 3

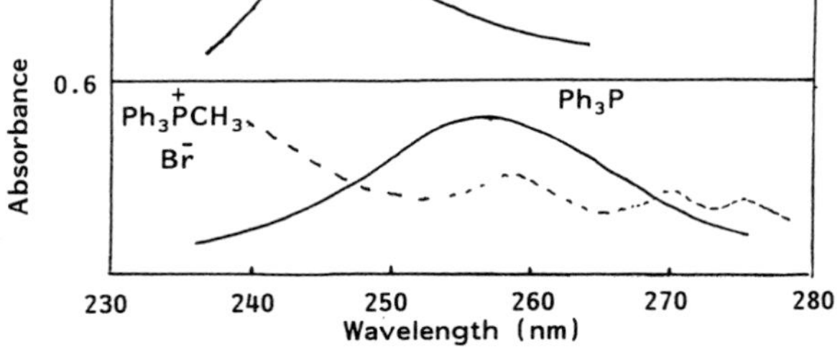

Fig.2 Evidence for zwitterion formation

Pepper has postulated that initiation by trialkyl phosphines can be described by a simple rate equation 1[6]

$$R_3P + CH_2=C\begin{smallmatrix}CN\\COOR\end{smallmatrix} \xrightarrow{k_i} R_3\overset{+}{P}-CH_2-\overset{-}{C}\begin{smallmatrix}CN\\COOR\end{smallmatrix} \qquad 1$$

In the case of triphenyl phosphine the rate constant for this addition is much smaller than for subsequent addition of monomer. When amines, including pyridine and derivatives of aliphatic amines, are employed as initiator several different kinetic features can be observed:

Higher molecular weights;
Slower rates of polymerisation;
An anomalous temperature dependence (faster at LOWER temperatures)

These observations were explained on the basis of Slow Initiation No Termination (SINT) kinetics combined with a multi-stage initiation process (Scheme 4)

$$
\begin{array}{ccc}
Py + M & \underset{K_{-I}}{\overset{K_I}{\rightleftharpoons}} & {}^+Z_1{}^- \\
{}^+Z_1{}^- + M & \xrightarrow{\phantom{K_{-I}}} & {}^+Z_2{}^-
\end{array}
\quad \text{Initiation}
$$

$$
\begin{array}{ccc}
{}^+Z_2{}^- + M & \longrightarrow & {}^+Z_3{}^- \\
{}^+Z_n{}^- + M & \longrightarrow & {}^+Z_{n+1}{}^-
\end{array}
\quad \text{Propagation}
$$

Scheme 4

This scheme indicates that the initiating zwitterion is only formed after two additions to monomers. This would of course explain the negative effect of temperature in that the concentration of initiating species would be controlled by an exothermic equilibrium constant $K=k_1/k_{-1}$. As the temperature is increased the equilibrium will move to the left. This type of behaviour has been confirmed from kinetic studies in the presence of strong acid which show that the rate of initiation decreases with increasing temperature.

In the presence of acidic impurities termination reactions occur as follows.

Strong acid:

$$^+Z^- \quad + \quad HA \quad \longrightarrow \quad ^+Z^-H + A^-$$

Weak acid:

$$^+Z^- \quad + \quad HX \quad \rightleftharpoons \quad ^+Z^-H + X^-$$

The conjugate base of a strong acid is not capable of initiating a new polymer chain. Consequently strong acids inhibit polymerisation. With weak acids the conjugate base is often sufficiently basic to initiate polymerisation. Weak acids are therefore regarded as retarders of polymerisation.

Pepper has also produced evidence to suggest that polymerisation of alkyl-2-cyanoacrylates can be initiated by propylene carbonate via an oxonium-carbenium zwitterion[68].

PROPERTIES OF POLY(ALKYL-2-CYANOACRYLATES)

Poly(alkyl-2-cyanoacrylates) are thermoplastic materials closely resembling poly(methylmethacrylate) and poly(styrene). The major polymer properties of the most common polymers are shown in Table 3.

Table 3 Properties of Cyanoacrylate Polymers

Polymer	Tg ($^\circ$C)	Solubility parameter	Density	Refractive index	Dielct. const.
Methyl	165	11.8	1.25	1.49	3.98
Ethyl	135	11.4	1.20	1.45	3.98
n-Propyl		11.0		1.45	3.78
i-Propyl		11.4		1.45	3.80
n-Butyl	85	10.8		1.45	3.88
i-Butyl		11.0		1.47	4.20
Allyl	115	11.5			
EtOEthyl		11.1		1.45	3.50
THFuryl		11.3			

The most common monomer used commercially is ethyl-2-cyanoacrylate. As a consequence most of the physical data available refers to poly(ethyl-2-cyanoacrylate)(PECA). PECA is a brittle glassy polymer. The glass transition temperature as measured by Dynamic Mechanical Thermal Analysis is in the region of 150°C (cf Table 3). The DMTA spectrum can be seen in Fig.3. A secondary transition is present at 50°C. The major mechanical features of PECA can be seen in Table 4[8].

Table 4 Mechanical Properties of PECA

Young's Modulus (25°C)	2.2 GPa
Breaking Strength	46.0 MPa
% Elongation @ break	4.3
Fracture Toughness	1.1 MNm$^{-3/2}$

The polymer is soluble in several polar solvents including nitromethane, tetrahydrofuran, dimethylformamide acetonitrile and acetone. A more extensive list of solvents for other cyanoacrylate polymers is given by Pepper and Donnelly[9].

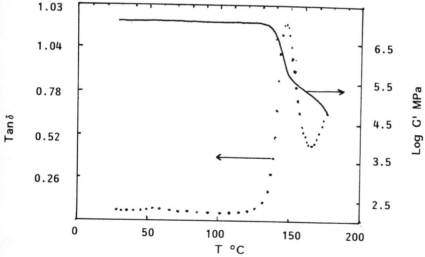

Fig.3 Dynamic mechanical spectrum of poly(ethyl cyanoacrylate).

Data on viscosities and unperturbed dimensions are also illustrated.

Very little data has been published regarding the microstructure of poly(alkyl-2-cyanoacrylates). Fawcett[10] et al have produced evidence to suggest that in certain solvent/amine initiating systems there is a tendency towards the formation of an isotactic polymer.

Poly(alkyl-2-cyanoacryaltes) are very sensitive to aqueous hydrolysis. This effect is not particularly marked at pH's less than 6. However in the presence of base the polymer breaks down,formaldehyde[11,12] being the main degradation product. A possible mechanism for the decomposition is given in Scheme 5[11].

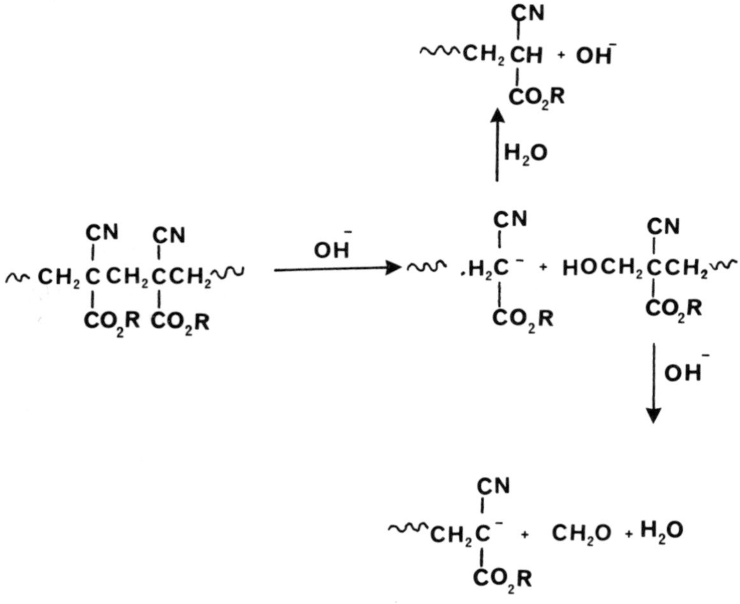

$$\sim\!\!\sim\!\!CH_2\underset{\underset{CO_2R}{|}}{\overset{\overset{CN}{|}}{CH}} + OH^-$$

$$\uparrow H_2O$$

$$\sim CH_2\underset{\underset{CO_2R}{|}}{\overset{\overset{CN}{|}}{C}}CH_2\underset{\underset{CO_2R}{|}}{\overset{\overset{CN}{|}}{C}}CH_2\sim \xrightarrow{\quad OH^-\quad} \sim\!\!\sim.H_2\underset{\underset{CO_2R}{|}}{\overset{\overset{CN}{|}}{C}}^- + HOCH_2\underset{\underset{CO_2R}{|}}{\overset{\overset{CN}{|}}{C}}CH_2\sim\!\!\sim$$

$$\downarrow OH^-$$

$$\sim\!\!\sim CH_2\underset{\underset{CO_2R}{|}}{\overset{\overset{CN}{|}}{C}}^- + CH_2O + H_2O$$

Scheme 5

Of particular interest to adhesives technology is the thermal stability of the polymer. When heated above 150°C the polymer decomposes to give monomer. It is believed that this involves an end initiated diffusion controlled retropolymerisation[13]. At least two groups have examined the thermal decomposition of cyanoacrylate polymers.Pepper and Burkinshaw[14] have postulated that the type of initiating species employed in the polymer synthesis has an effect on polymer stability. Polymers with pyridinium terminal groups were found to be significantly more stable than polymers with phosphonium terminal groups.Some of their results are shown in Table 5[14].

It can be clearly seen from Table 5 that the stability of the phosphine initiated polymer is less than that of the pyridine initiated material. This is not however reflected in the estimated values of activation energy. It was postulated that this increased stability was due to a thermal rearrangement of the terminal pyridinium group to produce a polymer chain with a more stable carbon-carbon terminal group. Some support for this proposal came from the observation that substituted pyridines such as lutidine and and 4-vinyl pyridine do not produce polymers with superior thermal resistance(Scheme 6).

Table 5 The effect of initiator on degradation
         characteristics

Initiator	Mn	$T^{o}C$ (2% Mass loss)	Rate @ 180°C ($\%s^{-1}$)	Ea (kJmol^{-1})
Ph$_3$P	1.07 x10^6	158	1.35 x10^{-1}	84
	3.07 x10^5	145	1.64 x10^{-1}	82
	2.44 x10^5	155	2.02 x10^{-1}	100
	7.00 x10^4	135	2.08 x10^{-1}	88
	4.40 x10^4	140	2.45 x10^{-1}	85
Pyridine	5.00 x10^6	185	7.43 x10^{-3}	104
	1.76 x10^6	217	6.11 x10^{-3}	61
	3.60 x10^5	200	5.00 x10^{-3}	60

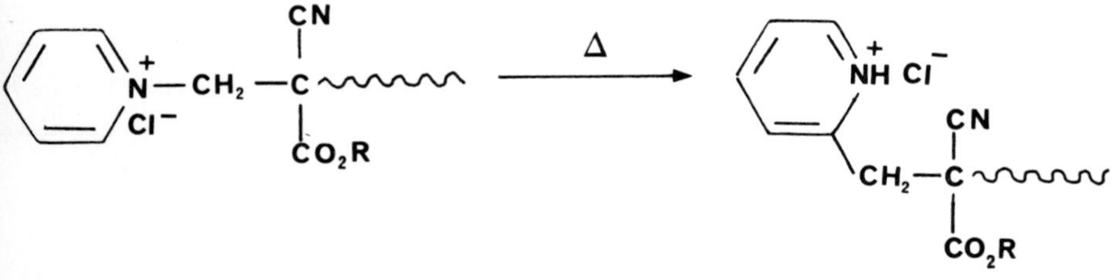

Scheme 6

These activation energies of thermal decomposition are lower than those reported for other polymers. This is a serious disadvantage which limits the application of cyanoacrylate adhesives. Negulesca[15] et al have determined the activation energies of decomposition for a range of poly(alkyl-2-cyanoacrylates). These can be seen in Table 6[15]. The value obtained for PECA is considerably higher that that quoted by Pepper[14]. The results in Table 5 were obtained by dynamic thermogravimetric analysis over a range of conversions and the final result estimated by extrapolation to zero conversion. Pepper's data was derived from isothermal thermogravimetry. The values of Pepper are similar to those obtained in the author's laboratory by isoconversional dynamic thermogravimetric analysis.

Table 6 Activation energies of decomposition of
           poly(alkyl-2-cyanoacrylates).

Polymer	$Ea(kJmol^{-1})$
Methyl CA	115.5
Ethyl CA	147.0
Propyl CA	163.8
Allyl CA	109.3,198.4

It should be noted that poly(allyl-2-cyanoacrylate) exhibits two distinct decomposition processes. The first is said to occur at low conversion of the polymer to volatiles. At higher conversions the polymer starts to crosslink by a radical mechanism at the allyl double bond. This results in a more thermally stable polymer.

CYANOACRYLATE ADHESIVES

Cyanoacrylate adhesives are often categorised by their viscosity: low, medium and high(variations in viscosity are achieved by the use of viscosity modifiers) and end use. The properties of a standard general purpose adhesive are shown in Table 7[16].

Table 7 ECA adhesive properties

Substrate	Bond Strength(MPa) (tensile shear)	Cure time(s)
Grit blasted steel	17	10-30
Etched aluminium	17	20-40
Polycarbonate	15	10-15
PVC	10	3-5
ABS	15	10-15
Butyl rubber	7	3-4
Polyethylene	0	>60
Paper	0	>60
Wood	0	>120

It can be seen from Table 7 that a wide range of substrates can be bonded with these adhesives. It should be noted that polyethylene, wood and paper are not bondable. Much industrial research has been directed to developing cyanoacrylate products for use with these "difficult" substrates.

Formulated adhesives consist of essentially pure monomer with relatively small amounts of property modifying additives. The polymerisation reaction is believed to be initiated by traces of basic material present on most surfaces, particularly in conjunction with surface moisture. These additives can classified as follows.

Stabilisers. The major difficulty associated with the the manufacture of cyanoacrylate adhesive is ensuring a balance between stability of the product and cure speed. The problem has been solved by careful choice of anionic polymerisation inhibitors. The materials employed are acidic compounds at levels between 5 and 1000 ppm. Table 8[8] shows a list of typical stabilisers.

Table 8 Anionic Polymerisation Inhibitors

Sulphur dioxide
Sulphur trioxide
Sulphonic acids
Sulphamides
Cationic exchange resins
Boric Acid Chelates

These function as described for the strong acids discussed earlier. They are all strongly acidic and prevent polymerisation by interaction with nucleophilic initiators or by terminating the zwitterion resulting from initiator-monomer addition. The level of stabiliser present is extremely critical. Too much stabiliser will give rise to a non active adhesive. Too little will result in an unstable formulation. It should also be noted that some of these stabilisers are volatile. In some cases this is considered advantageous to ensure stability in the vapour phase. In addition to anionic polymerisation inhibitors, free radical scavenging materials are also included to suppress the slow but significant tendency of cyanoacrylates to polymerise by a free radical mechanism. The materials used are invariably phenolic compounds such as hydroquinone or hindered phenols.

Accelerators. Accelerators increase the rate of
polymerisation. They should not be confused with
polymerisation initiators as they are not sufficiently
nucleophilic to induce polymerisation. The materials
described in the patent literature have one common feature
viz. they are all capable of sequestering alkali metal
cations. The mechanism by which accelerators function is
not clear, but it is believed to involve either increasing
ion separation at the growing chain end or activation of
anions on a substrate by cation sequestration to give so
called "naked" anions in the liquid adhesive. Examples of
compounds used as accelerators are crown ethers[17] (IV),
polyalkylene oxides (V)[18], podands (VI)[19] and calixarenes
(VII)[20]. These types of accelerator are particularly
effective on porous substrates such as wood and paper.
Until the advent of the use of accelerators it was not
possible to bond wood and paper with cyanoacrylate
adhesives.

    These developments have led to new applications for
cyanoacrylates and has extended their usefulness as a
consumer adhesive.

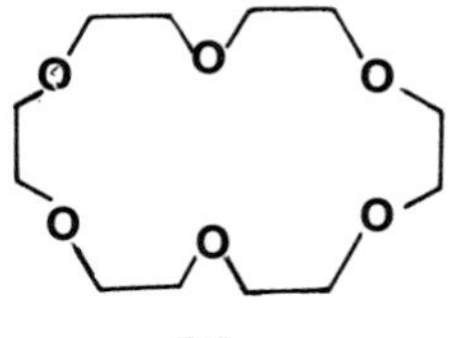

IV

$$RO\,CH_2\,CH_2\,O\left(CH_2\,CH_2\,O\right)_{\!n}^{-}\!H$$

V

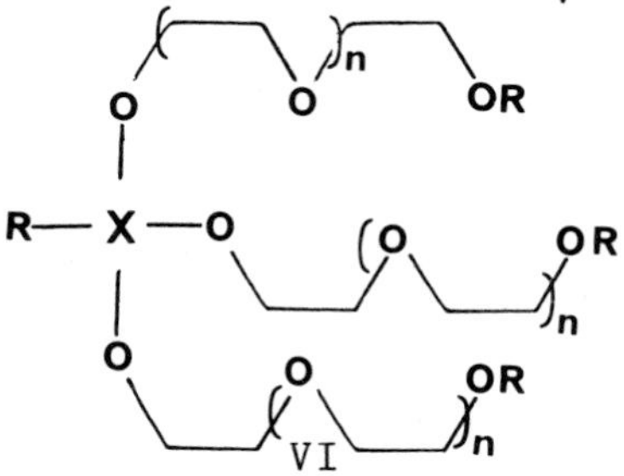

X= Multifunctional

VI

t-Bu

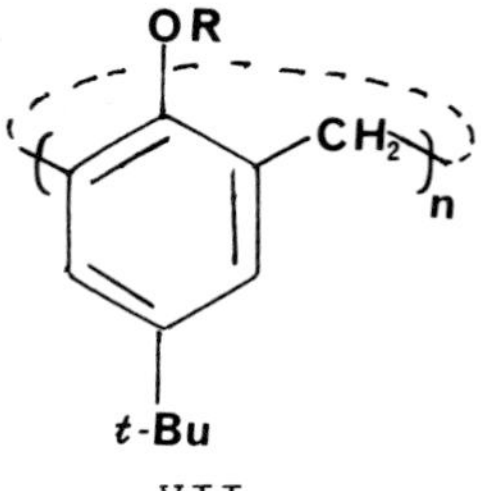

VII
n= 4, 6, 8

R = Alkyl

Adhesion Promoters. The patent literature describes the
use of unsaturated carboxylic acids and anhydrides as
adhesion promoters on metallic substrates. It is assumed
that the carboxylic acid group is capable of interacting
with the metal surface and that some degree of
copolymerisation takes place. There is however little or
no experimental evidence to substantiate copolymerisation.
The addition of these acidic species may also result in a
reduction of cure speed. Some indication of the
improvements in adhesion that can be achieved are shown in
Table 9[8].

Table 9 Adhesion promoters for cyanoacrylate adhesives

Adhesion promoter	% Level	Improvement of tensile shear strength(CRS)*
Itaconic acid	0.5-2.0	35%
Acetic acid	0.03-0.1	80%
Gallic acid/esters	0.02-0.2	30%
Polycarboxylic acids	0.1	300%
Polyhydroxybenzoic acids/polyether esters	0.2-10	1000%
Polyols,polyethers and aromatic polyols	0.05-20	500%
or acids	0.0001-0.5	30%

*CRS = Cold rolled steel

Plasticisers and Tougheners. Plasticisers are
required to reduce the inherent brittleness of the cured
adhesive. This is particularly marked for formulations
based on methyl and ethyl cyanoacrylate. Plasticisation
can be achieved by the use of both external and internal
plasticisers.

Toughness properties can be improved by the inclusion
of rubber toughening agents such as ABS or
MBS(methacrylate-butadiene-styrene). Plasticisation and
toughness are often only achieved at the expense of cure
speed. Table 10[8] gives a list of plasticisers and
toughening agents found in the patent literature.

An interesting approach to toughening is described in
a Japanese patent[21]. This involves formulating the
adhesive with an anionically polymerisable rubber forming
monomer. Some copolymerisation may also take place. The
adhesive is a mixture of ethyl cyanoacrylate and a
cyanopentadienoate (VIII)

Table 10 Plasticisers and tougheners for Cyanoacrylate
          adhesives.

Plasticisers (% Level)		Tougheners (% Level)	
Aliphatic esters	1-20	ABS	10-25
Glycerine triesters	1-20	VAC	10
Alkyl phthalates	1-20	MBS	20
Lactones	5-50	PBD-g-Styrene	1-15
Poly(n-Butyl ether)	3-15	Vamac B-124	10

VAC = vinylidene chloride-acrylonitrile
Vamac B-124 = ethylene-methyl acrylate

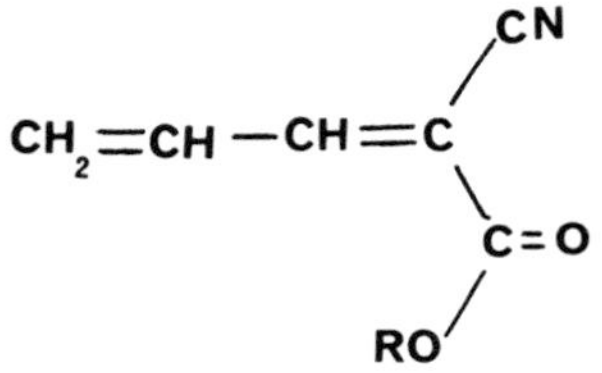

$$CH_2{=}CH{-}CH{=}C \begin{smallmatrix} CN \\ \\ C{=}O \\ RO \end{smallmatrix}$$

VIII

ENVIRONMENTAL PERFORMANCE

The durability of cyanoacrylate adhesive bonds is
reasonably good on rubbers and some polymer substrates.
However on glass and metals, both thermal and moisture
resistance are low. Heat resistance can be improved by
including additives in the formulation which give rise to
a more thermally stable polymer. Examples of this approach
are the use of biscyanoacrylates[22] (IX) and bismaleimides.

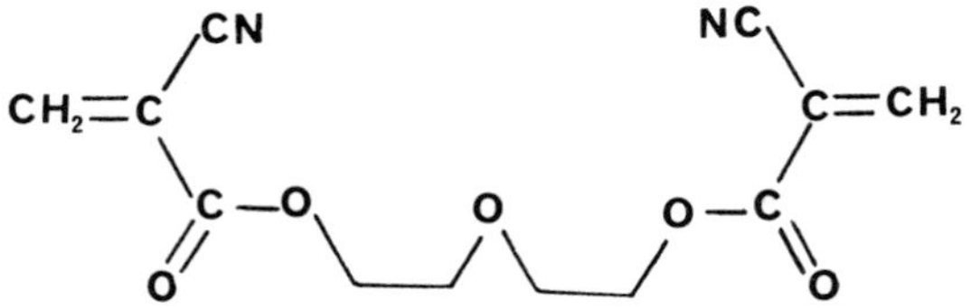

$$CH_2{=}C \begin{smallmatrix} CN \\ C \\ O \end{smallmatrix} {-}O \begin{smallmatrix} \\ O \\ \end{smallmatrix} O{-}C \begin{smallmatrix} NC \\ C{=}CH_2 \\ O \end{smallmatrix}$$

IX

Biscyanoacrylates are difficult and expensive to prepare
and the improvements obtained are only marginal. In the
case of bismaleimides it is difficult to see how they can
act as crosslinking agents. They probably function by
forming a more stable but separated network on heating.
The basicity of the imide link makes it difficult to

prepare formulations with a sufficient imide content to give significant improvements in thermal resistance.

Claims have been made that allyl-2-cyanoacrylate[23](X) and other cyanoacrylate monomers with side chain unsaturation can be used to improve heat resistance by virtue of radical induced crosslinking reactions. It is the authors experience that this is not the case. Crosslinking almost certainly occurs but the temperature required is so high that adhesion is lost before crosslinking occurs. It would appear however that lack of resistance to thermal ageing is not simply a manifestation of the thermal instability of the adhesive polymer.

$$CH_2 = C \begin{array}{c} CN \\ \\ C=O \\ \\ O \\ \\ CH_2 CH = CH_2 \end{array}$$

X

Exposure to heat gives rise to a "temperature induced loss of adhesion" rather than polymer degradation. The inclusion of carboxylic anhydrides appears to suppress this phenomena.

A similar approach has been taken to improving the moisture resistance of the adhesive bonds.Again the inclusion of anhydrides appears to have a beneficial effect on humidity resistance. Loss of adhesion in a moist environment is presumably related to the susceptibility of the polymer to hydrolysis. It has been suggested[12] that water diffuses along the metal oxide-polymer interface giving rise to polymer degradation and oxide growth. This new oxide layer and the hydrolytically weakened adhesive form a weak boundary layer.

Miscellaneous. Other modifications that can be made to cyanoacrylate adhesives include increasing the viscosity by the addition of thickeners such as poly(methylmethacrylate), cellulose esters or hydrophobic silicas. Colour can be imparted to the product by the use of selected dyes and pigments.

Polyolefin Bonding. The ability of cyanoacrylate adhesives
to bond a wide range of substrates with the exception of
polyethylene, polypropylene and PTFE has already been
mentioned. Recently products have been appearing on the
market in the form of a primer solution to be used in
conjunction with a cyanoacrylate adhesive. The primers
consist of dilute solutions of aluminium chelates[24,25,26],
titanium coupling compounds[27] or amines[28] in a volatile
solvent. In many cases the resultant bond is stronger than
can be achieved with surface treatment modifications such
as corona discharge and flame treatment. Some typical
results are shown in Table 11.

Medical Adhesive Applications. The ability of
cyanoacrylates to polymerise rapidly in the presence of
moisture makes them particularly attractive as tissue
bonding adhesives. In general the higher alkyl esters are
preferred for medical applications. Some typical medical
applications are shown in Table 12.

Table 11 Poly(olefin) bonding with cyanoacrylate adhesives

| | TENSILE SHEAR STRENGTH(MPa) | |
PRIMER	POLY(ETHYLENE)	POLY(PROPYLENE)
None	0	0.8
Aluminium Chelate(1)	2.0	9.3
Aluminium Chelate(2)	1.3	7.6
4-vinyl pyridine	4.1	9.3
Aliphatic amine	3.1	8.5
Loctite 757	6.0	8.0

Loctite 757 is a primer for use in conjunction with
cyanoacrylate adhesives on poly(olefins).

MISCELLANEOUS USES FOR POLY(ALKYL-2-CYANOACRYLATES)

Several non-adhesive uses can be found for
cyanoacrylate polymers both in patent and research
literature. Some of these are listed in Table 13.

HEALTH AND SAFETY FACTORS

Cyanoacrylate monomers have pungent odours and are
mildly lachrymator. Eye and skin contact should be
avoided. In the event of such contact the effected area
should be cleaned with water. In the case of eye contact
medical assistance should be sought.

Cyanoacrylate monomers are inflammable and should be isolated from sources of ignition. Basic materials such as amine and alcohols should be segregated from large quantities of monomer. These bases can initiate the rapid and highly exothermic polymerisation of alkyl-2-cyanoacrylates.

Table 12 Medical applications of cyanoacrylate adhesives

Tissue bonding:	butyl, i-butyl and other higher esters
Ear/eye/nose adhesives	cornea sealing, bone cementing in middle ear, stopping nasal bleeding
Dental adhesives	Hemostatic agent, dental filling
Suture adhesive	closing incisions, joining arteries and veins, treatment of ulcers, skin grafting male and female sterilisation
Nail bonding	Fixing of artificial nails

Table 13 Non-adhesive applications of Cyanoacrylate adhesives

Nano particles	Drug delivery
Photo resists	Solution and vapour deposition
Holography	Doped polymer films
Finger printing	Vapour deposition of polymer to form a finger print image.

REFERENCES
1. Advanced Engineering Materials in Europe, Vol.4-Structural Adhesives, IAL consultants Limited, 1986.
2. D.R.Smith, PhD Thesis, Northeasterr University, 1972.
3. J.M.Rooney, Polym.J., 1981, 13, 97
4. D.C.Pepper, J.Polym.Sci.,Polym.Symp., 1978, 62, 65.
5. D.S.Johnston and D.C.Pepper, Makromol.Chem., 1981, 182, 393.
6. D.C.Pepper, Polym.J., 1980, 12, 629.
7. I.C.Eromosele and D.C.Pepper, Makromol.Chem.,Rapid Comm., 1986, 7, 531.
8. G.H.Millet, Structural Adhesives, Chemistry and Technology, Ed. S.R.Hartshorn, Plenum Press, New York, 1986, Chapter 6, p249.

9. E.F.Donnelly and D.C.Pepper, <u>Makromol.Chem.,Rapid</u> <u>Comm.</u>,1981, <u>2</u>, 439.
10. A.H.Fawcett, J.Guthrie, M.S.Otterburn and D.Y.S.Szeto, <u>J.Polym.Sci.,Part C,Polym.Lett.</u>, 1988,<u>26</u>, 459.
11. F.Leonard, R.K.Kulkarni, G.Brandes, J.Nelson and J.J.Cameron, <u>J.Appl.Polym.Sci.</u>, 1966, <u>10</u>, 259.
12. K.F.Drain, J.Guthrie, C.L.Leung, F.R.Martin and M.S.Otterburn, <u>J.Adhes.</u> , 1984, <u>17</u>, 71.
13. J.M.Rooney, <u>Brit.Polym.J.</u>, 1981, <u>13</u>, 160.
14. C.Birkinshaw and D.C.Pepper, <u>Polym.Degrad.and Stab.</u>, 1986, <u>16</u>, 241.
15. I.I.Negulescu, E.M.Calugaru, C.Vasile and G.Dumitrescu, <u>J.Macromol.Sci.- Chem.</u>, 1987, <u>A24</u>, 75.
16. K.L.Shanty, S.Thennarasu and N.Krishnamurti, <u>J.Adhes.Sci.Technol.</u>, 1989, <u>3</u>, 237.
17. A.Motegi, E.Isowa and K.Kimura, US Patent 4,171,416, Toagosei Chemical Industry Company ,1979.
18. A.Motegi and K.Kimura,US Patent 4,170,585, Toagosei Chemical Industry Company 1979.
19. K.Reich and H.A.Tomasschek, US Patent 4,386,193, Teroson GmbH, 1983.
20. A.McKervey, US Patent 4,622,414, Loctite Limited, 1986.
21. T.Teramoto, N.Ijuin and T.Kotani, US Patent 4,313,865, Japan Synthetic Rubber Company, 1982.
22. C.J.Buck, US Patent 3,975,422, Johnson and Johnson, 1976.
23. D.L.Kotzev, T.C.Wight and D.W.Dwight, <u>J.Appl.Poly.Sci.</u> 1981, <u>26</u>, 1941.
24. Toagosei Chemical Industry Company, Jap.Patent 60,120,728, 1987.
25. Toagosei Chemical Industry Company, Jap.Patent 59,215,376, 1985.
26. Toagosei Chemical Industry Company, Eur.Patent 129,069, 1984.
27. Taoka Chemical Company Limited,Jap.Patent 61,136,567, 1986.
28. Alpha Giken Company, Jap.Patent 87,029,471, 1987.

Composites in Aerospace Applications

G. M. McNally

DEPARTMENT OF CHEMICAL ENGINEERING, THE QUEEN'S UNIVERSITY OF
BELFAST, BELFAST BT9 5A

1 INTRODUCTION

Polymer composite materials comprise high modulus fibres embedded in a high performance polymeric matrix. The benefits of using these materials as a replacement for metals and metal alloys for structural applications in the transport,[1] marine and aerospace industries are now generally well established and the growing awareness of their inherent advantages has lead to a steady increase in growth in the applications of these materials over the past 25 years. With some matrix resins however, a reduction in characteristic properties can arise, although this is usually as a result of improper materials handling procedures during fabrication of components and subsequent in-service environments. Some of these factors were examined, and are discussed in the latter sections of this presentation.

Table 1 Typical Mechanical Properties of Metals and
Composites

Material	Carbon Fibre UD epoxy	Carbon Fibre fabric/epoxy	Alloy Alluminium	Mild Steel
% Fibre	60	56		
Specific Gravity gms/cc	1.54	1.55	2.77	7.83
Specific Tensile modulus GPa/g/cc	89.61	39.59	26.14	26.44
Specific Tensile strength MPa/g/cc	985	436	159	194

Generally, composites have excellent mechanical properties and low specific gravities and so have outstanding weight performance characteristics that make them very attractive materials for structural applications, especially in the aerospace industry where weight/performance criteria are at a premium. Table 1 shows the typical properties of carbon fibre epoxy resin composites in comparison with metal alloys.

Composite materials are usually supplied in the form of a woven fibre fabric impregnated with partially reacted resin. These prepreg materials have good drape and tack characteristics that facilitate fabrication in laminate lay-up prior to autoclave curing. Prepregs also enable the fabricator to manufacture complex shapes at a much lower cost than with machine tooled metal alloys. There are also significant reductions in both job time and cost of tooling in the manufacture of small quantity components and prototypes. The prepreg can also be laid-up so that fibre orientation is aligned in the direction of in-service stresses, so that high structual efficiency can be built into the cured laminate during fabrication. Composite materials have also excellent fatigue life and very good environmental degradation and corrosion resistance and, because of their polymeric nature, have improved structural damping characteristics over metals.

One of the major concerns associated with materials used in aerospace applications is the effect impact damage may have on the static and fatigue strength of aircraft structures.[2] Such damage may occur as a result of bird impact, tool drop, or loose runway stones and can result in minor crazing, interlaminar shear delamination and subsequent fibre damage or in extreme cases complete puncture. In most cases carefully designed composite materials have improved impact properties over metal alloys and generally impact damage is not considered to be a major concern in the reduction of fatigue or static strength in these materials. On the other hand damaged composite structures are easily repaired by cut-out, implanting and recure using a heat blanket. Generally this type of repair is restricted to secondary structures.

Although composites have many commendable attributes in terms of ease of fabrication and property characteristics, their cost remains quite high in comparison to the metal alloys they are gradually replacing. Carbon fibre epoxy resin plain-weave type prepreg costs £24 per square metre and with a cured single ply thickness of around 0.2 mm the cost of this composite is approximately £70,000 per tonne. Fuselage and wing skin grade alluminium alloy is around £6,000-£9,000 per tonne and alluminium alloy used for Rolls Royce engine pods is £26,000 per tonne. However, associated with these alloys there is the added cost of carefully controlled chemical treatment of the manufactured components in large dip tanks, which is required before bonding or surface coating. On a volume basis the picture is more favourable, with costs of composites decreasing by up to 50% due to the low specific gravity of these materials. In-service energy conservation is now a decisive factor governing aerospace design and the attractive weight savings associated with the use of composites have led to these materials playing an ever increasing role in airframe technology. The Boeing 757 and 767 now use 1800Kg of carbon fibre epoxy resin composite per aircraft, a figure that is indicative of the progressive increase in the use of composites in civil aircraft.

On the AV-8B MK5 Harrier VTOL (Vertical Take-off and Landing) fighter aircraft, composite materials are now being used in many of the primary structures,[3] such as wing torque boxes and control surfaces, the horizontal tail and forward fuselage as well as many secondary structures. About 25% of the Harrier airframe is composite material and this results in a weight saving of 240Kgs per aircraft.

With the increasing demand for composites in aerospace applications and the continual development of new materials and major cost reductions associated with fabrication technology, the economics of using composite material will become even more attractive.

Table 2 Relative Cost of Structural Materials

Material	Cost £ per kilo
1. Aluminium Alloy-fuselage use	£ 6
2. Aluminium Alloy-wing skin use	£ 9
3. Aluminium Alloy-Rolls Royce Engine Cowls	£26
4. Polyester Resin	£26
5. Carbon fibre epoxy resin prepreg	£70
6. P.E.E.K. (poly-ether-ether-ketone)	£40-50
7. Kevlar (filament)	£10-100

2 THE DEVELOPMENT OF COMPOSITES IN AEROSPACE APPLICATIONS

The use of composites in aerospace applications dates back to the 1960s when glass fibre reinforced polyester resins were initially used in the manufacture of propellors and fuselages and subsequently in filament wound casings for the Polaris, Poseidon and Trident missiles. In order to broaden the use of composites in these high performance applications, new high modulus fibres were developed such as Boron and Carbon, and more recently, aramid fibres such as Kevlar. To derive maximum benefit from these high modulus fibres new resin systems have also been developed.

Although higher in price, epoxy resin based composites have much superior mechanical properties to polyester resins, and using appropriate hardener and accelerator systems the epoxies have also improved chemical and heat resistance. Therefore the epoxy resin laminates have played the most important role in current aircraft design. Recently, rubber modified epoxy resin systems have also become commercially available. These Toughened Epoxy systems were developed to enhance the impact tolerance of epoxy laminates.

With the rapid growth in the use of composites in aerospace applications, there has been a growing demand for composites with higher levels of thermal stability, especially in such applications as supersonic aircraft, when due to air friction, surface temperatures can reach 140°C during supersonic dashes. Even higher surface temperatures are experienced during VTOL, due to runway reflection of hot exhaust emissions.

In the mid 1970s polyimide resins became available. These materials have a polymeric backbone consisting essentially of a linear ring structure and so have high melting points of around 350°C and exhibit a high degree of thermal stability. Laminates of modified polyimide resins such as polybismaleimide can be used continuously up to 250°C and intermittently up to 400°C and so these resin systems have found particular use in supersonic and VTOL aircraft.[3] However these resin types are susceptible to hydrolysis above 100°C and are prone to cracking in water or steam above 100°C.

In 1977 ICI introduced a new thermoplastic resin PEEK (polyether-ether-ketone),[4] with high thermal stability and very low moisture absorption (0.1%) and excellent hydrolytic stability. Being a thermoplastic material PEEK has the ability to be thermoformed, albeit at high temperatures,[5] and so much shorter moulding times are required than the conventional autoclave curing of thermosets. These resins will also allow for remoulding and re-use of reject mouldings. The mechanical performance and impact behaviour of carbon fibre reinforced PEEK is also superior to similar carbon fibre epoxy laminates.[6]

However, the carbon/fibre epoxy resin based composites, fabricated using the autoclave moulding technique, are currently the most widely used composite materials in the aerospace industry for both primary and secondary airframe applications.

Autoclave Moulding of Epoxy Resin Laminates. Autoclave moulding is the most widely favoured fabrication technique employed in the manufacture of primary and secondary composite structures for aerospace applications.[7] The epoxy composite is supplied to the fabricator in the form of a B-stage Prepreg which is a partially reacted mixture of monomers impregnated into a woven fibre reinforcement.

Shapes are cut from the rolls of prepreg and these are then carefully layered on to the surface of a metal moulding platen (Fig. 1). The shaped prepreg is then overlaid with release film and a bleeder cloth which absorbs excess resin flow. This is then covered with a membrane which is sealed to the metal platen. A vacuum is then applied which exerts a pressure of approximately 1 Bar to the prepreg stack. This compacts the layers and removes any volatiles which may be evolved during curing. The whole assembly is then placed in an autoclave.

As the temperature of the autoclave increases the viscosity of the resin initially decreases and resin flow occurs.[8] With further increase in temperature the resin gels and cures to form a rigid matrix. The mould is held in the autoclave for some time at an elevated temperature according to the prepreg manufacturer's specifications. The autoclave is then cooled at a controlled rate to room temperature and the cured laminate is then removed from the mould and is ready for inspection, prior to assembly.

Provided the correct manufacturing technique is followed by the fabricator, high quality laminates are produced. However in the case of epoxy based composites,

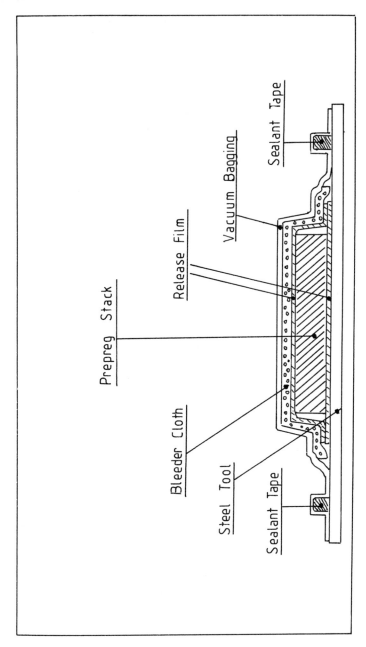

Figure1. Mould Lay-Up Conformation For Autoclave Curing of Prepregs

problems may arise due to inadequate materials handling control,[9,10] the curing procedure used and the subsequent in-service environments experienced by the laminate. The remainder of the present work investigates the detrimental effects such factors may have on the properties of the laminates as expressed in changes of glass transition temperatures (T_g).

(a) <u>Ageing of Prepreg</u>. In materials handling, the epoxy prepreg must be stored under refrigerated conditions, otherwise the reaction of the monomer mixture in such epoxy systems will continue under room temperature conditions. After some time the tack and drapability of the prepreg will deteriorate, as the prepreg was approaching the end of its useful life. Although stringent quality assurance specifications have to be met before a prepreg is accepted for use, a degree of uncertainty still exists with regard to the age at which a prepreg should be rejected, and if this age is exceeded what would be the consequences in terms of laminate properties. In this work the changes in glass transition temperatures of laminates made from grossly aged prepregs, is investigated.

(b) <u>Curing Conditions</u>. Manufacturers data normally states that the optimum properties will be achieved in the finished component by curing for a specified time at a particular temperature.[12] During the curing operation it is quite conceivable that either through operator error, or equipment malfunction, the prepreg is not cured according to these specified conditions. However, rarely is there any information given regarding the shortfall in properties which will arise if the actual cure cycle deviates from specification. The work presented here also investigates the extent to which the glass transition temperature of laminates cured under different conditions of temperature and time may vary.

(c) <u>Effect of Moisture</u>. It has been recognised that during their lifetime epoxy resin composite systems experience moisture absorption, and that this moisture ingress is detrimental to the mechanical properties of the laminate.[11,13] Many studies have been conducted to interpret and characterise the problem.[14,15] However, little is known regarding the moisture absorption characteristics of laminates manufactured under atypical conditions. The preliminary work presented here investigates some aspects of moisture absorption phenomena which are developed in atypically manufactured epoxy laminates.

Materials. The laminates used in these preliminary
investigations were manufactured from commercially
available carbon fibre epoxy resin based prepreg.
(a) Cycom 985 (3K-70-PW) prepreg, supplied by the
Cyanamid Corp. USA, were used for the ageing studies,
(b) Fiberite 934 (PW) prepreg, supplied by the Fiberite
Corp. USA, were used for the cure cycle studies.

Cycom 985 was left to age in atmospherically
controlled conditions (23°C 50% R.H.). At specific time
intervals, samples of the prepreg were removed and placed
in a freezer (-14°C) to retard further reaction of the
resin. At the end of an eight week period all the
samples were removed from the freezer and prepared for
lay-up. Specimen boards of the prepregs were cured in an
autoclave using the recommended cure cycle of 2°C/min up
to 179°C and held for 120 mins.

Specimen boards of Fiberite 934 were prepared from
fresh prepreg. These boards were cured under various
conditions of time and temperature according to Table 3.

Table 3   Cure Conditions for Laminate Samples

Sample	Cure Temperature (^{o}C)	Cure Time (mins)
A	180	45
B	180	60
C	180	120
D	140	120
E	150	120
F	160	120
G	170	120
H(D)	Recure 180°C	60

Experimental Techniques. Thermoanalytical techniques
are being increasingly employed for rapid and informative
quantitative characterisation of composite materials. In
recent published reports a close correlation tends to be
emerging between improved mechanical properties with
increasing T_g of the composite, as measured using thermal
analysis techniques. However the recorded glass
transition is very dependent on the test method and
experimental conditions used, and so it is important that
in comparative studies the analytical procedure is quoted
along with the T_g value. In this present work two
techniques were used, Dynamic Mechanical Thermal Analysis
and Thermomechanical Analysis.

(a) <u>Dynamic Mechanical Thermal Analysis</u>. Dynamic
Mechanical Thermal Analysis of the laminates prepared
from the aged Cycom 938 prepregs was studied using a
Rheovibron Dynamic Viscoelastometer. During the thermal
scan the samples were subjected to bending mode
oscillation at frequencies of 11, 35 and 110 Hz. The
recorded data was then processed using a Hewlett Packard
mini-computer to give thermograms of the damping factor
tan δ and dynamic modulii as functions of temperature.
The temperature at which tan δ max occurred was taken as
being the glass transition temperature.

(b) <u>Thermo Mechanical Analysis</u>. Thermo Mechanical
Analysis (T.M.A.) of laminates prepared from Fiberite 934
and cured under various conditions was studied using a
Perkin Elmer TMS 2 thermomechanical analyser, fitted with
a flexure probe arrangement. The T_g was recorded as
being the temperature of the first derivative peak on the
thermogram occurring directly after the onset of change
in expansion coefficient.

(c) <u>Moisture Absorption Studies</u>. To investigate the
moisture absorption characteristics of the composites,
preweighed samples of the various laminates were
subjected to cyclic immersion in water at elevated
temperature, using soxhlet apparatus. The temperature of
the water falling into the soxhlet thimble supporting the
samples was 85°C throughout the experimental
investigations. The samples were reweighed at various
time intervals to measure the extent of moisture uptake.

3 RESULTS AND DISCUSSION

The DMTA thermograms (Fig. 2) displayed the
characteristic viscoelastic transitions associated with
C/F epoxy resin systems,[16] with the tan δ peak, drop in
storage modulus and increase in loss modulus all
occurring in the region of the glass transition
temperature. Fig. 3 shows that the recorded T_gs of the
various laminates were generally found to be dependent on
the measurement frequency, i.e. as the frequency
increased so also did the recorded value of the T_g.
However this dependency of T_g with oscillation frequency
was found to diminish progressively with ageing of the
prepreg. Very little change in T_g was observed for
laminates made from grossly aged 56 day old prepreg.
Fig. 3 also shows that under fixed frequency conditions
the T_gs of the laminates show a progressive increase with
prepreg ageing. The magnitude of this recorded increase
was found to be more pronounced at the lower frequencies,

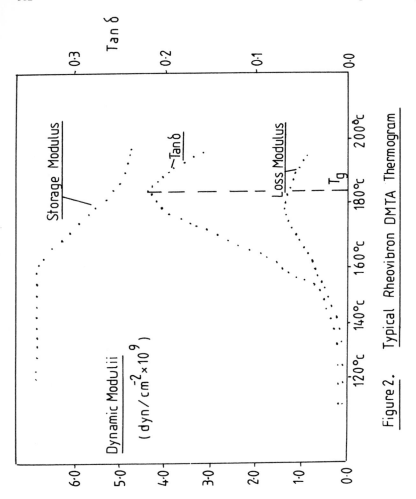

Figure 2.   Typical Rheovibron DMTA Thermogram

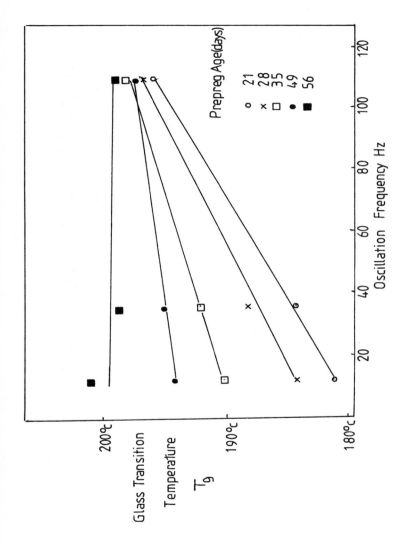

Figure 3.   Effect of DMTA Frequency on Recorded Value of T_g.

with increases of up to 20°C at 11 Hz but only 5°C at 110
Hz for laminates made from 21 day old and 56 day old
prepregs. These results showed that the differences in
viscoelastic properties were more distinguishable at
lower frequencies and so further DMTA analysis was
conducted with measurement frequency of 11 Hz.

Generally these findings indicate that the
characteristic properties of carbon fibre epoxy resin
based laminates would be enhanced by ageing the prepreg
prior to curing. However another associated factor
influencing the properties of these laminates is their
susceptibility to absorption of moisture. Absorbed
moisture acts as a plasticizing diluent in the resin
matrix leading to a reduction in mechanical properties.
The effect of absorbed moisture on the T_gs of the various
laminates is shown in Fig. 4. In almost all cases the
glass transition temperature was depressed with increase
in moisture content. However it is significant to note
that laminates with higher 'dry' T_gs, i.e. laminates
using grossly aged prepregs, exhibited a more acute
decrease in T_g with percentage moisture absorbed than did
those laminates prepared from fresher prepregs.

These investigations tend to show that the
properties of laminates having high T_gs will experience a
more pronounced detrimental response to moisture
absorption.

With regard to the implications on property
characteristics of laminates cured using atypical
conditions, Table 4 shows the effect of various cure
temperatures and cure times on the T_gs of laminates made
from fresh Fiberite 934 prepreg. The manufacturers
recommend autoclave curing at 180°C for 120 minutes and a
laminate cured under these conditions had a recorded T_g
of 225°C. The table shows that the recorded T_gs of
laminates cured at 180°C for 45 and 60 minutes were 222°C
and 223°C respectively, thus indicating that curing of
these particular samples at 180°C, increasing cure time
had very little effect on the T_g of the laminates.
However curing at lower temperatures gave laminates with
very much lower T_gs, thus indicating that there is a
progressively greater extent of undercure as the curing
temperature is lowered.

In practice components which had been erroneously
cured at lower temperatures would obviously be rejected
on the basis that they would not meet the required
property characteristics and so would be scrapped.

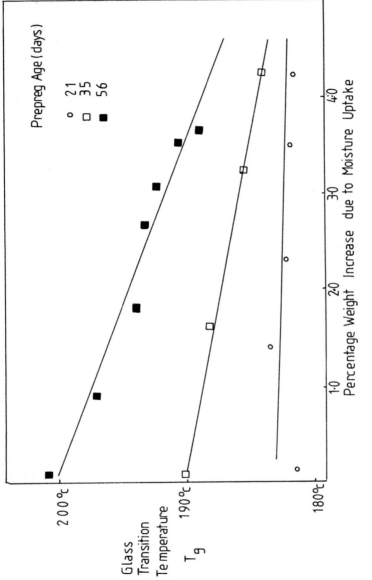

Figure 4. Effect of Moisture Content on T_g of Laminates Prepared from Aged Prepregs.

Table 4 The Effect of Cure Temperature and Time on T_g
       and Equilibrium Moisture Content

Cure Temperature (^{o}C)	Cure Time (Mins)	T_g (^{o}C)	Equilibrium Moisture Content (%)
180	45	222	3.70
180	60	223	3.6
180	120	225	3.6
140	120	139	5.87
150	120	192	4.88
160	120	208	4.35
170	120	218	3.75
D recured at 180	60	224	3.6

However it is interesting to note that the T_g of the
laminate cured at 140°C for 120 mins increased from 138°C
to 223°C on subsequent recuring at 180°C for 60 minutes.
This preliminary study would tend to imply that rather
than reject and scrap undercured components, these could
be recured at the correct temperature and so achieve the
required property characteristics. The salvaging of such
components would have significant cost saving advantages.

Accelerated moisture absorption studies using a
soxhlet apparatus (Fig. 5) shows that progressively more
moisture was absorbed by laminates cured at lower
temperatures. Under these accelerated test conditions
equilibrium moisture contents for the various laminates
varied from 5.8% to 3.6% equilibrium moisture content for
cure temperatures of 140°C and 180°C respectively.

4 CONCLUSIONS

In the DMTA studies of laminates made from aged prepregs
(Fig. 3), the general shift of the damping tan δ peak to
higher temperatures with increase in measurement
frequency, complies with the established theories
concerning material response to sinusoidal deformations
during Dynamic Mechanical Thermal Analysis.[17] This shift
is primarily related to the reduction in time available
for molecular motion to manifest itself at a particular
temperature, with the rapid imposition and removal of
stress encountered at higher measurement frequencies.
However it is interesting that as the prepreg age
increased, the degree of damping shift decreased and
hence the dependency of the recorded T_g on measurement

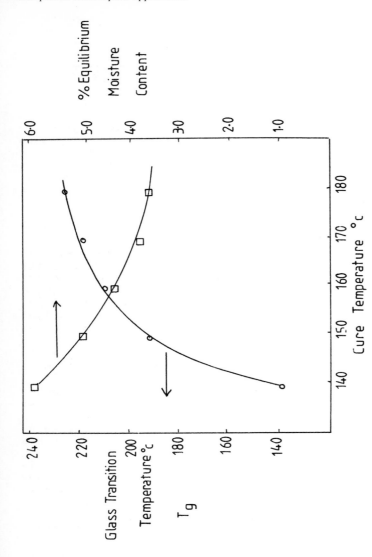

Figure 5. Effect of Cure Temperature on T_g and Equilibrium Moisture Content

frequency diminished. This observation would tend to suggest that the structure of the laminates made from grossly aged prepregs was progressively more rigid in character. The overall increase in T_g would also tend to imply a progressive increase in crosslink density being established in the laminates through ageing of the prepreg prior to curing. This proposed increase in crosslink density is in agreement with Chu[16] and Leckenby's[18] work on the relationship between network structure and the physical properties of amine cured epoxies. They observed that a molecular weight increase between crosslinks resulted in an overall lowering of the T_g, whereas increase in molecular weight through increase in crosslink density resulted in an overall increase in the observed T_g, due to more restricted chain movement in the highly crosslinked polymer network.

Hygrothermal exposure of the laminates with higher 'dry' T_gs showed a more extreme response to the plasticizing effects of moisture than did laminates with lower 'dry' T_gs. The adverse effects of moisture on the morphology of epoxy resins has been well documented[11,21] and can be characterised by various mechanisms and theoretical models based on free volume[14] and entropy.[15] In this present work the entropy model closely fits the trends observed in Fig. 4. This model theory is dependent on the presence of polar groups within the polymer network. In polymer networks of high crosslink density, chain movement will be restricted during polymerization and consequently it will be sterically difficult for unreacted functional groups to undergo reaction. Therefore as crosslink density increases, the effective concentration of unreacted hydrophilic sites increases.[19] Formation of intermolecular bonds between water molecules and the polymer hydrophilic sites occurs during sorption leading to the rupture of the intra molecular polymer matrix. The rupture of these bonds is manifested by a decrease in T_g with moisture content and increasing crosslink density.[9]

Very little change in T_g was observed for the laminate prepared from 21 day aged prepreg, indicating that moisture had been merely absorbed on to the free volume surface or microcavities of the matrix and so had very little plasticizing effect on the structure.

In the cure cycle studies, Fig. 5 shows that the T_g of the laminates decreases as cure temperature decreases. It is generally accepted[12,18] that the rate of cure of epoxy resin systems will be much slower at lower

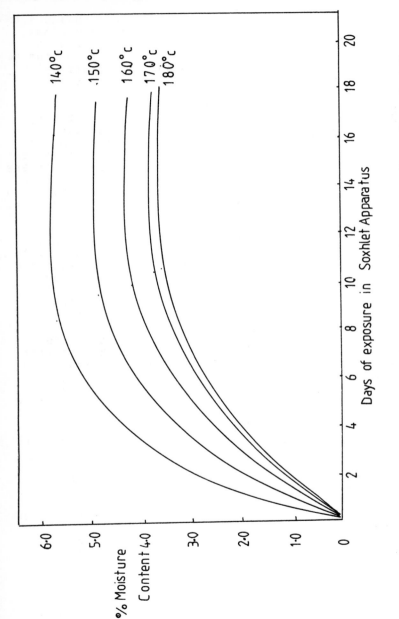

Figure 6.   Effect of Cure Temperature on Moisture Absorption Characteristics

temperatures and so the overall degree of cure after 120 mins will be smaller at lower cure temperatures. These undercured laminates will have a progressively lower crosslink density and will consequently have a greater degree of chain flexibility in the polymer network resulting in lower recorded T_gs.

Although the activation energy increases progressively as the curing reaction continues, these preliminary studies have shown that the glass transition temperature of undercured laminate D, increased on recuring to 224°C. This value is very close to that of laminate C, which was cured at the recommended conditions of 180°C for 120 mins, thus indicating that undercured components can achieve full-cure characteristics by reheating in an autoclave at the recommended temperature.

The moisture absorption studies (Fig. 6) show that undercured laminates have a higher moisture equilibrium content and generally a higher diffusion rate than fully cured laminates. The laminate cured at 140°C takes approximately 6 days to reach an equilibrium moisture content of 5.3%, whereas the laminate cured at 180°C takes approximately 11 days to reach an equilibrium moisture content of 3.5%. With the lower crosslink density[12,17] and more open network structure of undercured laminates moisture can diffuse rapidly into the matrix, and since the void volume in such laminates is greater, a larger capacity of moisture can therefore be accommodated.

REFERENCES

1. G.F. Smith, P.R.I. 3rd International Conference Fibre Reinforced Composites, Liverpool, March 1988.
2. A.A. Baker, R. Jones and R.J. Callinan, Composite Structures, 1985, 4, 15.
3. B.L. Riley, A.V.-8B/GR MK5 Airframe Composite Applications, Inst. Mech. Engineers Proceedings, 1986, 200, No. 50.
4. Aromatic Polymer Composites, ICI plc, Welwyn Garden City, 1982.
5. I. Brewster and J.B. Cattanach, SAMPE Conference Engineering with Composites, Paper 3.
6. S.M. Bishop, Composite Structures, 1985, 3, 295.
7. D. Purslow and R. Childs, Composites, April 1982, 17, No. 2, 127.
8. L.D. Laurer, SAMPE Quarterly, October 1983, 31.
9. F.N. Cogswell and D.C. Leach, Plastics and Rubber Processing and Applications, 1984, 4, 271.

10. S.N. Sanjana, <u>SAMPE Journal</u>, January 1980, 5.
11. W.W. Wright, <u>Composites</u>, July 1981, 201.
12. J.B. Johnston and C.N. Owston, <u>Composites</u>, May 1973, 111.
13. J.M. Barton and D.C. Greenfield, <u>British Polymer Journal</u>, 1986, <u>18</u>, No. 1, 51.
14. F.N. Kelley and F. Bueche, <u>Journal Polymer Science</u>, 1968, <u>50</u>, 2549.
15. H.G. Carter and K.G. Kibler, <u>Journal Composite Materials</u>, 1977, <u>11</u>, 265.
16. H.S. Chu and J.S. Sefaris, <u>Polymer Composites</u>, April 1984, <u>5</u>, No. 2, 124.
17. R.E. Wetton, "Developments in Polymer Characterisation", Polymer Labs. Loughborough.
18. P.S. Gill and J.N. Leckenby, <u>Composite Structures</u>, 1984, <u>2</u>, 235.
19. A. Apicella, L. Nicholais and C. de Cataldis, <u>Advances in Polymer Science</u>, 1985, <u>66</u>.
20. J. Migovic and H.T. Wang, <u>Journal of Applied Polymer Science</u>, 1989, <u>37</u>, 2661.
21. J.M. Barton, <u>British Polymer Journal</u>, 1986, <u>18</u>, No.1.

Kevlar® Aramid Fibers

J. A. Fitzgerald and R. S. Irwin[1]

FIBERS AND CENTRAL RESEARCH AND DEVELOPMENT DEPARTMENTS, E. I.
DU PONT DE NEMOURS & CO., INC., EXPERIMENTAL STATION,
WILMINGTON, DELAWARE 19880-0302, USA

1 INTRODUCTION

Kevlar® aramid fiber, based on poly(p-phenyleneterephthal-amide) (PPD-T) is the first high strength/high modulus fiber to be commercialized. Its outstanding tensile properties are the result of a high degree of alignment of fully extended polymer chains in the direction of the fiber axis, such that applied tensile loads are shared more or less equally by all the chains. In conventional fibers such as 66 nylon or PET, macromolecular orientation is achieved by drawing the as-spun fiber. Chain entanglements severely limit the degree of orientation thus achievable, so that fiber strength reaches only a rather small fraction of theoretical values for complete orientation. (In gel spinning, a fairly recent development, certain polymers such as polyethylene, of very high molecular weight, are drawn to very high degrees of orientation under conditions of minimal chain entanglement, to provide very much enhanced levels of tensile strength and modulus.) With Kevlar® high orientation is achieved in a totally different way. The relatively inflexible, extended PPD-T macromolecules in solution, at higher concentrations, form a liquid crystalline phase. Such phases are capable of very high macromolecular alignment under fiber spinning conditions and this structural arrangement can be captured in coagulation to fiber without need (or possibility) of further orientation by drawing.

Lyotropicity of p-aramids and its application to production of high-strength fibers was discovered at Du Pont in 1965 for poly(1,4- benzamide) (I) in alkylamide/salt solvents such as DMAc/LiCl (1,21). The results of a detailed study of these systems was subsequently applied to the less tractable but economically more attractive PPD-T (II) for the production of Kevlar®.

---

[1]To whom correspondence should be addressed.

(1)   (2)

Although liquid crystalline solution character of spinning solutions was the essence of Kevlar® fiber technology, certain other key problems had to be resolved in order to create a viable process. Thus a PPD-T molecular weight level three times higher than that described in the prior art was needed to provide optimum tensile property levels. This problem then had to be solved a second time when the polymerization system identified for initial manufacturing fell under suspicion as a human carcinogen. The need for a spinning solvent which would allow maximum capitalization of the potential of liquid spinning technology led to the recognition of pure sulfuric acid as far superior to ordinary "concentrated" sulfuric acid (96%). The trend towards higher fiber strength levels with increasing spin dope concentration was limited by apparent lack of solubility at concentrations above about 13%. A major breakthrough was the discovery that spinnable concentrations of 20% (w/w) could be achieved at 80°C without degradation of polymer or untoward corrosiveness hitherto anticipated with hot sulfuric acid. The air gap between the spinneret and the cold coagulant, made necessary to avoid freezing of spin dope within the spinneret assembly if it were immersed in the conventional way in the cold coagulant, produced an unexpected bonus in terms of exceptional fiber strength and ability to spin fine filament deniers at considerable throughput. The use of copious quantities of sulfuric acid as a spinning solvent posed a sizable problem with regard to its disposal in an environmentally acceptable manner, since recovery was hardly economical.

Kevlar® has been adapted to a variety of applications (2). It is best known as a high strength reinforcement fiber for rubber and plastics composites, and in ropes and cables. Its resistance to cutting and ballistic missile penetration has been the basis for use as protective gloves, clothing, and helmets. Its good thermooxidative and dimensional stability affords utility for high temperature applications such as asbestos replacement. It is also produced as fine, short fibrils or pulp which can be used to make papers. The tensile properties of Kevlar® 29 (a high strength fiber of somewhat higher toughness but lower modulus, especially useful for ropes and rubber reinforcement) and Kevlar® 49 (a stiffer but less tough version, having more highly developed crystallinity, for plastics reinforcement) are compared with those of high strength nylon in Table 1.

Table 1. Tensile Properties of Kevlar® Aramid Fibers

	Tenacity	Elongation	Modulus
Kevlar® 49	20 dN/tex (2800 MPa)	2.5%	860 dN/tex (124,000 MPa)
Kevlar® 29	20 dN/tex (2800 MPa)	4.0%	430 dN/tex (62,000 MPa)
66 Nylon	8.5 dN/tex (985 MPa)	18%	45 dN/tex (5,500 MPa)

In Figure 1 the specific stress-strain curves of Kevlar® fibers
is shown in relation to various other fibers used for reinforcement. The direct proportionality between stress and strain, and
absence of a discernible yield point is a consequence of high
macromolecular orientation. The contrast with the comparatively
poorly oriented all-meta isomer (Nomex® high temperature fiber)
is dramatic. Kevlar® is also distinguished from the other

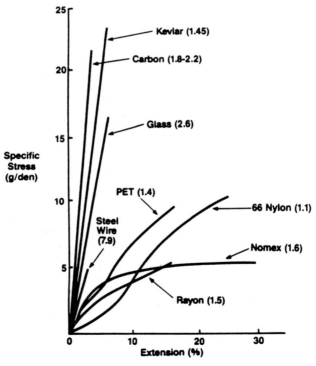

Figure 1. Stress-strain curves for reinforcing
fibers (density in g/cm³ in parentheses)

organic fibers of this group by its high thermooxidative stability; it does not soften or melt, and shows negligible pyrolytic weight loss below 500°C.

On an equal weight basis, Kevlar® is five times as strong as steel wire.

2 POLYMERIZATION

The preferred route to high molecular weight aramid polymers is usually the condensation of a diacyl chloride with a diamine, in an alkylamide solvent such as DMAc or NMP at temperatures in the range 0-25°C (3).

$$H_2N \cdot Ar \cdot NH_2 + ClOC \cdot Ar' \cdot COCl + 2B \rightarrow \{HN \cdot Ar \cdot NHOC \cdot Ar' \cdot CO\} + 2B \cdot HCl$$

The weak basicity of alkylamides enables the solvent to act also as an HCl acceptor and, because it is present in large excess, the condensation reaction is driven to completion, i.e., formation of high molecular weight polymer. Stronger HCl acceptors, such as tertiary amines, have been applied but the reaction is already very fast, so that they provide a definite advantage only if non-basic solvents are to be used. Low temperature solution polymerization has long been operated on a commercial scale by Du Pont for the production of poly(1,3-phenyleneisophthalamide) (MPD-I; the all-meta isomer of PPD-T) for the manufacture of Nomex® high temperature aramid fibers. In this process the solution, containing about 20% MPD-I in DMAc, is treated with a stoichiometric amount of CaO to neutralize byproduct HCl; the $CaCl_2$ so formed acts as a solubility enhancer and prevents precipitation of MPD-I. The neutral polymer solution is dry-spun, and the fiber drawn through hot water to simultaneously provide macromolecular orientation and extract $CaCl_2$ and residual solvent. After crystallization by high temperature drying, tensile strength and modulus are about 4 gpd and 100 gpd, respectively.

The relatively inflexible, extended-chain character of PPD-T drastically diminishes solubility in organic media compared with MPD-I, with profound consequences for its polymerizability. When terephthaloyl chloride and p-phenylenediamine are combined in the usual way in a solvent such as DMAc or NMP, the polymer precipitates at low molecular weight (D.P. about 20), whereupon chain growth ceases. Certain solvent combinations, notably NMP:HMPA (2:1) (HMPA is hexamethylphosphoramide) (4) were found to provide superior solvent power such that rather higher molecular weights could be attained before solubility was lost, in which case the stiffness of the macromolecules prevents rearrangement to form a crystalline precipitate, and the combination of polymer and solvent forms a gel. Within the gel matrix there is sufficient mobility to enable chain growth to continue up to the molecular weight levels necessary for optimum fiber properties. Polymerization is expedited by applying agitation, such as extruder mixing, to the gel; this tends to break it up into a crumb-like consistency. It was later found possible to operate a commercial polymerization process with HMPA alone, rather than a mixed solvent. Later it was determined that HMPA is carcinogenic towards small mammals so that, as a possible human carcinogen, it became necessary to replace it by a safer solvent.

Extrusion of the gelled polymerizate into a coagulant obviously does not provide a worthwhile polymer product directly. For high strength fibers it is necessary to isolate the polymer by precipitation in water, and redissolve the dried material in a second solvent which can provide fluidity necessary for spinning. No suitable organic solvent system is known. However, PPD-T is soluble in 100% sulfuric acid up to a level of 20% (w/w) and though this is a solid at room temperature, it has satisfactory fluidity above 70°C to make possible spinning to high strength fiber.

Since a single-solvent process would be economically more desirable than a two-solvent process for PPD-T fibers, considerable effort has been directed towards achieving this, but without real success. Sulfuric acid is not a suitable polymerization solvent because molecular weights are low and some degree of ring sulfonation is difficult to avoid. PPD-T can be prepared in the absence of a solvent either (a) by reaction of p-phenylenediamine with terephthaloyl chloride in the vapor state (5) or (b) by reaction of p-phenylenediamine with diphenyl terephthalate in the solid state for brief periods at about 500°C (6); although high molecular weights were obtained in both cases, it was not possible to diminish a concurrent chain branching side-reaction to a level where it would not have deleterious consequences for solution viscosity and spinnability.

Table 2 shows some of the macromolecular characteristics of PPD-T as used for the production of Kevlar®, along with comparative data for 66 nylon.

Table 2. Macromolecular Properties of PPD-T (Kevlar®)

	PPD-T	66 Nylon
$\overline{M}w$	50,000	40,000
$\overline{M}n$	20,000	20,000
Dispersity	2.5	2.0
Degree of Polymerization	84	93
Chain Length (nm)	108	150
Inherent Viscosity (H_2SO_4)	6.0	–

3 PPD-T SOLUTIONS

The conformation and organization of the relatively stiff, extended PPD-T macromolecules in solution dominates PPD-T technology. Consequences with regard to high polymer formation have already been discussed. The viscosity of sulfuric acid spin dopes can undergo profound changes with increasing concentration and is very sensitive to shear rate. Most important of all, the orientation and conformation of the solute molecules in the extrudate from the spinneret is captured and reflected in the coagulated fiber and thenceforward determines fibers property levels.

Unlike conventional fibers, molecular orientation of PPD-T in Kevlar® cannot be altered by high temperature drawing, and the only change possible in the as-spun fiber structure is further crystal growth under tension at high temperature which is manifested by an increase in modulus but not in tenacity (Table 1).

In PPD-T the amide bridges act in rigid, four-atom, planar units because the N(H)-C(O) bond has considerable double bond character and therefore presents a very high barrier (~20 Kcal/mole) to rotation.

Rotation about the Aryl-N(H) and Aryl-C(O) bonds is, on the other hand, far less restricted; the energy barrier between stable conformations is less than 2 Kcal/ mole. Fig. 2 shows the energy barriers for mutual rotation, Θ, between the two amide groups in the model compound N,N'-dimethylterephthalamide (7).

Fig. 2 also indicates that the stabilizing energy for the most stable form versus the least stable form is small.

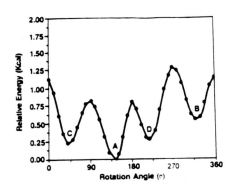

A - Most Stable Form

B - Least Stable Form

C,D - Intermediate Forms

<u>Figure 2</u>.   Rotational energy barriers for Aryl-C(O) bonds in N,N'-dimethylterephthalamide

Nevertheless, although energy differences between different conformations in PPD-T in dilute solution are undoubtedly also small, there is a high preference for the most stable form, which coincides with the most extended, most linear conformation (Fig. 3).   This conformation is in dynamic equilibrium with higher energy less linear forms.   The effect of introducing a higher energy repeat unit conformation into an otherwise lowest energy polymer chain segment is dramatized in Fig. 3.

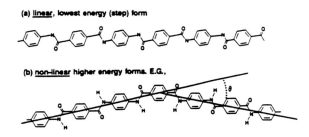

(a) <u>linear</u>, lowest energy (step) form

(b) <u>non-linear</u> higher energy forms. E.G.,

<u>Figure 3</u>.   PPD-T conformations in dilute solution

The macromolecular shape may be likened to a mildly curving worm.   A measure of molecular straightness is persistence length as shown for PPD-T and contrasting polymers in Table 3.

Table 3. Conformation in Dilute Solution

Polymer	Persistence Length	% of Length of Macromolecules	Comment
PPD-T	24 nm	20%	Relatively extended and stiff.
MPD-I	2.0 nm	2	Meta-isomer of PPD-T Flexible, coiled.
66 Nylon	0.54 nm	<1	Highly flexible coil.
PBZT*	--	Approaches 100%	Rigid rod.

*Polybenzthiazole

The macromolecular shape may be likened to a mildly curving worm. A measure of molecular straightness is persistence length as shown for PPD-T and contrasting polymers in Table 3.

The relatively stiff, extended nature of PPD-T macromolecules is reflected in a variety of polymer properties; in Table 4 comparisons are made with the inherently far more flexible and coiled meta-isomer of PPD-T (MPD-I). Segmental movement in PPD-T is severely restricted so that thermal transitions, including decompositions, occur at higher temperatures.

Liquid Crystallinity

The ability of certain elongated organic molecules to organize themselves in solution (lyotropicity) or in the melt (thermotropic) in such a way that they exhibit the optical properties of a crystal while retaining liquid-like fluidity has been known for many years. Two common modes of organization

Table 4. Polymer Properties Dominated by Chain Stiffness

	PPD-T	MPD-I (meta isomer of PPD-T)
Solubility (organics)	No fluid solutions	Soluble
Glass Transition	425°C	265°C
Crystallite Melting	550°C (est.)	415°C (with decomp.)
Decomposition (TGA)		
− Initial	' 500°C	410°C
− Fast	550°C	450°C
Crystallizability	Very High	Moderate
Lyotropicity	Very High	Absent

Nematic — One-dimensional order
— Low viscosity

Smectic — Two-dimensional order
— Very viscous

Figure 4. Polymer liquid crystal phases

(Fig. 4) are the nematic state (in which the molecules lie parallel to one another but with no particular registry in the transverse direction) and the smectic state (molecules again aligned but the ends are in register).

In 1955 Flory predicted that rigid, rod-like polymer molecules, in solution above certain critical concentrations, would spontaneously organize as a liquid crystalline phase for maximization of packing efficiency. At that time, no suitable soluble rod-like polymers were known. In 1965 at Du Pont it was observed (8) that poly(1,4-benzamide), although hardly a rigid, rod-like polymer as envisaged by Flory, dissolved in amide-salt solvents such as DMAc/LiCl to form lyotropic solutions. Like PPD-T this has a substantial persistence length and may be similarly described as an extended, low-amplitude worm in dilute solution. As shown in Fig. 5, above a certain critical composition the polymer chains reorganize as domains or groups of mutually parallel, fully extended macromolecules with directionality varying randomly from domain to domain.

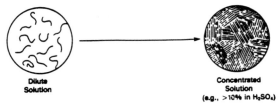

Dilute
Solution

Concentrated
Solution
(e.g., >10% in H₂SO₄)

Figure 5. Reorganization of Solution
Structure with Increased
Concentration

The same behavior is perceived for PPD-T in sulfuric acid; above
a critical concentration of about 12%, a lyotropic phase appears.
It is not known whether the macromolecules progressively extend
more completely as concentration is increased, below the critical
concentration, or whether macromolecular straightening occurs
simultaneously with the formation of the liquid crystalline
phase.

When lyotropic solutions, e.g., of PPD-T in sulfuric acid,
are subjected to shear, as in extrusion through a spinneret, or
to elongational flow, as during spin-stretching in the air-gap
just ahead of the coagulant, the domains and their constituent
PPD-T macromolecules all become oriented in the flow direction
(Fig. 6). Domains merge and lose their individual identities.
Coagulation of PPD-T solutions in this oriented condition is the
basis of high macromolecular alignment in as-spun fibers. If the
shearing or extensional force is removed from an oriented PPD-T
solution, it slowly relaxes to the original, randomly-oriented
domain structure. Domain structure in PPD-T/H_2SO_4 solutions may
be observed directly in a polarizing microscope with the specimen
between crossed polarizers. A bright field with a myriad of
colors is indicative of diffraction and depolarization of light
by domains randomly oriented in three dimensions. With an isotropic solution under the same conditions the field of view is
dark. When the specimen is subjected to shear by moving the
microscope slides in opposite directions, the orientational
effect on the domain structure becomes readily visible.

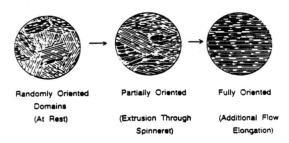

Figure 6. Alignment and Coalescence
of Domains

Another demonstration of the alignment of domains is the
beautiful pearly or opalescent appearance of such solutions when
stirred; at rest they are slightly cloudy. Elongation of
domains, of suitable dimensions, by stirring causes light to be
preferentially reflected and refracted in a common direction.

The transition from isotropic to lyotropic phase in PPD-T/H_2SO_4 solutions is accompanied by a profound change in solution viscosity (Fig. 7). In the isotropic region viscosity builds up rapidly as concentration increases as might be expected for stiff, extended macromolecules. At the critical concentration, lyotropic phase starts to appear and there is a decrease in viscosity as the proportion of lyotropic phase builds up. Ultimately viscosity passes through a minimum where almost all solution is lyotropic phase and beyond this point viscosity increases again. The buildup of viscosity in isotropic phases may be considered to be largely due to chain entanglements providing resistance to flow. This is represented for a random coil polymer in Fig. 8.

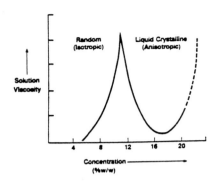

Figure 7.   Viscosity of PPD-T Solutions

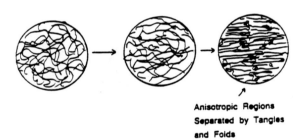

Anisotropic Regions
Separated by Tangles
and Folds

Figure 8.   Drawing of Conventional, Flexible Chain Polymers

In liquid crystalline phases, entanglements are almost absent so that there is much less resistance to flow. This viscosity change with entry into the lyotropic phase has the great practical consequence of making possible polymer spin dopes of much higher concentration, and therefore far better economics, than might otherwise be possible in the absence of liquid crystalline phases.

The creation of high macromolecular alignment in the process of spinning lyotropic PPD-T solutions (9) is represented in Fig. 9. When the lyotropic solution experiences shear in extrusion through a spinneret hole, there is substantial orientation of macromolecules in the flow direction. If the extrudate is not subjected to any extensional force, viscoelasticity will cause a bulge in the emergent extrudate which acts to disorient the macromolecules. Disorientation will be increased further down the threadline by natural relaxation forces. The threadline under free fall conditions moves more slowly than the liquid flow within the spinneret hole. When the extrudate

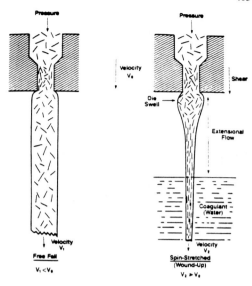

Figure 9.   Development of Orientation in Fiber Spinning

is subjected to extensional flow (as caused by winding of a coagulated fiber with positive stretching of the fluid in the air gap region), although disorientation due to bulging is not prevented, orientation is recovered and even enhanced by extensional flow forces; these obviously inhibit any relaxational deorientation. The result is that a high degree of orientation is captured by coagulation. For good orientation the spin stretch need not be higher than a factor of about three, because domains are readily oriented, although for optimal properties higher stretching is preferred. In ordinary wet spinning, where the spinneret is immersed in the coagulant, coagulation begins in the bulge region so that subsequent draw-down could not possibly achieve as complete orientation as with the air gap.

The air gap serves two additional very important functions. It allows spin dope at a considerably higher temperature, e.g., 80°C, to be extruded into a cold coagulant. Difficulties are obvious if a hot spinneret has to be immersed in a cold coagulant. Additionally, it facilitates a higher output per spinning position whereby a larger amount of dope can be extruded through each hole, in a given period, if the final low filament denier can be achieved by draw-down and faster windup. Experience has shown that fiber homogeneity and property levels are optimal if the fiber gauge or denier is kept to a low level of about 2 dpf.

There is a limit to the amount of draw-down and windup rate, at a point where ultimately fiber tensile properties diminish to an unacceptable level.

For formation of liquid crystal solutions the concentration of sulfuric acid is quite critical. As the phase diagram (Fig. 10) shows, isotropic solutions only are formed with sulfuric acid at ambient temperature. Polymer solubility and lyotropicity increase markedly as sulfuric acid approaches 100%. Beyond the 100% level the solutions contain free SO_3 and there is a tendency for partial ring sulfonation of the polymer. Such a sulfonated product, unless neutralized, has diminished chemical stability because the acid groups autocatalyze the fission of amide groups in various chemical environments.

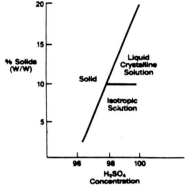

Figure 10. Criticality of 100% Sulfuric Acid as Spinning Solvent

Solutions of PPD-T in 100% sulfuric acid in the 12-20% (w/w) concentration range are actually crystalline solid complexes at ambient temperatures. It has been estimated that 20% solids approaches a pure complex containing quite few, free sulfuric acid molecules. In a sense, a spin dope of PPD-T in sulfuric acid at the preferred concentration of about 20% (w/w) may be regarded as a molten complex rather than a true solution. The relationship between solution melting point and concentration is shown in Fig. 11. Chemical stability of PPD-T in H_2SO_4 is satisfactory in the preferred range of ~80°C but in the region of a ~120°C degradation, as signified by M.W. loss, is unacceptable.

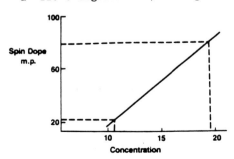

Figure 11. Relationship of m.p. to concentration for PPD-T (D.P. ~100) in 100% sulfuric acid

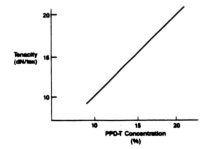

Figure 12. Relationship of fiber tensile strength to dope concentration

Spin dope concentration has a significant influence on the tenacity of as-spun PPD-T, as shown in Fig. 12. This is un-doubtedly associated with the fact that in the lyotropic concen-tration range there is a progressive increase in the content of liquid crystalline phase and this no doubt is associated with the degree of orientation in the as-spun fiber.

4 FIBER STRUCTURE

PPD-T macromolecules are generally well oriented in the axial direction and essentially fully extended. There is substantial crystallinity, even in the as-spun fiber, wherein the unit cell is monoclinic (pseudo orthorhombic) (Fig. 13). Crystallites are about 5 nm and 7 nm, respectively, across in the as-spun and heat-treated fibers.

Within the crystalline regions the consecutive amide groups may be regarded as coplanar with the intervening p-phenylene rings alternately tilted at about $\pm20°$ to the amide groups. The amide groups can form hydrogen bonds with these in adjacent macromolecules to form hydrogen bonded sheets as shown in Fig. 14. The arrangement of crystallites within the fiber is such that hydrogen-bonded planes are radial with respect to the fiber axis. The radial crystallite arrangement has been demonstrated by Hagege et al. (10) with Ag_2S-stained fiber specimens via dark field transmission electron microscopy. Meridional and equator-ial lattice pages can actually be observed in high reduction micrographs of PPD-T fibers, fragmented by ultrasonic radiation, by Dobb et al. (11).

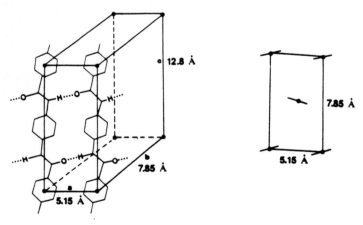

Figure 13. Unit cell of PPD-T

Superimposed on the radial crystallite arrangement in the fiber is a pleat morphology (Fig. 15)(12). The pleats have a periodicity of about 250 nm and a pleat angle of 170° in the as-spun fiber. Dobb et al. (13) demonstrated the existence of pleats by dark field electron microscopy, using longitudinal fiber sections. Pleats are the consequence of slight shrinkage compression in the interior of the fiber during drying and consolidation. The pleat structure is diminished by heat treatment, especially with superheated steam, under tension to increase crystalline size and this is reflected in higher initial modulus. In tensile testing of as-spun PPD-T fibers, the initial modulus is depressed to an extent by the presence of the pleats but as stress is applied progressively, the pleats are pulled out, modulus increases (stress-strain curve is slightly concave upwards), and tenacity at break reflects the material with pleats removed. Pleating also accounts for the higher elongation to break and toughness of as-spun, compared to heat-treated fibers.

PPD-T, in common with other very highly oriented organic fibers, characteristically fibrillates under excessive stress. Thus Hagege et al. (14) demonstrated that a surface layer on the fiber is easily peeled away to reveal an underlying fibrillar structure. Fibrillation may also be easily demonstrated by excessively twisting PPD-T fibers (14) or by simply breaking them under tension. It is not certain whether the appearance of fibrils as described is the result of fibrillar separation along lines of weakness, inherent to the fiber structure from the method of spinning, or whether the well-oriented polymer cleaves easily along normal crystal planes. Etching experiments, e.g., by HCl vapors (15) suggest that flowed regions may separate incipient fibrillar regions. In any case, the ease with which the fibers fibrillate illustrates the inherently weak transverse strength whereby macromolecules are held together merely by Van der Waals forces with some hydrogen bonding.

Hydrogen Bonded Sheet   **Sheets Stack Together**

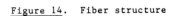

Figure 14.   Fiber structure   Figure 15.   Pleated
                                                 sheets

Associated with low transverse strength of PPD-T fibers is low compressive strength, e.g., at least six times lower than carbon fibers. Under compressive forces, which cannot be relieved by fiber bonding faults or kink bands are propagated diagonally across the fiber. This has been well described and illustrated by Dobb et al. (16). Thus Kevlar® is not suitable for structures

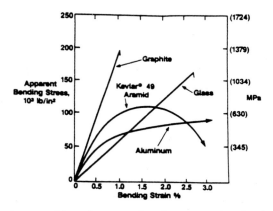

Figure 16. Compressional strengths (unidirectional composite bending stress/strain curves [in epoxy resins])

subjected to compression in use, although it provides excellent tensile reinforcement. It is possible, in some cases, to provide composite of carbon and Kevlar® fibers where carbon meets compressional forces at one location and Kevlar® meets extensional forces at another. The outstanding advantage of Kevlar® as a plastics reinforcement is that, unlike carbon or glass, it does not undergo catastrophic failure as illustrated in Fig. 16. Even heavily kink-banded Kevlar® retains considerable tensile strength.

5 MISCELLANEOUS FIBER PROPERTIES

The excellent retention of strength of Kevlar® fibers after exposure to elevated temperatures up to at least 200°C, is illustrated in Fig. 17. The breaking strength of Kevlar® fibers, measured at elevated temperatures, is shown in Fig. 18.

While most organic solvents have no effect on Kevlar® fibers, it is attacked by strong acids or bases at high temperatures, at high concentrations or elevated temperatures. Kevlar® is also affected deleteriously by ultraviolet light but the affected surfaces are self-screening so that thick fabrics do not lose much strength, although thin fabrics may lose 50% of their strength on exposure to Florida sunlight for 5 weeks. Kevlar®-reinforced composites are, of course, unaffected.

Dimensional stability is excellent. Kevlar® shows negligible dimensional change in dry air at 160°C, in boiling water, or at high relative humidity, even though moisture pickup may be as high as 7.5%.

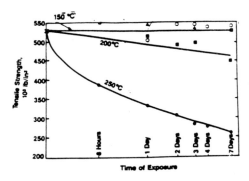

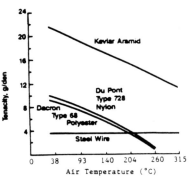

Fig. 17. Effect of elevated temperature on the tensile strength of
Kevlar® 49/epoxy

Fig. 18. Breaking tenacity
of industrial filament
yarns in air at elevated
temperature

The tensile properties of Kevlar® are compared with competitive materials as shown in Fig. 19.

The ultimate strength of PPD-T has been predicted to be 120 gpd, with a modulus of 1700 gpd, for perfectly aligned macromolecules of infinite molecular weight, in a flow-free structure. This is, of course, not a realistic goal. At the molecular weight levels of commercial Kevlar® predicted tenacity is 60 gpd and modulus is 1500 gpd. Current limitations of strength in commercial Kevlar® to less than 50% of the theoretical is attributed to imperfect alignment (including pleat structure) and flaws. In modulus Kevlar® approaches the theoretical values more closely. The theoretical limitation is inherent to the actual chemical structure which is not as stiff as certain other rod-like, fused ring structures such as polybenzthiazoles.

6   SCALE-UP TO A MARKET DEVELOPMENT PLANT

The discovery of air-gap spinning of liquid crystalline PPD-T/sulfuric acid complexes was an invention worthy of scaleup. The product was certainly unique, the process looked scalable and the economics looked satisfactory. The next step in the process was to scale-up to a market development plant.

This became a particularly challenging task. It involved moving the development from the laboratory to a plant site. At this point it was particularly important to coalesce a multidisciplinary team.

Translation of a laboratory discovery to a practical, scalable, commercializable process is one of the hardest tasks faced by a technology-driven industry. In the case of Kevlar®, a task force of dozens of scientists and engineers of many disciplines was assembled. The task was to develop the manufacturing basic data and tackle scale-up. In less than two years from the laboratory discoveries of Blades, Kevlar® was being shipped from a 1 million pound per year market development plant. This was a timetable, considering the complexity of the process, that was unprecedented in the Du Pont Co. Fibers Department. Five years later commercial product was being produced in a 15 million pound per year market development plant.

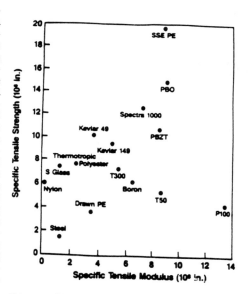

Figure 19. Plot of specific strength vs. specific modulus for various reinforcing fibers

During the scale-up stage several difficult hurdles were encountered, some quite unusual and unexpected. A complex laboratory process had to be translated into a workable plant operation. The ingredients were corrosive, there were environmental concerns such as waste disposal and the polymerization solvent had an unexpected level of toxicity. Two of these hurdles are described below.

An unusual environmental problem arose involving waste disposal. The spinning solution is based on a complex containing five moles of sulfuric acid per PPD-T amide bond, or 4 pounds of sulfuric acid per pound of polymer. The problem was to dispose of the spent acid after spinning. The best option turned out to be conversion of the sulfuric acid into calcium sulfate (gypsum). For every pound of fiber seven pounds of gypsum is generated. Each year the Kevlar® plant generates a lot of gypsum – quite a storage problem! As a matter of interest, a crystalline, pure gypsum is produced that is attractive to both wallboard and cement manufacturers, and eventually the gypsum will be recycled.

The second hurdle deals with toxic materials. In the Kevlar® process, two solvents are used. The spinning solvent is sulfuric acid, discussed above, and the polymerization solvent was hexamethylphosphoramide, or HMPA (III).

$$(CH_3)_2N - \overset{\overset{\displaystyle O}{\|}}{\underset{\underset{\displaystyle N(CH_3)_2}{|}}{P}} - N(CH_3)_2$$

(III)

$$\begin{array}{c} H_2C - CH_2 \\ H_2C \qquad C=O \\ \diagdown N \diagup \\ CH_3 \end{array}$$

(IV)

The Kevlar® business was in the scale-up stage on a fast track heading toward full commercialization. Suddenly it came to a self-imposed barrier - potential toxicity of the HMPA polymerization solvent.

We all pay much attention to toxicity and handling of hazardous materials. Use of HMPA on a large scale was new, but the material was well known for many years and there were no highly unusual toxic effects reported. To be completely sure about HMPA toxicity, a lifetime exposure study with rats was initiated by Du Pont to determine any possible carcinogenicity potential [17]. This study was one of the first of its kind, and done purely as a precautionary measure. The results of that study showed that HMPA was an experimental animal carcinogen. Immediate steps were taken in the handling of HMPA to be certain that there was no hazard to the workers, the community or the customers.

However, it is clearly preferable not to use a potentially hazardous material if a safer alternative can be found. Therefore, a crash technical program was mounted to find an HMPA replacement. The task was to find an acceptable polymerization solvent of low toxicity; one that would give polymer yielding fiber properties identical to those already introduced to the trade, as well as fitting into the process and the expensive equipment layout designed for polymerization in HMPA. The chemistry turned out to be relatively straightforward [18]. The combination of N-methyl pyrrolidone (IV) and calcium chloride was selected as the solvent of choice. The engineering and product proveout were formidable challenges, however.

Polymerization in NMP/CaCl$_2$ proceeded smoothly. High inherent viscosity polymer was obtained with no difficulty. However, the polymer could not be spun to the equivalent tenacity of the same inherent viscosity polymer prepared in HMPA.

The problem turned out to be molecular weight distribution. The data showed that at equal inherent viscosity, the number average molecular weight of polymer made in the NMP/ CaCl$_2$ system was lower than in the HMPA system. Using a gel permeation chromatography technique developed for this program, this difference was traced to the presence of a large low molecular weight fraction in the NMP/CaCl₂ system (Fig. 20). Further analysis showed that this resulted from precipitation of oligomers from the polymerization mixture. These were not present in the HMPA system because of it's superior solvent power.

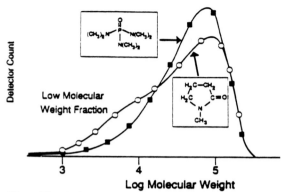

Fig. 20. Gel Permeation Elution Curves for PPD-T Prepared in HMPA and in HMP/CaCl₂

The problem was solved through design of a reactor system to eliminate early nucleation and precipitation of low molecular weight polymer. This is another example where solving a problem required multi-disciplinary chemical and engineering skills. Extensive internal and trade testing clearly demonstrated that the Kevlar® fibers made in the two polymerization systems were indistinguishable. The program for an alternate solvent took an estimated 40 man-years of technical effort.

7   FULL COMMERCIALIZATION

Thus far, the development of a sound scientific understanding of this new product, and the new processes to make it, and a few of the process development hurdles in scale-up to a market development plant have been described. Perhaps an even greater challenge was to demonstrate the market potential of Kevlar®. This was necessary to justify the final step in the innovation process - a full scale commercial plant requiring a $400,000,000 investment. Hence, throughout the development there was intensive parallel effort to find practical applications for this new fiber. For Kevlar® to be commercial, sufficient value in use had to be found versus incumbent fibers like nylon, steel, fiberglass and carbon to warrant a pricing structure that made economic sense.

8   SYSTEMS APPROACH TO APPLICATIONS

Early in the development, it was recognized that Kevlar® was a unique fiber that wouldn't automatically fit into existing applications. This became apparent from initial evaluation in tires, ballistics, composites, ropes, cables, etc. Each application had to be looked at as a "system" requiring a systems approach. Early partnerships with customers were vital to success.

It was necessary to creatively integrate base technologies of polymer chemistry, fiber science, engineering and application development to give an "advanced structure". The process for doing this, or "systems" engineering, has been vital to Kevlar® product development. It requires the combined talents of professionals in many disciplines. A final advanced structure usually involves much iteration between each base technology and sometimes multiple products organized into an integrated total system.

To illustrate the systems approach, two examples have been selected: Kevlar® in tires and in ropes and cables.

The key Kevlar® properties of high tensile strength and modulus quickly suggested potential application as a reinforcement for rubber in radial tires.

In the belt of a radial tire, modulus, or stiffness, is particularly important for tread life, cornering stability, and handling characteristics. In the carcass, or sidewall, strength is important for overall tire durability. Fiber characteristics beyond modulus and strength which are important include tensile and compressive fatigue resistance.

The major technical challenge in order to succeed as the radial tire belt reinforcement was to develop a "systems" technology which minimized fiber strength loss from compressive fatigue.

The forces on a tire have been mathematically modelled by both Du Pont and tire manufacturers. These studies show the edge of the normal cut belt, as highlighted in the circle in Figure

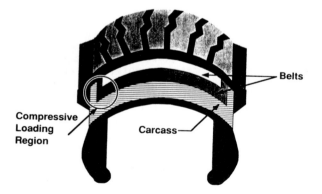

Fig. 21.   Radial tire cut-away showing compressive
           loading

21, can undergo severe compressive cycling during cornering. Du Pont experiments had shown that the major mechanism of strength loss for Kevlar® in the belt of a tire will be compressive fatigue, unless special care is taken in the design.

Figure 22 shows the effect of repeated compressional loading on the tensile strength of Kevlar®. Single filaments were subjected to repeated compressive strains of 1.0%, 0.8%, and 0.5%, and then residual strength was measured after various numbers of cycles. No strength loss was seen after several hundred thousand cycles at 0.5% strain, but at higher strain levels, strength loss began to appear at some threshold and increased thereafter with continued cycling.

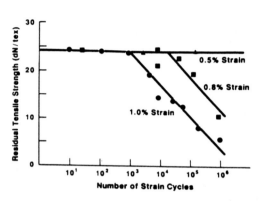

Fig. 22. Repeated Compression Loading of Kevlar® Filaments

The reason for the compressive failure we hypothesize is due to the molecular structure of PPD-T. A compressive stress results in formation of structural defects called kink bands and eventual brittle failure.

At a compressive strain of about 0.5%, a yield is observed. It is quite possible that this corresponds to a compressive buckling of the PPD-T molecules by molecular rotation around the amide carbon-nitrogen bond. This can occur by a transition from the normal extended trans configuration to a kinked cis configuration. This results in a bend in the chain which propagates across the unit cell, the micro-structure, and finally results in a kink band across the entire fiber.

It was necessary, therefore, to design a reinforcement system to keep compressive strains under 0.5% in order to obtain satisfactory fatigue performance in the tire and to compete effectively with steel tire cords.

Two approaches to acceptable fatigue performance were taken.

The first was to optimize the cord structure used in the belt. Tire cords are normally made by twisting individual yarns, then combining these yarns by further twisting into a cord. It

is well known that com-
pressive fatigue resist-
ance is proportional to
twist level. At the same
time, however, cord ten-
sile strength and modulus
decrease very fast with
increasing twist (Figure
23).

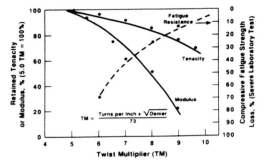

Therefore, the cord
was designed with a twist
level selected to achieve
the best balance of ten-
sile properties, on the
one hand, and fatigue re-
sistance, on the other.

Figure 23. Cord properties as a
function of twist level

The second approach was to redesign the belt
to place more cord in the critical belt edge region, which
reduced the compressive stress on individual filaments by
distributing the stress load over more filaments. This was done
by folding the bottom belt around the edge of the top belt, as
highlighted in the circle of Figure 24, which provided 50% more
fiber in the critical region than the normal cut belt.

The benefits of these design changes were confirmed in
a 38,000 kilometer fleet performance test. At a given twist
level, the folded belt is far superior to the normal cut belt,
and is also proportional to twist level, as shown in Figure 25.

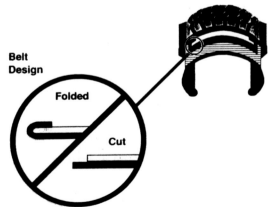

Fig. 24.   Cut-away of tires showing normal and
          folded belts

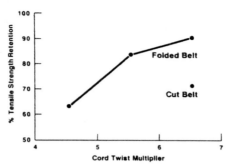

Figure 25. Fleet test results showing effect of cut vs. folded belts

These tests also convinced us that satisfactory performance
can be obtained with one kilogram of Kevlar® replacing five
kilograms of steel cord. Radial passenger car tires reinforced
with belts of Kevlar® are now available in over 30 tire brands
produced by 10 major tire companies.

In the next application, the main driving force for Kevlar®
in ropes and cables is the high strength per unit weight, inherent low elongation and low creep. In air, the specific
strength is seven times steel. In sea water, the specific
strength is more than 20 times steel (Figure 26). This means one
can use smaller, lighter more easily handled lines. In long
lengths where the self-weight of steel becomes critical, Kevlar®
can offer more payload.

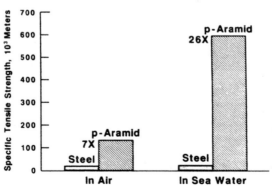

Figure 26. Specific tensile strength
of Kevlar® vs. steel in air and water

The particular application to be described is that of a line. The systems technology involved a specially engineered rope design to reduce internal stresses to increase wear life. Riser-tensioner lines are used on floating offshore oil drilling plat-forms (Fig. 27). The purpose of the riser-tensioner line is to keep the rise pipe or outer drill casing at a constant elevation and under uniform tension while the vessel surges with the waves. These are normally 44 mm diameter steel wire ropes which experience considerable cycling over pulleys as the platform moves.

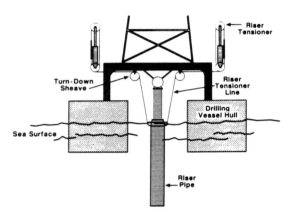

Figure 27. Deep water oil drilling platform with riser-tensioners

Laboratory studies had shown that small diameter ropes of Kevlar® made in low twist stranded wire rope constructions could far surpass steel in cycling performance over pulleys. However, scale-up of the best small constructions to 44mm diameter sur-prisingly gave rope lifetimes that were only 5-10% that of steel wire rope. Analysis showed that internal forces in the twisted rope rise rapidly with increasing diameter [19,20]. These internal loads stem from radial squeezing forces which increase rapidly with increasing twist levels (Fig. 28). This brings pressure against the pulleys and leads to high frictional heating, high internal abrasion, and shear fatigue failure of yarn as rope elements move.

Several design changes were made that improved the lifetime of ropes of Kevlar® over 50-fold as shown in Figure 29. On the left is the former construction, and on the right the redesigned rope. One change was to increase the number of strand sizes from one to three. The purpose was to minimize cross-overs of inner and outer strand layers by nesting the outer strands in the valleys

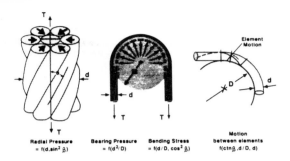

Figure 28.   Forces on a rope

of the inner layer. This compacted the structure and spread the
lateral loads uniformly over a greater area giving a five-fold
improvement in lifetime. A second change was to lubricate the
strands by jacketing each strand with a braid impregnated with
fluorocarbons. This reduced friction, heat buildup, abrasion and
internal shear stresses.   The result was a six-fold
improvement.   The third approach was to optimize twist helix
angle to minimize radial squeezing forces without seriously
affecting other rope properties.   This gave an additional
two-fold improvement. The result was a rope having more than
three times the life of steel in severe laboratory tests and more
than five times in service. More than a dozen oil rigs have
utilized riser-tensioner lines of Kevlar®, including those from
major drilling contractors including Santa Fe International,
Western Oceanic, Zapata and Sedco.

Limited space permits
description of only two
examples   of   Kevlar®
applications   research.
There   are   numerous
others.   In   the   mid-
seventies   Kevlar®   was
participating in only 10
market segments and less
than 50 specific appli-
cations.   Today, Kevlar®
is in more than 20 market
segments,   serving   more
than   200   applications.
Continued   growth   is
anticipated.   Kevlar® is

Former Construction
- Strands same size
- Unlubricated
- Layers at same helix angle
- High helix angle

Redesigned Rope
- Three strand sizes
- Lubricated (5.9x)
- Strands nested (5.x)
- Lower helix angle (1.7x)

Figure 29.   Rope constructions

produced in Richmond, VA, where we have a 45MM lb. plant; in
1988, a second plant was started up in Northern Ireland and plans
for a third plant in Japan have been announced.

9 SUMMARY

The Kevlar® story exemplifies the kind of basic product structure/property relationships, scale-up challenges, and market development research involved in bringing a laboratory discovery to commercial reality.

REFERENCES

1. Tanner, D., Fitzgerald, J. A., and Phillips, B. R., *Angew. Chem.* Adv. Mater. (1989) 101, 665.

2. Du Pont Fibers Technical Information Bulletins (a) "Properties of Industrial Filament Yarns of Kevlar® Aramid Fiber for Tires, Hose, and Rubberized Belting", (K4; August 1979); (b) "Characteristics and Uses of Kevlar® 49 Aramid High Modulus Organic Fiber" (K5; September 1981).

3. Morgan, P. W., "Condensation Polymers by Interfacial and Solution Methods", Interscience, New York (1965).

4. Bair, T. I., and Morgan, P. W., U.S. Patent 3,673,143, 1973.

5. Shin, H., U.S. Patent 4,009,153, 1977.

6. Fitzgerald, J. A., Irwin, R. S., Likhyani, K. K., and Memeger, W., Preprints, First Pacific Polymer Conference - December 12-15 (1989), p. 441-2.

7. Vorpagel, E., Du Pont, private communication, 1989.

8. Kwolek, S. L., U.S. Patents 3,600,350, 1971, and 3,671,542, 1972.

9. Blades, H., U.S. Patents 3,767,756, 1973, and 3,869,429, 1975.

10. Hagege, R., Jarrin, M., and Sotton, M., J. Microscopy, (1979) 115, 65-72.

11. Dobb, M. G., Johnson, D. J., and Saville, B. P., J. Poly. Sci., (1977), Poly. Symp. 58, 237-51.

12. Dobb, M. G., Johnson, D. J., and Saville, B. P., J. Poly. Sci. (1977), Phy. Ed., 15, 2201-2211.

13. Dobb, M. G., Johnson, D. J., and Saville, B. P., Phil. Trans. R. Soc. London (1980), A294, 483-5.

14. Jaquenart, J., and Hagege, R., Journal de Microscopic et de Spectroscopie Electroniques (1980), 3, 427-438.

15. Li, L. S., Allard, L. F., and Bigelow, W. C., J. Macromol. Sci. - Physics (1983) B22(2), 269-90.

16. Dobb, M. G., Johnson, D. J., and Saville, B. P., Polymer, (1981), 22 (7), 960.

17. K. P. Lee and H. J. Trochimowicz, "Pulmonary Response to Inhaled Hexamethylphosphoramide in Rats", J. Toxicol. & Appl. Pharm. (1982), 62, 90-103.

18. W. B. Black and J. Preston, "Fiber-Forming Aromatic Polyamides" in Man-Made Fibers, (H. Mark, editor), pp 297-301, Interscience Publishers, New York (1968) Vol 2.

19. P. T. Gibson, "Analytical and Experimental Investigation of Aircraft Arresting Gear Purchase Cable", NTIS Report AD 904263, Battelle Memorial Institute, Columbus, OH, July 3, 1967.

20. P. T. Gibson, "Continuation of Analytical and Experimental Investigation of Aircraft Arresting Gear Purchase Cable", NTIS Report AD 869092, Battelle Memorial Institute, Columbus, OH, April 8, 1969.

21. D. Tanner, J. A. Fitzgerald, and P. G. Riewald, Chapter 2, Handbook of Fiber Science and Technology: Vol. III. High Technology Fibers, Part B, ed. M. Lewis and J. Preston, Marcel Dekker, Inc., New York and Basel (1989).

Carbon Fibre Production

J. E. Wilson

COURTAULDS RESEARCH, P.O. BOX 111, LOCKHURST LANE, COVENTRY,
CV6 5RS

1 INTRODUCTION

Carbon Fibres have come a long way since Thomas Edison
first produced them by carbonisation of either viscose
rayon or cotton in the nineteenth century. These fibres
were used as filaments for early light bulbs but as they
were extremely brittle soon gave way to the advent of
tungsten. Tungsten offered enhanced mechanical properties
whilst maintaining the ability to conduct electrical
current and become incandescent.

Over the course of the 20th Century, major steps have
been made in the production of synthetic fibres, and in
1957, Courtaulds commenced production of its polyacrylonitrile (PAN) fibre "Courtelle". This material is used
extensively in a wide range of apparel end uses and provides the chemical basis for production of carbon fibres
by the polyacrylonitrile route.

In the early 60's, developments at Farnborough and
in Japan led to initial production of carbon fibre from
polyacrylonitrile as we now know it. A few years later,
a Parliamentary Committee recommended that there should
be a U.K. manufacturer of carbon fibre, and in 1968,
Courtaulds began work on carbon fibre under licence from
the NRDC.

Initially, carbon fibre was made by a batch process,
starting with a spool of polyacrylonitrile fibre. This
precursor fibre was wound onto a frame, put into an oven
and oxidised in air for about five hours at 225°C. Oxidised fibre was then cut from the frame into one metre
lengths and placed into a furnace containing an inert

nitrogen atmosphere, at over 1000°C. Finally, the fibre had a surface treatment to improve resin adhesion before yielding the final carbon product in metre long lengths. Obviously, at this time, this process had significant scope for development.

2 PRODUCTION ROUTES

Apart from a small amount of carbon fibre produced from viscose rayon for ablative end uses, where mechanical properties are not critical, the majority of carbon fibre is currently produced either from pitch based precursors or polyacrylonitirile.

In this review, I will concentrate on the process for producing carbon fibre from polyacrylonitrile. The production of carbon fibre by this route involves three basic stages;

1. Production of the polymer.
2. Spinning the polymer solution (or dope) into precursor fibre.
3. Conversion of precursor to carbon fibre.

The production of polymer is carried out by polymerisation of a monomer mix of over 90% w/w acrylonitrile. In practice, most acrylic and precursor manufacturers also incorporate one or two other comonomers such as methyl acrylate and methacrylic acid. These are added to overcome solubility problems associated with the homopolymer and to impart enough plasticity to enable the polymer to be stretched in the precursor production stage. Overall, the monomers are polymerised to give a product with the following basic structure at a molecular weight of about 100,000.

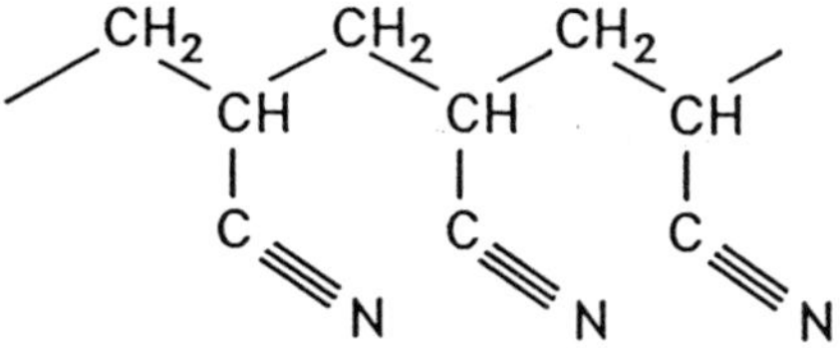

Scheme 1   Polyacrylonitrile Polymer

One typical polymerisation route is to utilise a free radical process starting mainly with acrylonitrile and involving the well known steps of initiation, propagation and termination. Under the action of heat, the initiator splits to form a free radical. This then reacts with the acrylonitrile monomer to form a larger radical which can then react with another monomer molecule, and so on, with the build up of a growing polymer chain. Termination can occur by reaction of two of the free radicals. However, in practice, on a production basis, most of the chain termination is controlled by addition of a chain transfer agent (CTA) to the reagents mix. Compounds with readily abstractable hydrogen atoms are most suited to this role. In the industrial process, the key requirements are to control the molecular weight of the polymer and the % conversion of monomer to polymer.

The polymerisation process is a continuous one based on the preparation of a reagents mix from four ingredients; the initiator, the various monomers (in the correct ratio), an aqueous solvent and finally a liquid product. The resultant product is then continuously metered forward to a stirred reactor vessel, where under the action of heat, reaction takes place to form the polymer. The % conversion of monomers to polymer is controlled by the level of initiator, the pump rate and the temperature of the reactor. The unreacted monomers are removed in a vacuum demonomerisation process before the product is passed forward for fibre spinning. The final product is a viscous solution of the polymer and is referred to as "spinning dope".

Having produced the polymer/dope, the next stage of the process is to spin it into a precursor fibre. After filtration and deaereation, the first stage of the process involves the coagulation of filaments by extrusion of the dope through fine holes into an aqueous spin bath. The resulting filaments are very weak and are then treated to a series of stretching processes. This aligns the polymer molecules within the filaments and considerably increases the fibre strength. The fibre is washed to remove residual solvent, before being treated with a lubricating size, dried, and collected.

The resultant product is available in 12,000, 6,000 and 3,000 (12, 6 & 3k) filament tow sizes with individual filaments being about 12 μ in diameter.

The third and final stage of the process involves processing the PAN precursor to carbon fibre. The process is a continuous on-line one with the first stage involving

passing the PAN precursor as a warp sheet into an air oxidation oven maintained at a temperature of 200-300°C. The
chemistry which takes place at this stage is complex[1] and
can be summarised as follows;

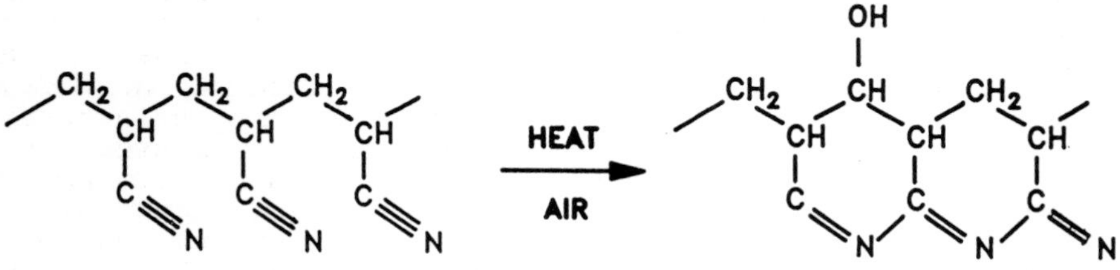

Scheme 2   Cyclysation/Stabilisation of Pan Product

From here, the resultant stabilised product is passed
forward to a pre-carbonisation furnace where, at a temperature of 500-900°C, under an inert gas, the fibre experiences a 30% weight loss with the evolution of considerable
amounts of gas and tar. The material is then passed forward to the carbonisation furnace where, in an inert atmosphere, at a temperature of 1300-3000°C, the final structure
and properties are formed.

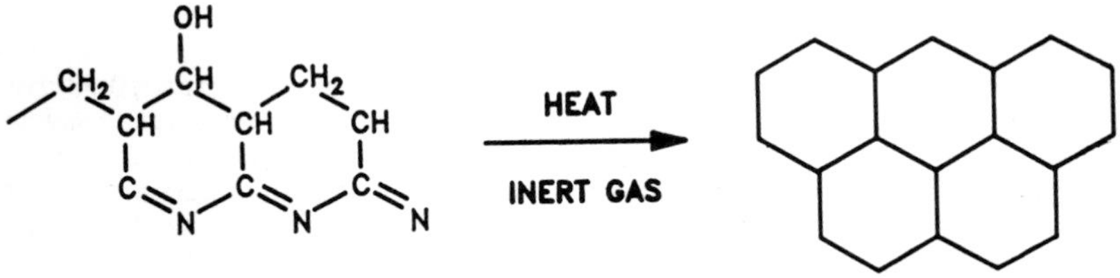

Scheme 3   Carbonisation of Oxidised Product

The next stage is to give the fibre an oxidative
surface treatment which is vital to achieving good interfacial bonding with resins in composite. It is then given
a surface finish/size application before finally being
collected. Over the process as a whole, there is about a
50% weight loss with the final carbon fibre product (12k,
6k or 3k) having an individual filament diameter of about
7μ.

The oxidation stage is the rate limiting step and has
a crucial bearing on the quality and properties of the
final product. In order to achieve the density necessary
for satisfactory carbonisation, a laboratory treatment time
of over three hours would be required. Commercial oxidation is now carried out at much shorter times. However,
on a production basis, a balance has to be achieved between
maximising temperatures, to increase speed/productivity,
and running the risk of the PAN fibre bursting into flames.

As I have already shown, the oxidation process is
quite complex involving a cyclisation reaction, together
with a diffusion controlled penetration of oxygen into the
body of the filaments. With a poor polymer composition,
the wrong temperature conditions, or variable/large diameter precursor filaments, it is relatively easy to generate
unevenly oxidised products which will produce very inferior
carbon fibres.

Carbon fibres offer very high strengths and moduli,
but with the added attraction of low densities. As a
result, carbon fibre/resin composites are finding increased
usage as substitutes for metals across a broad range of
areas including aerospace, industrial, and sporting goods
applications. Courtaulds Grafil essentially produce a
range of three products; a high strain, an intermediate
and a high modulus product, with properties ranging from
580 ksi (4.0 GPa) strength at 33.5 msi (230 GPa) modulus
to 380 ksi (2.6 GPa) at 51 msi (350 GPa).

The properties of the final carbon product are determined by a number of factors. The base orientation of the
polymer chains, incorporation of defects, and ultimate
firing temperature, all have significant impact on fibre
performance. Over the years, since carbon fibre was developed, strength and modulus values have increased significantly as the production process has been refined. In
particular, considerable emphasis has been placed on the
removal of defects from the fibre structure.[2,3,4,5]
Such defects can take the form of particulates or voids,
both of which can act as stress concentrators, reducing
fibre strengths.

3 APPLICATIONS

In order to realise the high strength/stiffness of advanced
fibre materials, incorporation into a matrix is necessary.
The matrix material supports the reinforcement transferring
loads into the fibres, which are oriented in such a way as
to make good use of their specific properties. In general,

thermosetting epoxy resins are used in carbon composites which are typically produced to a 60:40 fibre to resin ratio. Increasingly, however, thermoplastics are used as matrix materials. Carbon itself can also be used as a matrix material in carbon-carbon components.

A significant number of processes have developed to incorporate advanced fibres into matrix materials in the correct orientation. In many ways, it is these processes that control the growth of the carbon fibre market. In order to fully exploit the fibre properties, processes such as "Prepregging" and "Filament Winding" are necessary.

In Prepreg manufacture, carbon fibre tows are aligned and placed adjacent to each other before being impregnated with a resin matrix. A schematic of the hot melt prepreg process is shown in Figure 1. Impregnated fibres are sandwiched between release papers before being consolidated between compaction rollers and collected.

This prepreg product can then be used as a supply of oriented resin coated fibre which is then laminated and cured to form a composite product.

Filament Winding (Figure 2), as the name suggests, is a process that involves winding resin impregnated fibres onto a mandrel in order to form a composite shape. The technique has developed to employ a large number of basic winding patterns in order to make best use of the fibre reinforcement. Advanced techniques can also be used to make quite complex shapes. Filament Winding is used extensively to produce a wide range of products such as gas cylinder canisters and drive shafts.

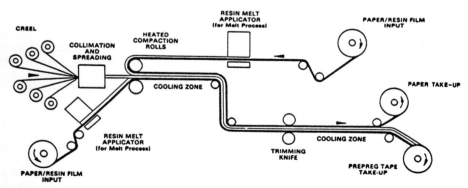

<u>Figure 1</u>   Prepreg Manufacture - Melt/Film Process

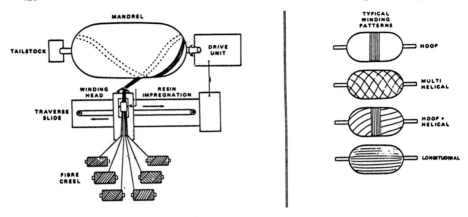

<u>Figure 2</u>   Filament Winding

Significant enhancement of properties can be achieved
however by far simpler routes such as Injection Moulding
which uses discontinuous fibres in a thermoplastic matrix.

The applications of carbon fibre reinforced composites
have become increasingly diverse in line with the range
of processing technologies. Initially, aerospace designers
were the main customers prepared to pay the prices required
(£150/kg) for carbon based products. More recently, prices
for standard products have reduced considerably ($\approx$ £20/kg)
due to improved process efficiencies which have opened up
completely new markets and applications.

In general, it is the strength and stiffness of carbon
fibres in both tension and compression that justifies its
use in a wide range of lightweight structures. Increasingly
however, other unique properties determine the choice of
reinforcing materials. Low coefficients of thermal expansion, electrical conductivity, good vibrational damping or
biocompatibility are just some of the properties which may
justify selection for a composite product. Of course, a
much wider range of reinforcing fibres are now available
to the composite designer. Hybrid systems, utilising
mixtures of reinforcing fibres, are used extensively.
These combine the advantages of different fibre types and
result in tailored products for specific market requirements.

The choice of a wide range of reinforcing materials,
coupled with a large number of processing routes and matrix
types result in something of a double edged sword. On one

hand, the facility exists to design a composite to meet some very specific requirements, whilst on the other, the increased choice results in a highly complicated picture when compared with conventional isotropic materials. Undoubtedly, the difficulties involved in designing with composite materials have restricted the rate at which this market has developed. As a greater number of products and processes have emerged to utilise the properties of these materials, confidence has grown, and completely new market areas have appeared. This trend is expected to continue with composites in general and carbon fibre in particular.

4 SUMMARY

In summary, carbon fibres have developed significantly since Edison first used them for light bulb filaments. Technical properties have increased by 200% over the last 20 years and an increase of approximately 3500% has been made in fibre sales. Process economies have resulted in significant reductions in cost for these materials, opening up an increased range of commercial processing routes and applications. Predictions for the future vary enormously depending to a great extent on the ability to continue to reduce the costs of the carbon fibre process, whilst improving the materials' properties.

REFERENCES

1. Jean-Baptiste Donnet and Roop Chand Bansal, 'Carbon Fibres', Marcel Dekker Inc., 1984, Pgs 14-23.
2. R. Moreton, Fibre Science and Technology, 1968, 1, 273.
3. J.W. Johnson, Applied Polymer Symposium, 1969, 9, 229.
4. J.W. Johnson and D.J. Thorne, Carbon, 1969, 7, 659.
5. R. Moreton and W. Watt, Nature, 1974, 247, 360.

Main Chain Thermotropic Liquid Crystal Polymers

W. A. MacDonald

ICI WILTON MATERIALS RESEARCH CENTRE, P.O. BOX 90,
MIDDLESBOROUGH

1 INTRODUCTION

Low molecular weight liquid crystalline compounds have
been known for about 100 years but main chain thermotropic
liquid crystal polymers have attained prominence only in
the last 15 years. The first well characterised
description of a polymer exhibiting thermotropic behaviour
appeared in the mid 1970's when Jackson described a series
of copolymers prepared by the acidolysis of poly(ethylene
terephthalate) with p-acetoxybenzoic acid which exhibited
the phenomenon of opaque melts, low melt viscosities and
anisotropic properties[1]. Since then there has been
considerable research in both industry and academia and a
recent survey of the liquid crystal polymer (LCP) field
estimated that 40 companies are known to be active in LCP
research internationally with further work being carried
out in smaller specialised companies[2].

2 DESIGN OF THERMOTROPIC LIQUID CRYSTAL POLYMERS

Completely rigid rod-like molecules such as
poly(4-oxybenzoyl) or poly (p-phenylene terephthalate)
tend to be highly crystalline and intractable with melting
points above the decomposition temperature of the polymers
(>450°C). The problem of thermotropic LCP design is to
disrupt the regularity of the intractable para-linked
aromatic polymers to the point at which mesomorphic
behaviour is manifest below the decomposition temperature
and the materials can be processed in fluid yet ordered
states. The disruption must not be taken to the stage
however, where conventional isotropic fluid behaviour is
preferred. These requirements that the

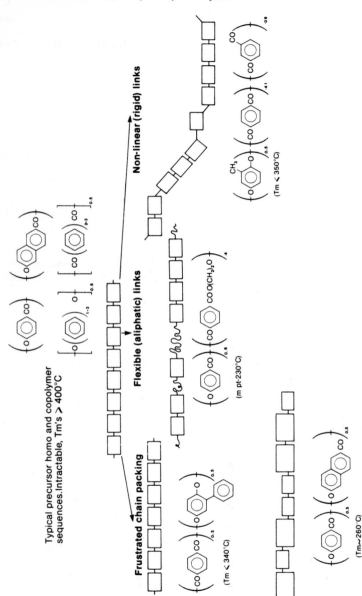

Figure 1 Control of Thermotropy in polyarylates

polymer must retain some rod like nature but at the same time be melt processable below 400-450°C have limited thermotropic LCP's mainly to polymers based on the linear ester or ester/amide bonds.   With polyester/polyesteramides, disruption is normally achieved by the three copolymerisation techniques outlined in Fig 1[3] ie
(i) frustrated chain packing which refers to the mechanism which whilst maintaining essential linearity and chain stiffness makes close and regular correlation into a 3D lattice difficult eg the incorporation of naphthalene, biphenyl or substituted phenyl moieties into the backbone.
(ii) Flexible spacers where aliphatic spacer units are incorporated.
(iii) Non linear links where meta or ortho substituted phenyls or groups incorporating 'kinked' bonds within the monomer units are included in the polymer backbone.

Fig 2[4] illustrates the typical monomer unit types that are used to lower Tm and Table 1 shows a range of typical thermotropic LCP's developed by industry which illustrate the various strategies for lowering Tm.

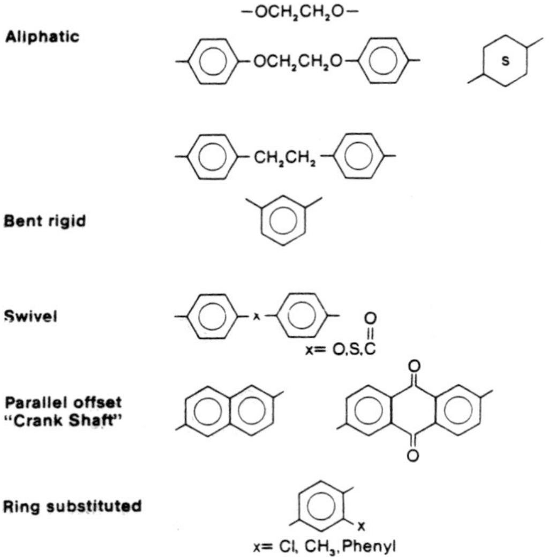

Figure 2 Strategies for lowering T_m in LCPs.

Celanese — Ref 5

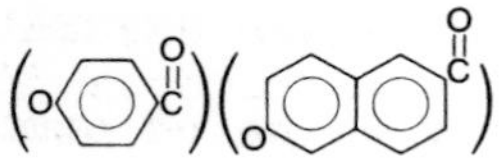

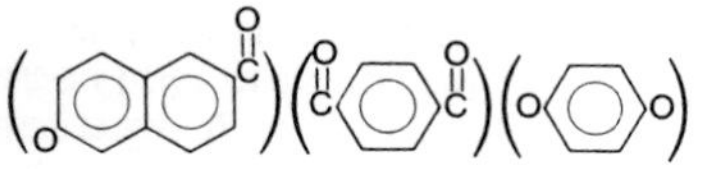

6

7

Dupont — 8 , 9 , 10

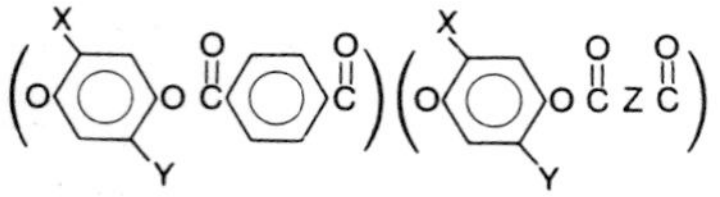

X and Y = -H, -Cl, -Br, -CH₃

Z = ...

Dupont — 11

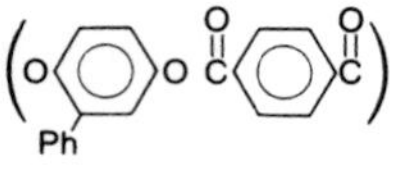

Owens Corning (Montedison) — 12

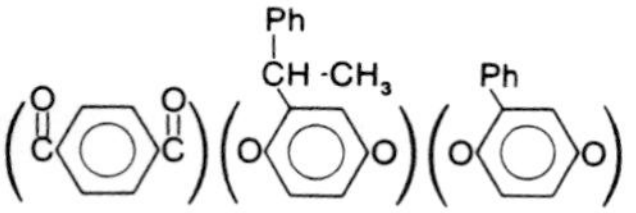

Eastman Kodak — 13 , 14

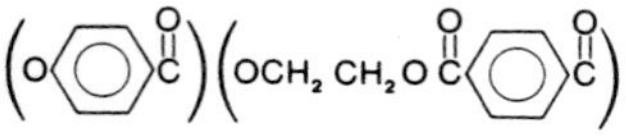

Carborundum (Amoco) — 15

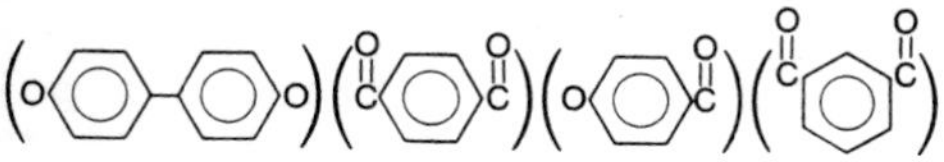

ICI — 16

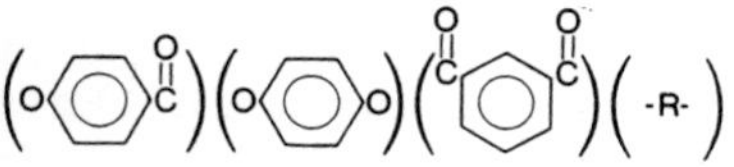

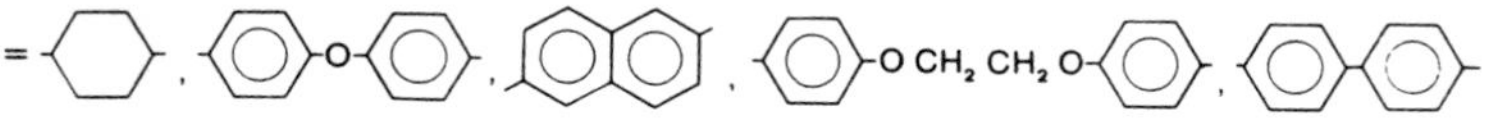

R - aromatic diol, carboxylic acid, hydroxy acid (less than 2%)

Table 1   LCP's developed by industry illustrating the various stategies for controlling Tm

3 SYNTHESIS OF THERMOTROPIC LIQUID CRYSTAL POLYMERS

The majority of LCP's, especially those of commercial significance are prepared by an ester exchange reaction between acetoxyaryl groups and the carboxylic acid group with the elimination of acetic acid, at temperatures above the Tm of the polymer produced. A typical melt acidolysis reaction is outlined in Fig 3[4].

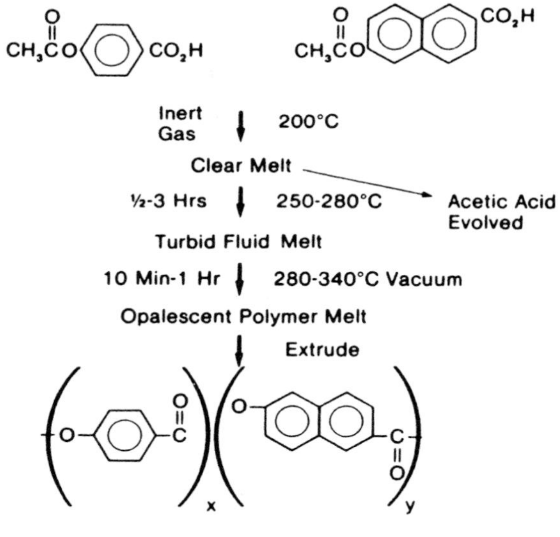

Figure 3   Typical LCP melt polymerisation synthetic scheme.

However this polymerisation technique is limited by the viscosity of the melt and the upper limit therefore for the molecular weight of LCP's prepared by this route is determined by the ability to extrude the polymer from the melt autoclave. This limit becomes more severe as the value of Tm for the polymer rises over 300°C. One route taken to get around this problem has been to prepare a low molecular weight prepolymer by this route and then to polymerise this prepolymer in the solid state[1]. More recently a novel high temperature non aqueous dispersion polymerisation route to LCP's has been developed at ICI[17]. Polymerisation is carried out as outlined in Fig 4 in an inert heat transfer medium such as liquid paraffin or Santotherm 66[R] in the presence of polymeric and or hydrophobic inorganic stabilisers.

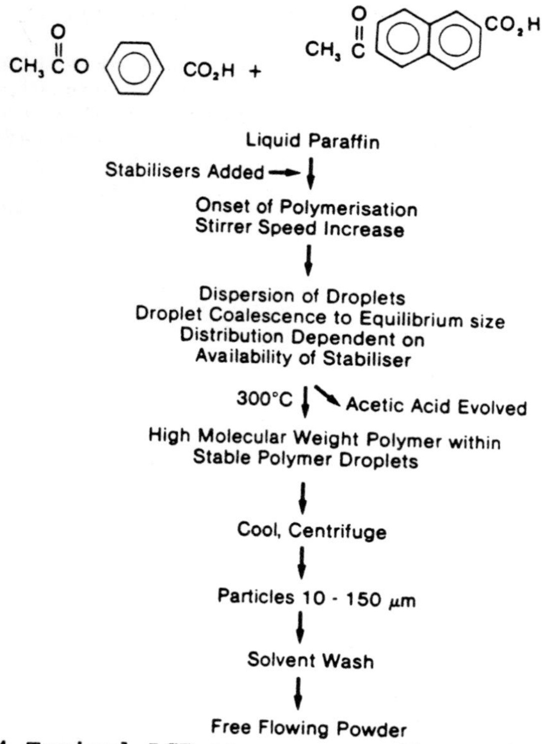

Liquid Paraffin

Stabilisers Added →

Onset of Polymerisation
Stirrer Speed Increase

Dispersion of Droplets
Droplet Coalescence to Equilibrium size
Distribution Dependent on
Availability of Stabiliser

300°C — Acetic Acid Evolved

High Molecular Weight Polymer within
Stable Polymer Droplets

Cool, Centrifuge

Particles 10 - 150 μm

Solvent Wash

Free Flowing Powder

Figure 4 Typical LCP non aqueous dispersion
polymerisation synthetic scheme.

These stabilisers sit at the surface of the droplet
and prevent flocculation occurring during the
polymerisation process. Unlike the melt polymerisation
process the polymerisation can be carried out at
relatively low polymerisation temperatures, in some cases
below the Tm of the polymer and no vacuum is required
because of the efficient removal of the acetic acid
by-product from the small polymer droplets. Furthermore
the polymerisation process is no longer viscosity limited
and very high molecular weight and/or intractable polymers
can be prepared by this route. The LCP's prepared were
found to have significantly improved mechanical
properties, particularly with respect to tensile strength
and impact strength when compared to analogues prepared by
the traditional melt acidolysis route (unpublished work
from this laboratory).

4 SOLUTION AND MELT CHARACTERISATION

LCP's are insoluble in common laboratory solvents and characterisation in solution requires the use of relatively exotic solvents or mixtures of solvents. This has restricted analysis of the solution properties especially with the fully aromatic LCP's.

MLCP melts are characterised by being opaque and exhibiting the characteristic shear whitening effect associated with anisotropic melts. Under cross polars the melts exhibit birefringence and the typical threaded schlieren textures associated with domain texture in areas of high local orientation. This domain texture is equivalent to the textures seen with small molecule nematic liquid crystals. Combining hot stage microscopy and DSC analysis allows fairly complete phase diagrams to be constructed for polymers of the type shown below which exhibit an isotropic and liquid crystalline phase as shown in Fig 5[18].

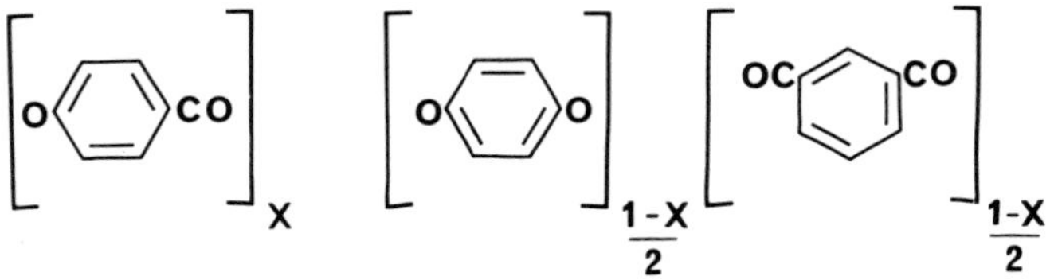

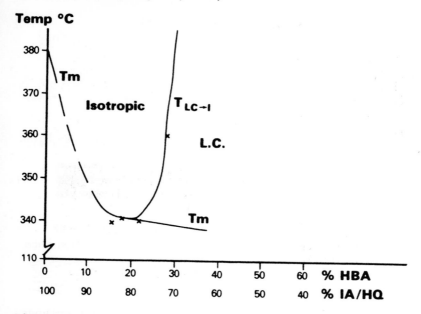

Figure 5 Transition temperature versus composition for LCP's based on HBA/IA/HQ

5 MORPHOLOGY AND STRUCTURE

 The hierachy of structure in LCP's can be divided into three main categories as illustrated in Fig 6 ie macro, micro, and molecular morphology.

 Macromorpholgy is taken to include the structure of fabricated articles and the macroscopic morphology of injection moulded bars consists of three macrolayers -two highly orientated skin layers with a less well ordered core in between [19,20]. The highly ordered skin, result from elongational/extensional flow giving rise to high levels of molecular orientation in the direction of flow at the surface. The core on the other hand is subjected only to shear flow which produces significantly less molecular orientation which may be perpendicular to the flow direction representing localised flow patterns. The micromorphology ie the domain and banded texture of LCP's is probably the least well understood of the hierarchy of structure and will not be discussed in this chapter. The molecular morphology ie the way LCP chains pack and crystallise and the conformational changes that the

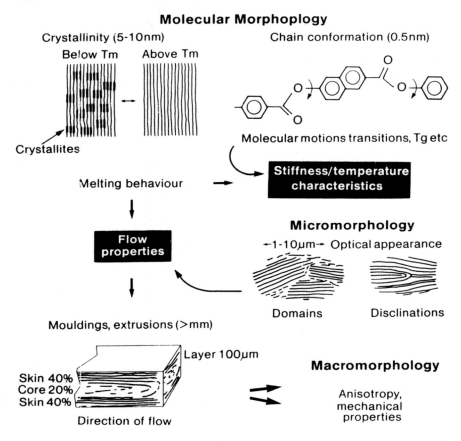

Figure 6   Hierarchy of structures in LCP's

polymer backbone undergoes as the temperature is raised is the underlying key to the behaviour of LCP's and dictates to a large extent the macro-and micromorphological behaviour of LCP's.

The nature of crystallites in solidified rigid chain LCP's of the type shown below was examined by Blundell.

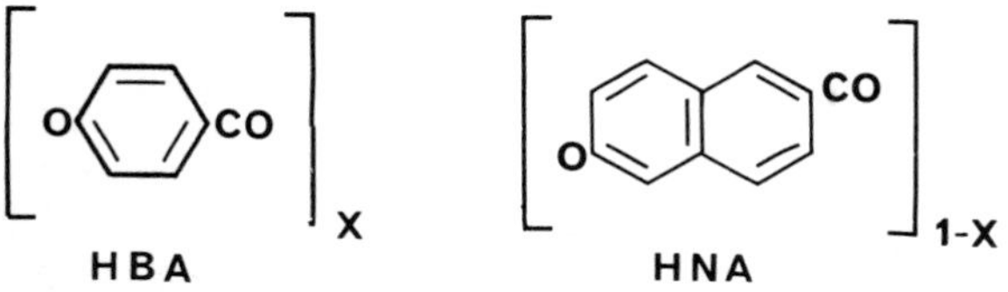

HBA   HNA   II

X-ray diffraction and thermal analysis showed[21] that identical heats of fusion were observed for slow cooled specimens with well developed crystals and for quenched specimens with almost amorphous X-ray patterns. This could only be explained by the presence of microcrystals that were much smaller than those possible in conventional polymers. If the surface energy per unit area were of the same value as that usually found in chain folded lamellar crystals then from Equation 1

$$\Delta h = \Delta h_\infty - \frac{A}{V} \cdot \gamma \qquad \text{Eqn 1}$$

where Δh_∞ = heat of fusion per unit volume of an infinite crystal; A = surface area surrounding regions of three dimensional crystals; X = surface energy per unit area; the term $\frac{A}{V} \cdot \gamma$, would be so large that the crystal would be unstable and would spontaneously melt. It was proposed by Blundell that the surface energy of LCP's was small and a direct consequence of the molecular morphology of the nematic LCP liquid crystal state. Unlike the vast disorder experienced when a conventional lamellar crystal melts as illustrated in Fig 7 it was envisaged that in a nematic LCP there was little change in the general configuration of the molecules before and after melting.

Below T_m Above T_m

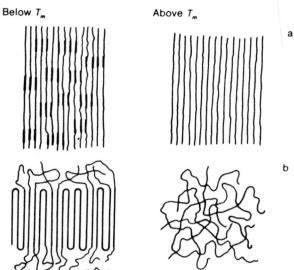

a

b

<u>Figure 7</u> Blundell's schematic diagram of the morphologies
above and below the crystal melting point for (a)
rigid chain nematic polymer, and (b) conventional
polymer with chain folded lamellar crystals. The
thicker parts of the lines represent regions
where chains form 3-D crystal lattices[69].

This has important implications for warpage and
shrinkage in moulding and will be discussed later.

It was also observed in the same paper that the heat
of fusion (ΔH_f) and entropy of fusion (ΔS_f) of the LCP's
studied were significantly lower than that of a
corresponding isotropic polymer such as polyethylene
terephthalate. This has important implications in the
design of LCP's. The low ΔH_f is directly related to the
lower level of molecular cohesion within the crystallites
which results from chain irregularities, whereas the low
ΔS_f is a direct consequence of the chain stiffness. Chain
irregularities and chain stiffness are two distinct
properties that can be separately designed into the LCP's.
In the polymers of type II it automatically follows that
there will be a low entropy change since there is little
or no change in overall configuration on crystallisation.
From Eqn 2.

$$Tm = \frac{\Delta Hf}{\Delta Sf} \qquad \text{Eqn 2}$$

given that ΔSf is low then the ΔHf must also be low otherwise Tm would be too high and the system intractable. Thus in order to make processable LCP's based on frustrated chain packing, irregularities must be introduced into the chain in order to limit the effective bonding of the crystals. This then gives rise to the disordered crystals of poor 3D order.

There has been considerable research directed at elucidating the nature of the crystallites in LCP's of the above type in particularly from the laboratories of Windle and Blackwell and the author recommends that interested parties investigate the references provided [21-28].

The situation where the LCP is based on a kinked structure such as polymer I (see section 4) crystals with good 3D order are observed. The deduced values of ΔSf and ΔHf are considerably higher and are more akin to the values observed in conventional isotropic polymers such as PET[18] Table 2. This can be rationalised in terms of the meta links of the isophthalic acid giving rise to potential flexibility in the chain and hence enabling greater conformational freedom. If ΔSf is larger then from equation 2 an acceptable Tm can be achieved with crystals of higher order and better chain packing ie ΔHf can be larger.

Further evidence for the higher order in LCP's containing kinks comes from the effect of annealing on

<u>Table 2</u> Crystallinity Data of LCP's With and Without Kinks
Compared to Polyethylene terephthalate

| Composition (Annealed Mouldings) | DSC | | X-Ray Fractional Crystallinity | ΔH_f kJ/kg | ΔS_f kJ/kg/°C |
	Tm °C	Peak Area kJ/kg			
33.3 HBA/ 33.3IA/33.3 HQ	320	18.8	0.16	117	0.2
40 HBA/30 IA/30 HQ	320	12.7	0.11	115	0.2
73 HBA/27 HNA	290	–	0.25	20	0.04
PET	250	–	0.5	135	0.26

the density of LCP's based on frustrated chain packing
versus LCP's based on kinked structures. It has been
shown that LCP's based on polymer II show no change in
density on annealing despite the crystallinity
increasing[29]. Unpublished work from our laboratories
agrees with the above result, but shows however that an
LCP based on polymer I which contains isophthalic acid
does increase its density as the crystallinity increases
(as measured by wide angle X-ray diffraction) Table 3, ie
the LCP containing kinks is achieving close regular
packing in the crystalline regions and is behaving like a
conventional isotropic polymer.

    The implication of these observations is that the
polymer architecture has a significant effect on the
crystalline nature and this in turn has major implications
for bulk polymer properties such as warpage and shrinkage
(see later).

Table 3 Density Versus Crystallinity Data on Liquid Crystal Polymers Containing Kinks or Crankshaft Units

	Density g/cm^3	% Crystallinity
36HBA/32 IA/32 HQ Control	1.390	<5
Quenched	1.391	<5
Annealed 250°C for 24 hrs	1.422	12.6
73 HBA/27 HNA Control	1.401	15
Annealed 250°C for 24 hrs	1.401	25

The ordered regions in LCP's appear to play a similar role to crystals in conventional polymers ie they tie the molecules together and impart molecular rigidity. However the level of crystallinity in LCP's is low, often under 20% for an unannealed sample and the crystallites in LCP's only account for a minor proportion of the total material. The remaining material although conforming to a general nematic configuration possesses no three dimensional correlation between segments of adjacent molecules. The variation of stiffness with temperature in the region below the melting point of the crystallite depends very much on the molecular segment in the non crystalline region. Taking LCP's based on frustrated packing such as polymer II, the dynamical mechanical properties are typified by an appreciable fall off in modulus with temperature which is particularly significant over the temperature range 0-120°C[30,31,32]. Three separate molecular processes have been identified in LCP's of this type. The upper α process at about 110°C appears to be related to the Tg transition of a conventional polymer although the Tg is surprisingly low for a fully aromatic polyester. The prominent β process in the region of 50°C is associated with the motion of the naphthyl moiety and the shallow process at about -40°C is attributed to similar motions associated with the phenyl unit[4].

LCP's based on kinks rather than frustrated chain

packing do not exhibit the β loss process associated with
the naphthyl moieties with the result that LCP's based on
kinks eg polymer I (X = 0.36) have lost 50% of the 0°C DMA
modulus at 120°C, whereas the naphthalene containing LCP's
have lost 50% of the 0°C DMA modulus at 80°C[31]. Thus
again changes in the chemistry of the LCP backbone can
have significant effects on bulk mechanical properties,
this time with respect to modulus retention at high
temperatures. However although the LCP's containing kinks
have superior modulus retention with temperature compared
to the naphthalene containing LCP's they still have poor
modulus retention with temperature when compared to
isotropic aromatic polyester analogues (Fig 8[31].)

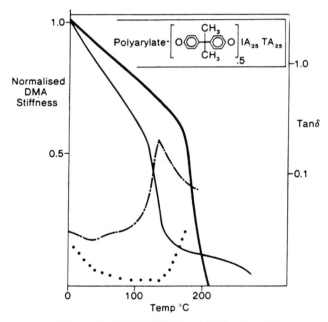

— Normalised DMA Stiffness of HBA$_{36}$ IA$_{32}$ HQ$_{32}$
--- Tanδ Loss process of HBA$_{36}$ IA$_{32}$ HQ$_{32}$
— Normalised DMA Stiffness of Polyarylate
••• Tanδ Loss process of polyarylate

<u>Figure 8</u> DMA of polymer I (X=0.36) compared to an
           isotropic aromatic polyester

    This rapid fall off in modulus in the 0-120°C
temperature range appears to be common to most LCP's.

The reason for this behaviour is not clearly understood but may be related to large range conformational motions occurring in the temperature regime 0-120°C arising from the rod like nature of the LCP's.

6 PROPERTIES OF THERMOTROPIC LIQUID CRYSTAL POLYMERS

6.1 Rheology

The rheology of LCP's is complex and will not be discussed in detail in this paper. In general, the shear viscosity of LCP's is much lower than that of conventional polymers at a comparable molecular weight and the transition from the isotropic state to the liquid crystalline state is generally accompanied by a significant decrease in melt viscosity. At the onset of nematic behaviour, the melt viscosity of the LCP's is three decades less order of magnitude than that of a similar but non mesogenic polymer[3].

Molecular orientation occurs readily during melt flow and it has been demonstrated that elongational/extensional flow is primarily responsible for orientation of LCP's during melt processing [20,33]. Elongational flow is introduced by drawing down during fibre spinning and is also present in the fountain flow at the advancing front of an injection moulding process. LCP's melts have very long relaxation times[34] retaining this orientation after flow has ceased and producing anisotropic articles[35].

From a commercial point of view the low melt viscosity of LCP's at high shear rates relative to conventional polymers is one of the key features [36] of LCP's and enables their use:
(a) to injection mould component with long or complex flow paths and thin sections.
(b) at very high filler loadings (up to 70% by weight)
(c) as a processing aid (bulk lubricant) for conventional thermoplastics (providing the processing temperatures of the two polymers overlap). In general the addition of 10% weight/weight of an LCP to a conventional thermoplastic approximately halves the melt viscosity of the polymer so that it becomes twice as easy to process[37]. This drop in viscosity is greater than that observed with isotropic immiscible blends and the reason for it occurring is not fully understood.

6.2 Mechanical

Mechanical properties, particularly tensile strengths and stiffness depend upon the degree of orientation achieved. This is limited to some extend by the fabrication method and type of article produced as shown schematically in Fig 9.

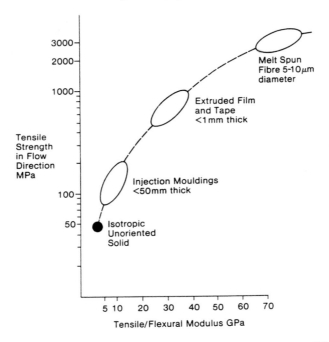

Figure 9 Schematic illustration of how tensile strength and stiffness depend upon the degree of orientation achieved.

Thus a compression moulded unoriented LCP has mechanical properties similar to that of a conventional isotropic polymer. On injection moulding, tensile bars of LCP's generally show superior mechanical moduli to that of conventional glass fibre reinforced isotropic thermoplastics with the highly oriented skin making the major contribution to the stiffness of the moulding[36]. As the level of elongational flow increases the mechanical properties increase to the very high mechanical moduli and tensile strengths demonstrated by LCP fibres[4,38].

As a simplification, the layered structure of an injection moulding can be considered to be microcomposites made up of layers in which the direction of reinforcement changes from layer to layer. The observed mechanical modulus of the overall moulding will reflect the integrated effect of the component layers. The important factors governing the modulus would be expected to be the thickness of the component layers and the direction and degree of orientation of the polymer chains within the layers. These factors will largely depend upon the rheological conditions and on the intrinsic response of the polymer chains to the flow conditions. Variations in the mechanical stiffness can also be achieved by altering the formulation. The phase diagram Fig 5 of the polymer series based on polymer I indicates a change from isotropic to liquid crystalline melt at around 20% HBA[18,39,40]. Studies on the flexural modulus of injection moulded tensile bars over the isotropic to liquid crystalline range 15%<HBA <36% showed a smooth systematic increase in modulus rather than a pronounced increase in modulus as the polymer formed an L.C. mesophase with increasing content of HBA (Fig 10)[18].

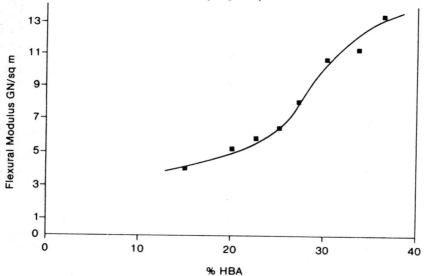

<u>Figure 10</u> Flexural modulus versus composition for the series of formulations based on polymer I.

Three selected examples of moulded bars from this series were examined by X-ray diffraction to quantify the chain orientation function as a function of depth of the mouldings [40]. Although skin core morphology was observed in these mouldings it was shown that the increase in modulus observed across the series was mainly a consequence of the increase in molecular orientation within the layers. The conclusion was drawn that in order to design a system with high self reinforcing modulus, it is not sufficient just to ensure that the polymer forms a liquid crystalline mesophase. It is also necessary that the polymer mesophase is capable of becoming highly oriented along the flow field so that there is a high global orientation especially within the critical skin regions ie the molecular linearity of the chain in addition to the rheological effects discussed in Section 6.1 can have a pronounced effect on bulk mechanical properties.

LCP's are tough materials and a benign failure is experienced on impact, similar to that exhibited by long fibre reinforced polymers or natural wood, ie the failure is neither ductile nor brittle and the mouldings generally do not shatter.

As discussed in Section 5 LCP's exhibit a significant fall off in modulus with temperature and although this can be influenced by control of the component parts of the LCP this is obviously undesirable in an engineering resin. However this is partly offset by the exceptionally high stiffness and strength of LCP's at room temperature and useful mechanical properties are retained at high temperatures (>200°C).

One of the major problem areas in injection moulding LCP's is their poor weld line strength because flow fronts do not knit easily together. Careful control over mould design to move weld lines into areas where they least effect properties is important. This is an inherent fault of LCP's and may prove to be a major 'Achilles Heel'.

Fabricated LCP articles are anisotropic and the anisotropy ratio ie the difference in properties along and across the flow direction increases with the degree of orientation and so is highest in fibres. Injection mouldings exhibit anisotropy ratios of between 4:1 to 10:1 depending upon mould thickness[41]. Anisotropy increases with decreasing thickness as the proportion of skin to core increases[1]. Interestingly, whereas the introduction of fillers tends to increase the anisotropy ratio of conventional isotropic thermoplastics, the introduction of fillers to LCP's disrupts the alignment of the LCP molecules and reduces the anisotropy ratio [41]. This has the benefit of improving the properties in the across flow direction.

LCP's absorb very low levels of moisture (typically less than 0.2% on immersion in water) and therefore the change in dimension of mouldings of the LCP's due to moisture absorption is very low. The coefficient of linear thermal expansion of LCP's is much lower than that for conventional polymers (even when glass fibre reinforced) and is comparable to those for metals as shown in Fig 11.

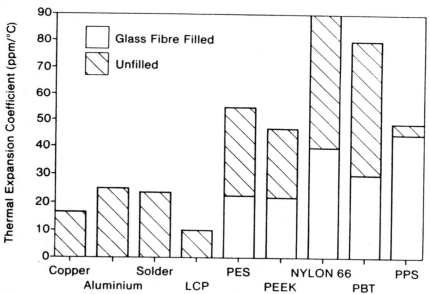

Figure 11 Linear thermal expansion coefficient of thermoplastics.

LCP's also exhibit very low mould shrinkage and minimal sinkage and warpage compared to conventional isotropic polymers permitting precision moulding of components. This is largely related to the unusual molecular morphology of LCP's where there is very little change in the general configuration of the molecules before and after melting. With isotropic crystalline thermoplastics on the other hand the formation of chain folded lamellae can give rise to warpage/shrinkage problems. However as mentioned in Section 5, as 'kinks'

gains more flexibility and crystals of relatively high
perfection are observed. This may have an adverse effect
on the low warpage/shrinkage associated with LCP's and is
the likely explanation for why glass filled mouldings of
polymer I (LCP-1 in Fig 12) exhibit higher degrees of
warpage than glass filled mouldings of polymer II (LCP-2)
N.B The warpage behaviour is still considerably better in
the case of polymer I than that of nylon 6/6 etc.

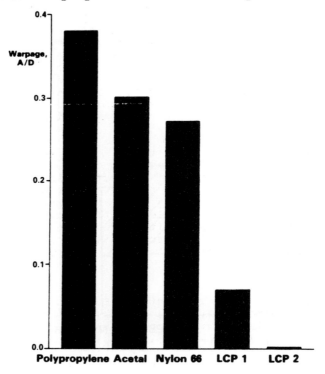

<u>Figure 12</u> Warpage of 30% glass fibre filled
           thermoplastics (100mm diam x1.6mm thick disc)

     Thus the presence of flexibility in the LCP backbone
in addition to affecting the molecular linearity may also
have a significant effect on warpage and shrinkage by
altering the crystallisation behaviour of LCP's. These
factors must be taken into account when designing LCP's
for high performance precision moulding applications.

6.3 Miscellaneous Properties

Gas transport studies on the LCP based on polymer II have shown that this polymer has excellent barrier properties and at 35°C the permeability coefficients for He, H_2, O_2, Ar, N_2 and CO_2 in this polymer are comparable to or smaller than those for polyacrylonitrile which is one of the least permeable polymers known [42]. This low permeability is believed to be due to the liquid crystalline order. Significantly, initial unpublished studies carried out in our laboratories on LCP's containing 'kinks' of the type illustrated by polymer I indicate that although the barrier properties are still good, they are not as good as the rod like polymers based on frustrated packing discussed above. This is presumably related to the ability of the polymer chains to pack, and again indicates that the inherent rod like nature of the polymer backbone can have a significant effect on bulk properties.

LCP's have excellent resistance to a wide range of organic solvents and exhibit very good hydrolysis resistance and the retention of properties in both acidic and basic environments is very good [4]. As with gas barrier properties it is likely that the liquid crystalline order is responsible for this excellent chemical resistance and again it is found that the chemical resistance of LCP's decrease if kinks are introduced into the backbone[4].

7 APPLICATIONS

The differences in behaviour between LCPs and conventional isotropic polymers and the unique set of properties resulting from the behaviour of LCP's are summarised in Fig 13 and the main application for LCP's will be in areas that exploit combinations of the key properties such as strength, easy flow, excellent dimensional stability, the ability to incorporate high levels of fillers and excellent chemical resistance. Examples of these can be seen from applications in the electronic industry such as surface mount units and printed wiring boards where the similarity in thermal expansion for metal and LCP's is expected to result in good component integrity and minimal strain when components containing metals (eg solder) and LCP's are in contact and are subject to thermal cycling or shock. In addition to this the easy flow and low warpage will allow precision moulding of complicated components and the ability to withstand strong solvents, increasingly used to clean between even smaller places, will be increasingly important.

Applications outside the electronic industry include LCP tower packing saddles for eg formic acid plants which replace ceramics.   Here, the better chemical resistance and breaking strengths are the critical properties.   It is interesting to note that the main application areas identified so far exploit the thermal and chemical properties rather than mechanical strength, the original property that LCP's was pushed for.

One of the major brakes on the LCP market expansion has been the relatively high cost of the polymers (currently >£12/kg) resulting from expensive feedstocks, and there is currently considerable research activity in industry aimed at lowering monomer costs.   The other major 'Achilles Heel' of LCP's is the weld line problem discussed earlier.

The LCP blends area is an area attracting great interest at present and there are three main themes to the research.
(1) The addition of a small amount of an LCP to another polymer to improve flow.
(2) Blending LCP's with cheaper polymers to give LCP properties at lower cost and also to overcome some of the anisotropy problems.   An example of this is an LCP/nylon blend which exhibits a very low coefficient of thermal expansion[43].
(3) Blending LCP's to high performance polymers to impart LCP properties to higher added value polymers.

   8 CONCLUSION

The key properties of LCP's outlined in Fig 13 can be directly related to the unique molecular morphology of LCP's which in turn controls the micro- and macromorphology.   The molecular morphology is in turn directly related to the polymer architecture and it has been shown in this paper that changing the polymer architecture can have a significant effect on the properties on fabricated articles.

The exceptional balance of properties with the ability to tailor the properties of LCP's for particular end uses is creating new opportunities for designers and material specifiers.   The current LCP world plant capacity is around 3500 metric tons and is dominated by Amoco and Hoescht Celanese.   It is predicted that this

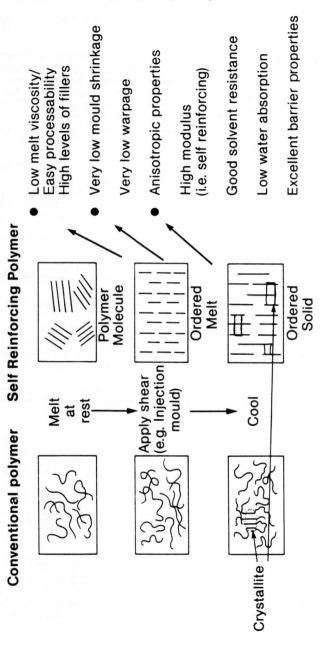

Figure 13 Differences in behaviour between LCP's and isotropic polymers and summary of the LCP key properties

will rise to over 10,000 metric tons in the early 90's.
However the estimated world sales of LCP for engineering
plastic type applications is only about 500 metric tons
and forecasts of growth depend heavily on price. For a
price greater than £12/kg a 20% per annum growth rate
increasing to 40% if the polymer price comes down to
£6.5/kg, as predicted[2]. Thus LCP's are slowly becoming
established in the market place, but although they offer
an exciting combination of properties the major
determining factor in the future success of these
materials will be success in bringing down LCP monomer
costs.

REFERENCES

1   Jackson, W.J. and Kuhfuss H.F., <u>J. Polym. Sci. Chem.
     Ed.</u>, 1976, <u>14</u>, 2043.

2   Technology Catalyst Inc, Liquid Crystal Polymer
     Technology Technical Review 1988.

3   Griffin, B.P. and Cox, M.K., <u>Brit. Polym. J.</u>, 1980,
     <u>12</u>, 147

4   Calundann, G.W. and Jaffe, M. 'Robert A Welch
     Conferences in Chemical Research Proc. Synth.
     Polymers' 1982, 247.

5   Calundann, G.W., U.S. Patent 4161470, 1979,
     (Celanese Co).

6   Calundann, G.W., U.S. Patent 4184996 1980 (Celanese
     Co).

7   Calundann, G.W., Charbonneau, L.F. and East, A.J.,
     US Patent. 4330457, 1982 (Celanese Corp).

8   Pletcher, T.C., U.S.Patent 3991013 and 3991014, 1976
     (E.I. DuPont de Nemours and Co).

9   Kleinschuster, J.J., Pletcher, T.C., Schaefgen, J.R.
     and Luise, R.R., Ger. Offen 2520819 and 2520820,
     1975 (E.I. DuPont de Nemours and Co).

10   Elliot, S.P., Ger Offen 2751585, 1978 (E.I. Dupont
     de Nemours and Co).

11   Payet, C.R., Ger. Offen, 2751653, 1978, (E.I.DuPont
     de Nemours and Co).

12 Boudreaux, E., Lee, D.M., Hutchings, D.A., Sieloff, G.M. and Willard, G.F., US Patent. 4668760, 1984 (Owens-Corning Fibreglass).

13 Kuhfuss, H.F. and Jackson, W.J., US Patent. 3778410, 1973 (Eastman Kodak Co).

14 Kuhfuss, H.F. and Jackson, W.J., US Patent. 3804805, 1974 (Eastman Kodak Co).

15 Cottis, S.G. Economy, J. and Nowak, B.E. U.S. Patent 3975486, 1973 (Carborundum Co).

16 MacDonald, W.A., and Ryan, T.G., Eur. Patent. 275164, 1988, (Imperial Chemical Industries PLC).

17 Carter, N., Griffin, B.P., MacDonald, W.A. and Ryan, T.G., Eur. Patent. 275163, 1988, (Imperial Chemical Industries PLC).

18 Blundell, D.J., MacDonald, W.A. and Chives, R.A., <u>High Perf. Polyms.</u>, 1989, <u>1</u>, 97.

19 Weng, T., Hiltner, A and Baer, E., J. Mat. Sci., 1986, <u>21</u>, 744.

20 Ide, Y. and Ophir, Z., <u>Polym. Engng. Sci</u>., 1983, <u>23</u>, 261.

21 Blundell, D.J., <u>Polymer</u>, 1982, <u>23</u>, 359.

22 Chivers, R.A., Blackwell, J., Gutierrez, G.A., Stomotoff, J.B. and Yoon, H., in 'Polymeric Liquid Crystals' Blumstein, A., (Ed.), Plenum Press, new York, 1985.

23 Mitchell, G.R., and Windle, A.H., <u>Colloid and Polym. Sci</u>., 1985, 263, 230.

24 Biswas, A. and Blackwell, J., <u>Macromolecules</u>, 1988, <u>21</u>, 3152.

25 Hanna, S. and Windle, A.H., <u>Polymer</u>, 1988, <u>29</u>, 223.

26 Windle, A.H., Viney, C., Golombok, R., Donald, A.M. and Mitchell, G.R., <u>Faraday Disc Chem. Soc</u>., 1985, 79.

27 Golombok, R., Hanna, S. and Windle, A.H., <u>Mol. Cryst. Lq. Cryst</u>, 1988, <u>155</u>, 281.

28   Spontak, R. and Windle A.H., work to be published.

29   Butzbach, G.D., Wendorff, J.H. ad Zimmermann, H.J.,
      Makromol. Chem. Rapid Commun., 1985, 6, 821.

30   Blundell, D.J. and Buckingham, K.A., Polymer, 1985,
      26, 1623.

31   MacDonald, W.A., Mol. Cryst. Liq. Cryst., 1987, 153,
      311.

32   Troughton, M.J., Unwin, A.P., Davies, G.R. and Ward
      I.M., Polymer, 1988, 29, 1389.

33   Viola, G.G., Baird, D.G. and Wilkes, G.L., Polymer
      Engng. Sci., 1985, 25, 888.

34   Ide, Y. and Chung, T.S., J. Macromol. Sci., 1984/85,
      B23, 497.

35   Huynh-ba, G., ACS Polymeric Mat. Sci. and Engng.,
      1984, 50, 141.

36   Cox, M.K., Mol. Cryst. Liq. Cryst., 1987, 153, 415.

37   Cogswell, F.N., Griffin, B.P. and Rose, J.B. Eur.
      Patent 30417 (1981).

38   Prevorsek, D.C., in 'Polymer Liquid Crystals',
      Ciferri, A., Krigbaum, W.R. and Meyer, R.B. (Eds),
      Academic Press, 1982.

39   Erdermir, A.B., Johnson, D.J. and Tomka, J.G.,
      Polymer, 1986, 27, 441.

40   Blundell, D.J., Chivers, R.A., Curson, A.D., Love,
      J.C. and MacDonald, W.A., Polymer, 1988, 29, 1459.

41   Naitove, M.N., Plast. Technol, 1985, 31, 35.

42   Chiou, J.S., and Paul, D.R., 'J. Polym. Sci. Part B
      Polym. Phys., 1987, 25, 1699.

43   Chung, T-S, Plastics Engineering, 1987, 43, No. 10,
      39.

Polyimides: Their Synthesis, Characterization, Properties, and Uses

J. N. Hay

KOBE STEEL EUROPE LTD., THE NODUS CENTRE, UNIVERSITY OF SURREY,
GUILDFORD GU2 5TB

1 INTRODUCTION

Polymers containing heterocyclic rings in the main chain are
becoming increasingly important industrially, due to the
attractive combination of properties conferred by the presence
of the heterocyclic units. In particular, such polymers
frequently exhibit excellent properties at elevated
temperatures, thus qualifying them for use in applications
experiencing high service temperatures. Such temperatures
are often encountered in applications as diverse as aero
engines and microelectronic devices, where the advantages
offered by these polymers enable them to command a relatively
high price.

Types of polyheterocycles attracting commercial attention
include polyphenylquinoxalines, polybenzimidazoles,
polybenzothiazoles and polyimides. By far the most important
of these are the polyimides (PIs). At present, only the
cyclic imide polymers are important in a commercial context
and, indeed, few acyclic polyimides are known. The generic
structure of the cyclic polyimides is shown in Figure 1.
Polyimides were first discovered in 1908 [1] but it was not
until the late 1950's that Du Pont in the USA made commercial
exploitation of these materials a reality. Since then, the
number of polyimides synthesised has proliferated, as has
the number of companies active in the field.

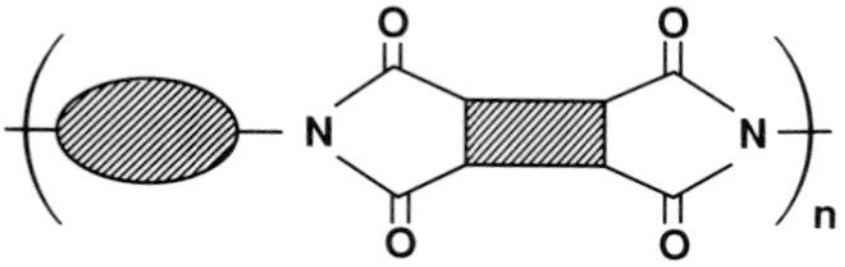

<u>Figure 1</u> Generic structure of polyimides

Polyimides - Market Perspective

Figure 2 illustrates the conclusions of a recent market survey carried out by Kline and Company.[2] The survey of high temperature polymers shows that sales of polyimides in 1988

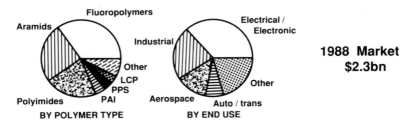

1988 Market
$2.3bn

Figure 2 Worldwide market for high temperature polymers (1988)

were exceeded only by those of fluoropolymers (e.g. PTFE) and aramids (e.g. Kevlar). Approximately 22% of the 1988 market of $2.3 billion was accounted for by polyimides. In addition, while the overall market for high temperature polymers is expected to grow by over 8% per annum to $5 billion by 1998, growth of polyimide sales is forecast at over 12% per annum. This clearly illustrates the present and future commercial significance of this class of polymers.

The following sections will discuss in more detail the synthesis, characterisation, properties and applications of the polyimides and will highlight the importance of a thorough understanding of the chemistry to the successful commercialisation of these polymers.

2   SYNTHESIS OF POLYIMIDES

Polyimide Types

In some texts, the classification of polyimide type is unnecessarily complex and confusing. Chemically, only two types of polymer need be considered and these are illustrated in Figure 3. If the material is not crosslinked (and no chain branching occurs), a linear, or thermoplastic, polyimide results. In some cases, true thermoplasticity is never observed because the polymer decomposes below its softening point. This is true of Kapton, for example. The polymers can be either crystalline or amorphous, depending on structure. If the polymer is crosslinked, a thermoset polyimide is produced. Crosslinking is normally induced by thermal decomposition or polymerisation of a reactive end-group. A variety of such groups can be used, such as norbornene, acetylene and benzocyclobutene. The type of end-group influences both the processing and properties of the final product.

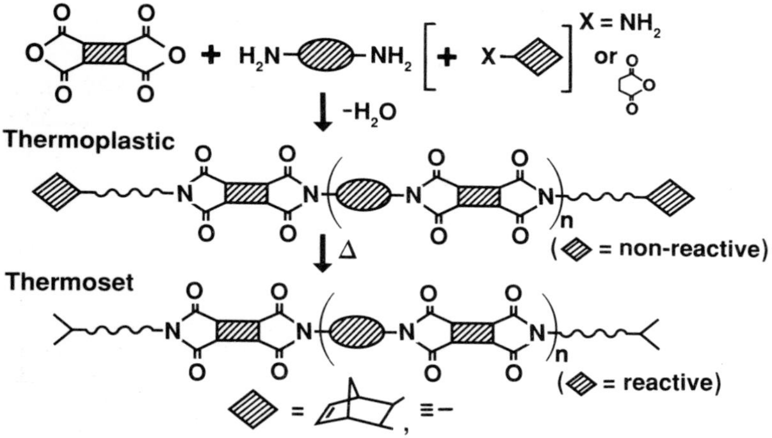

Figure 3 Chemistry of polyimides

General Features of Polyimide Synthesis

The polyimide formation reaction is a classical polycondensation reaction and is therefore characterised by the elimination of a small molecule or evolution of a volatile by-product. This can sometimes lead to difficulties during the processing of these materials, where efficient removal of volatiles is often essential. High purity monomers need to be used to achieve high polymer molecular weight (MW). Difunctional monomers should contain as little as possible monofunctional impurity which would act as a 'chain stopper'. Similarly, precise control of the stoichiometry is essential for formation of high MW polymer.

Synthetic Routes to Polyimides

Linear Polyimides . By far the most common route to polyimides is the reaction of a dianhydride with a diamine, either in solution or in the melt. The synthesis is illustrated in Figure 3 and schematically in Figure 4, showing the two steps of the reaction more clearly. This reaction is considered in more detail later, when the

kinetic and mechanistic features of the reaction are discussed.
Variations on the diamine/dianhydride reaction can sometimes
offer advantages over the more conventional synthesis.

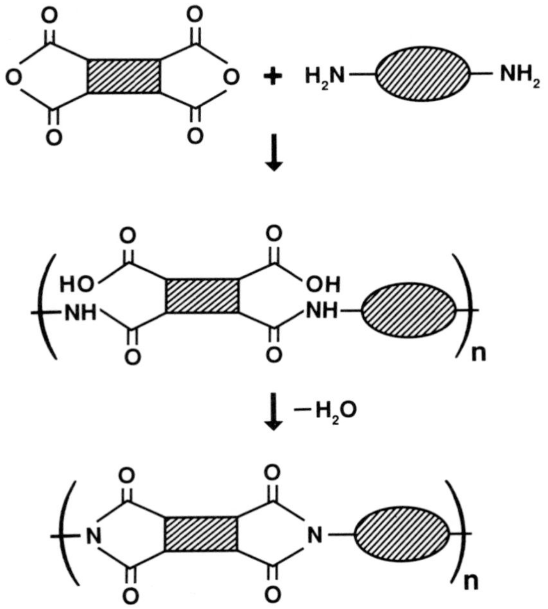

<u>Figure 4</u> Diamine/dianhydride reaction scheme

    The reaction of diisocyanates with dianhydrides is very
facile, [3] but the resultant polymer is often of low molecular
weight. The relative reactivity of diisocyanates towards
protic impurities such as water may be problematical. High
molecular weight polymer can be obtained by reaction of
dianhydrides with N-silylated diamines (e.g. Me$_3$SiHN-Ar-
NHSiMe$_3$).[3] Thioanhydrides will react with diamines under
relatively mild conditions to give high molecular weight
polymer,[3] but the evolution of hydrogen sulphide as a by-
product can pose problems on an industrial scale! Another
dianhydride variant involves the use of diacid-diester
derivatives. This approach is used in PMR polyimides (see
later).

    Another synthetic route to linear polyimides is the
activated aromatic nucleophilic substitution reaction of
nitroaromatics by phenoxide dianions (Figure 5).[4]   The
imide ring activates the nitro group sufficiently for
nucleophilic substitution to be fairly facile. The classical

route to polyether imides such as General Electric's Ultem

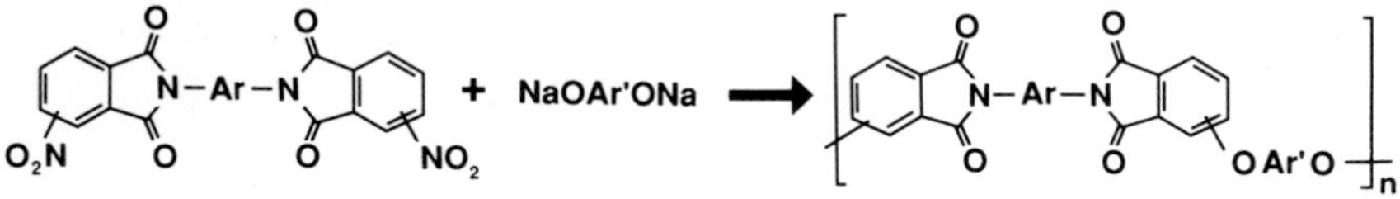

Figure 5 Synthesis of polyimides by aromatic nucleophilic substitution

is shown in Figure 6. Although the polymer-forming step is a conventional diamine/dianhydride reaction, the key step is the aromatic nucleophilic substitution reaction leading ultimately to the dianhydride. An increasingly wide range of bisphenols is available commercially, greatly increasing the applicability of these synthetic routes.

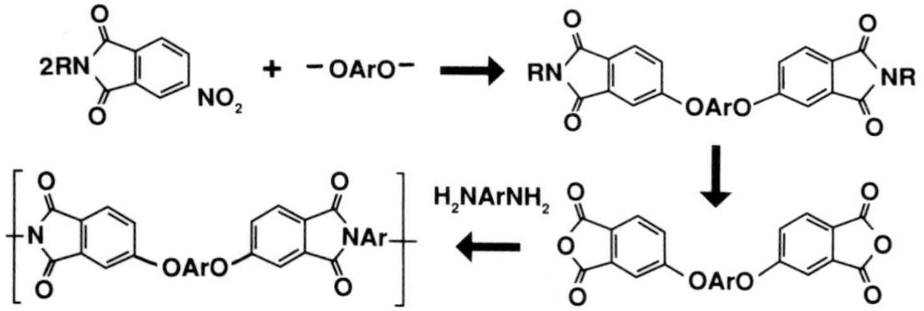

Figure 6 Synthesis of polyether imides

Finally, in addition to polyether imides, a variety of 'mixed' polyimides can be synthesised by the methods outlined above. These include polyamide imides, polyester imides and polysiloxane imides.

Crosslinked Polyimides. The PMR approach is illustrated in Figure 7 for the aerospace polyimide, PMR-15,[5] which is discussed in more detail later. PMR stands for 'Polymerisation of Monomeric Reactants' and refers to the way in which the material is polymerised in-situ (usually impregnated on carbon or glass fibre) during the processing of a final component. The '-15' designates the molecular weight of the first-formed oligomer (1500). In this case, the diamine (MDA) is mixed with the half-ester derivative of the dianhydride, this monomer (BTDE) being used to facilitate dissolution in a low boiling solvent such as methanol and prevent polymerisation occurring at room temperature. A reactive end-cap is provided via a norbornene half-ester (NE). The reaction is conducted in the molten state and proceeds via an intermediate thermoplastic oligomer which is then crosslinked under heat (> 290°C) and pressure (> 200 psi)

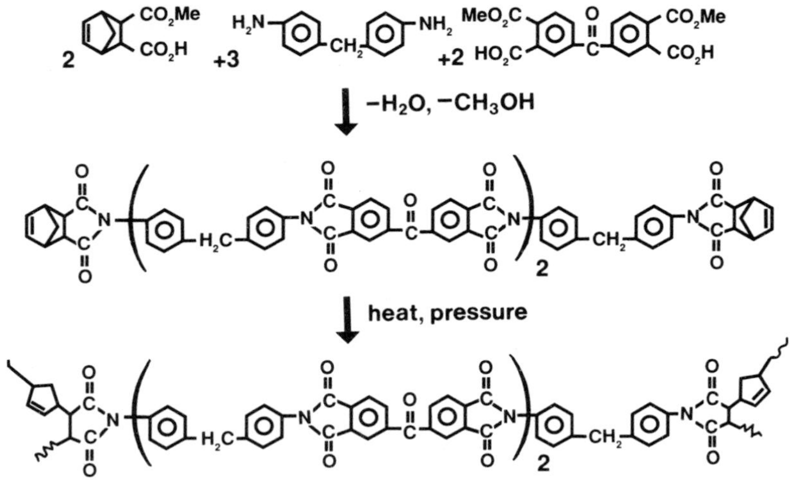

Figure 7 PMR-15 reaction scheme

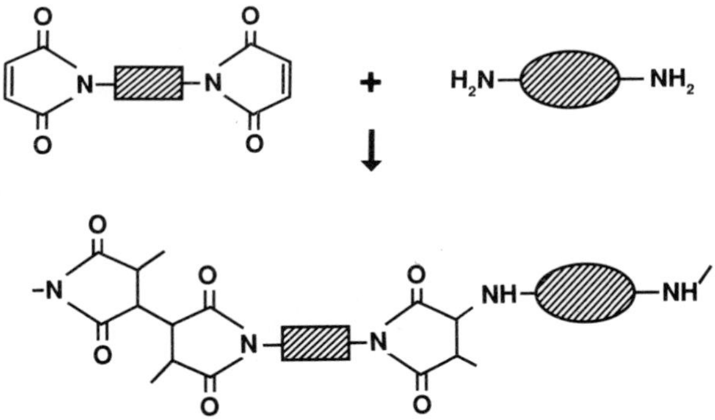

Figure 8 BMI chemistry

through a retro Diels-Alder reaction and subsequent
polymerisation of the norbornene end-group. The tetraacid
derivative of the dianhydride may also be used, but this
would normally require the use of a more polar, higher
boiling point solvent such as N-methylpyrrolidone which can
prove difficult to remove during processing.

A further route to thermosetting polyimides involves the
crosslinking of reactive bis-imides, such as bismaleimides
(BMIs) and, more recently, bis(benzocyclobutene imides) (BCBs).
The BMI monomers are prepared by reaction of aromatic diamines

with maleic anhydride. Polymers can be formed purely by thermal radical crosslinking of one or more BMI monomers, but more commonly comonomers are added to improve the properties (e.g. toughness) of the final product. Comonomers used include aromatic diamines, allylphenyl and propenyl compounds.[6] Figure 8 shows both the radical homopolymerisation of a BMI and the Michael addition of a diamine to the maleimide double bond. Allylphenyl monomers coreact with BMIs via an ene reaction followed by a Diels-Alder addition while the propenyl compounds react initially via a Diels-Alder reaction.

BCB monomers were first invented by the US Air Force[7] and have since been developed commercially by Dow. BCBs are synthesised by reaction of 4-aminobenzocyclobutene with an aromatic dianhydride. Polymerisation of the BCB proceeds via thermally induced opening of the benzocyclobutene ring to give an o-quinodimethane intermediate which then polymerises (Figure 9). BCBs can also be copolymerised with BMIs.[6]

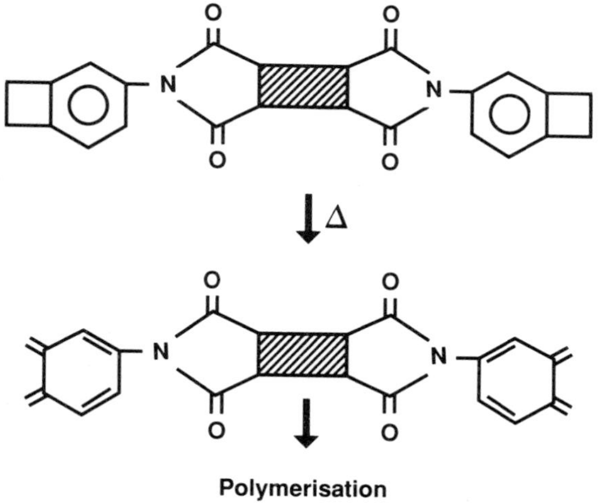

Polymerisation

<u>Figure 9</u> Polymerisation of bis(benzocyclobutene imides)

Thermosetting polyimides can also be prepared from thermoplastic oligomers which have been terminated with BCB groups (cf. PMR-15). Another type of reactive end-group commonly used is the acetylene group,[8] which polymerises via a linear radical process and not trimerisation to a benzene ring as originally thought. The processing of these materials has been improved by synthesising the isoimide - containing oligomer by dicyclohexylcarbodiimide cyclodehydration of the amic acid oligomer.[9] The isoimide has better flow

characteristics than the imide but rearranges to the imide
during elevated temperature processing.

Mechanistic and Kinetic Features of Diamine/dianhydride Reaction

Despite its role as the most generally applicable route
to polyimides, the reaction of diamines with dianhydrides has
been little understood until relatively recently. Much of the
improvement in the understanding of the mechanisms operative
in this reaction is attributable to Russian workers.[10] The
reaction can be carried out either in two steps or one step.
Each is considered below.

Two Step Synthesis - Polyamic Acid Formation. The two
step synthesis of polyimides involves initial formation of a
polyamic acid solution, followed by chemical or thermal
cyclodehydration (see Figure 4). The kinetics and mechanisms
of polyamic acid formation have been studied in some detail.
The reaction is reversible, but in polar aprotic solvents, the
reverse reaction is hindered due to hydrogen bonding effects.
The equilibrium constant is therefore large,[11] leading to
high molecular weight polymer. The propagation reaction is
also exothermic at room temperature. Decreasing the reaction
temperature should therefore shift the equilibrium to the
right and increase the polyamic acid molecular weight.[12]
This effect is, however, mitigated by the high value of the
equilibrium constant. Concentration effects have also been
observed in the reaction due to the forward reaction being
bimolecular and the reverse unimolecular.

Despite the efforts of many workers, the precise reaction
kinetics are still unclear. The polymerisation rate is
solvent dependent, increasing in more polar and more basic
solvents. Changing the solvent can change the kinetics
followed from irreversible second order [13] to reversible
autocatalytic.[14] The rate of the dianhydride - diamine
reaction has been found to be dependent on the electron
affinity of the dianhydride and the basicity (pKa) of the
diamine.

Polyamic Acid Formation - Practical Considerations.
Some practical aspects of the diamine-dianhydride reaction
are worthy of note. The first is that the use of solid
dianhydride leads to high values of $\bar{M}w$. In this case, the
reaction is effectively an interfacial polymerisation and
becomes diffusion controlled.[15] In addition to high $\bar{M}w$, the
polyamic acid has a broad, often bimodal, molecular weight
distribution. Another result of this effect is that the
polyamic acid solution can re-equilibrate on storage, resulting

in the commonly observed reduction in solution viscosity.

Two potential side reactions may be significant. Any water present will result in a competition between the propagation reaction and hydrolysis of the anhydride. Amine impurities are often present in the amide solvents commonly used. This can lead to chain termination during polymerisation.

Two Step Synthesis - Polyamic Acid Cyclisation . The cyclisation (imidisation) of polyamic acids can be brought about either by thermal or chemical means.

Thermal imidisation is the preferred industrial route. Generally, cyclisation is induced by heating the polyamic acid to 250-400 ° C in the solid state. Thin films are easier to imidise than powders or fibres. Cyclisation can lead to either imide or isoimide structures, but if any of the latter is formed, it probably thermally rearranges to the imide. A further reaction which usually occurs to some extent is cyclisation with resultant chain scission to regenerate shorter-chain anhydride- and amine-terminated polymers or oligomers.[16] These moieties can recombine later in the reaction. The chain scission can be avoided by using amine salts of the polyamic acid.

The imidisation reaction is accelerated by the presence of small amounts of residual amide solvent. The kinetics are complex. Imidisation is initially fast but as reaction proceeds becomes much slower. Explanations for this behaviour include loss of solvent as imidisation proceeds[17] and reduction in molecular mobility due to the polymer Tg rising above the reaction temperature.[18]

Chemical imidisation of the polyamic acid is normally carried out by treatment with an aliphatic anhydride and a tertiary amine. Common reagents are acetic anhydride with triethylamine or pyridine. Amide solvents such as NMP or DMF are normally used. The reaction is carried out in dilute solution (ca. 20% w/v); despite this, precipitation of the forming polyimide often occurs. This necessitates subsequent heat treatment of the solid polymer to complete the cyclisation. This also serves to rearrange any residual isoimide, which is formed in greater amounts during chemical imidisation than thermal cyclisation. The molecular weight of the polymer changes very little on chemical imidisation, in contrast to the decrease in molecular weight often observed during thermal treatment. The mechanism of chemical imidisation is shown in Figure 10.[19] The first step involves acylation of the acid to the corresponding anhydride, followed by ring closure to either the kinetic product (isoimide) or the thermodynamic product (imide). Rearrangement of the isoimide to the imide is catalysed by the acetate ion.

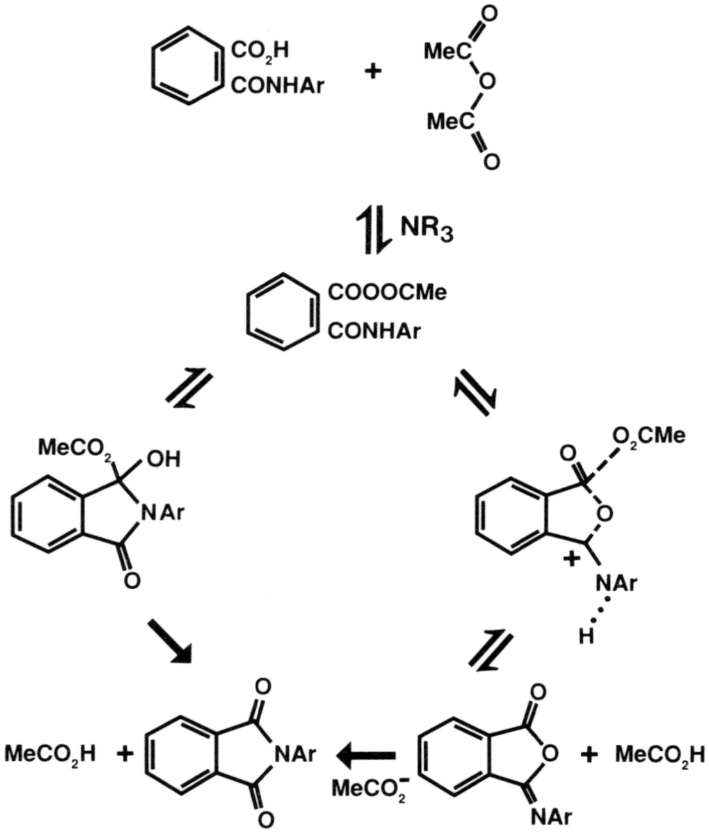

<u>Figure 10</u> Mechanism of chemical imidisation

One Step Polyimide Synthesis. This method can be used for the synthesis of soluble polyimides and involves heating the monomers in a high boiling solvent such as NMP or m-cresol. The water generated is removed by using an azeotroping cosolvent such as toluene. This method is sometimes useful for polymerising unreactive monomers. A more crystalline product can be obtained than in the two step process. The rate determining step in the one step reaction is the reaction between the anhydride and the amine.[20] Subsequent imidisation is very fast.

Recent and Future Directions in Polyimide Synthesis

Some recent trends in the development of improved synthetic routes to polyimides have included the use of new azeotroping cosolvents such as N-cyclohexylpyrrolidone and the use of electromagnetic, or microwave, curing. Further advances in improving imidisation efficiency will be required in future. There is a need for new reactive end-caps for thermosetting polyimides to provide easier processing or increased thermal stability of the crosslinks. NASA, for example, have developed novel cyclophane end-caps. New polyimides will continue to be synthesised to provide specific property improvements in particular applications. Finally, and perhaps most importantly, there will be an increasing driving force to reduce the cost of these polymers, for example by synthesis of novel, cheaper monomers.

3 POLYIMIDE CHARACTERISATION

For convenience, the characterisation methods used for polyimides are classified below into five categories, although the distinction between these is not rigid.

Physical Characterisation

A wide range of techniques is used to characterise the physical properties of the polyimides, whether in the solid state, the melt or in solution. The most fundamental property which is usually measured is the glass transition temperature (Tg). The Tg is measured by thermal or dynamic mechanical methods such as differential scanning calorimetry (DSC), dynamic mechanical (thermal) analysis (DM(T)A), thermomechanical analysis (TMA) and Rheometrics dynamic spectroscopy (RDS). The molecular weight can be determined by gel permeation chromatography (GPC) (relative method) or light scattering (in principle, absolute method). Solution viscometry is commonly used to determine the intrinsic viscosity of polyamic acids, but complications arise with the polyimides, perhaps as a result of solvent complexation. The rheological behaviour of the polyimides above Tg can be studied by RDS. The morphology of polyimide specimens can often be observed by scanning or transmission electron microscopy (SEM or TEM). Crystallinity or ordering in the polymers can be detected by X-ray scattering, DSC or even UV-visible fluorescence spectroscopy. Another commonly quoted parameter is (thermo-oxidative) stability, as represented by the thermogravimetric analysis (TGA) curve of the polymer.

Chemical Characterisation

The chemical structure of the polyimides can be studied by conventional NMR and IR spectroscopic techniques, which are particularly sensitive to the chemical environment of the carbonyl group (e.g. amide, imide, anhydride).[21,22] For intractable solid polymers, solid state NMR and pyrolysis GC-MS methods may be applied, although the information gained is more limited. Even traditional 'wet' methods have their uses and hydrazinolysis has been used to good effect to cleave polyimides into more easily analysed fragments.

Mechanistic Characterisation

This can best be defined as the application of the techniques described above, plus some additional methods, to the study of the mechanism of polyimide formation.   Such mechanistic information is frequently invaluable in developing cure cycles and in understanding the behaviour of the final product.

Thermal transitions occurring during cure are identified by techniques such as DSC,[23] DMTA or RDS.   These transitions may represent either chemical (e.g imidisation) or physical changes (e.g. Tg).   The chemical transformations can be followed more clearly by IR spectroscopy, using a heated cell in the spectrometer.   The various carbonyl functionalities lend themselves ideally to this purpose.   The effect of temperature on the band frequencies has been reported.[24] In principle, Raman spectroscopy could also be used; however, interference due to fluorescence is often a problem.   FT Raman may provide a solution.   The industrial relevance of the IR studies is illustrated by the involvement of such companies as IBM, NASA, Bell and BP.   A technique which can provide information on both chemical and physical changes is dielectrometry, which studies the variation of dielectric loss and permittivity with temperature.[25] The use of various combined analytical techniques has been reported, for example TGA-GC-MS by Toray and TGA-DTA-MS by BP.[26]

Cure Monitoring

Mechanistic characterisation can be extended to allow in-situ real time monitoring of resin cure.   This is a relatively recent development which should find application as a process control method in polyimide fabrication.   The main methods used are spectroscopic and dielectric. Potentially useful spectroscopic techniques include near infra-red (NIR), mid infra-red, Raman and fluorescence spectroscopy. These are all characterised by the need to use fibre optics to link the curing resin with the spectrometer. Dynamic dielectric analysis employs ceramic dielectric sensors to follow the cure. These two methods complement each other and are likely to be best used in tandem.

Quality Control (QC)

Many of the methods described previously can be used for QC of the polyimide products. Strict QC of the monomer precursors is particularly necessary where the product is intractable. This is true of PMR-15, for example, where companies such as BP, GE, Rohr and NASA have developed HPLC methods for the analysis of monomers and monomer mixtures.[27,28]

Miscellaneous Characterisation Methods

Surface analysis techniques such as XPS and Auger can be used to characterise the surface of polyimide films. Other properties which are measured, depending on the final application, include electrical properties (e.g. dielectric constant), gas sorption and tribological behaviour.

Future Needs

It is possible to identify areas where advances in analytical techniques would greatly benefit polyimide technology. Improved methods are needed for the analysis of intractable solid polymers. Solid state NMR has a role to play here. Better combined analytical techniques would lead to a better understanding of cure mechanisms. A relatively routine method for the determination of absolute polyimide molecular weight would allow a greater insight into the effect of structure on properties. Finally, improvements are needed in fibre optics before some of the spectroscopic cure monitoring methods can be used in a production environment.

4 POLYIMIDE PROPERTIES

The properties of the polyimides are strongly dependent on the detailed polymer structure. Since a wide range of structures are accessible via the synthetic routes described in Section 2, a wide range of properties can be obtained leading to a wide range of applications. Unfortunately, there is often a trade-off between the desired properties. For example, while it is possible to synthesise polyimides with a high Tg, these are normally very difficult to process due to high melt viscosity.

In general, polyimides have good thermo-oxidative

stability and can be used at temperatures up to 300-350 °C. Mechanical properties are good and the polymers adhere well to metals such as aluminium or copper.[29] Other useful properties include low dielectric constant, the ability to form films and fibres and sorption of various gases.

A wide range of structure-property relationships has been studied, including structure-Tg, structure-solubility, structure-thermal expansion coefficient, structure-permeability and structure-processability.

Effect of Structure on Tg

The glass transition temperature of polyimides is strongly influenced by both intramolecular (chain mobility) and intermolecular (chain packing) factors. Depending on the polymer structure, Tg values can range from less than 200 °C to more than 400 °C. A number of empirical rules have been formulated to relate Tg to structure. Attempts have also been made to predict Tg values using computer modelling,[30] but this can sometimes be complicated by the occurrence of charge transfer complexation between polymer chains.

In general, aromatic structures have a higher Tg than aliphatic, while para linkages tend to raise the Tg relative to meta. The Tg is also raised by hexafluoroisopropylidene connecting groups between aromatic rings, due to restricted rotation. Polymers containing directly linked aryl groups such as biphenyl have higher Tg's than those with heteroatom linkages such as ether.

While high Tg polyimide structures are accessible by following the general rules outlined above, the very factors which lead to high Tg tend also to result in high melt viscosities and therefore poor processability. One possible solution to this problem is to use an oligomer approach (see later).

Thermo-Oxidative Stability

The thermo-oxidative stability of a polymer is one aspect of its durability, viz its continued integrity under actual service conditions. Stability can be improved by incorporation of predominantly 'stable' groups such as aromatic or trifluoromethyl groups. Once again, however, such an approach tends to introduce a conflict with easy processability.

Processability

Processability is often the Achilles' heel of high temperature polyimides. As discussed above, the factors which make a polymer suitable for use at elevated temperatures also often lead to difficult or impossible processing. The PMR approach was one early attempt to solve this problem. Deliberate use of isoimides, which thermally rearrange to imides at high temperatures, has also had some success with acetylene-terminated oligomers in particular.[9] Use of copolymers such as imide-siloxanes has also resulted in improved processability.[31] New approaches are undoubtedly needed, however, and chemistry has an important role to play here.

One exciting possibility concerns the development of new coupling reactions which lead to stable linking groups, without the evolution of volatile by-products. This would then allow processable functionally - terminated oligomers to be used in component fabrication, allowing easy processing followed by chain extension and/or crosslinking to give stable, high MW products with good mechanical properties. Success in this area would no doubt lead to far more widespread use of polyimides than at present, including the application of newer processing techniques such as resin transfer moulding (RTM). The reality may be another matter!

5 APPLICATION OF POLYIMIDES

As discussed earlier, the polyimides find widespread application due to the wide range of chemistry and properties accessible. Their commercial importance is illustrated by the fact that at least 2300 patents have been published on polyimides in the last five years. This represents an average increase of 16% per annum. These figures probably grossly underestimate the patent activity in Japan. Many major multinational chemical and manufacturing companies are actively involved in polyimides either as suppliers or end-users or both. These include Du Pont, IBM, Shell, ICI, BP, General Electric, Hitachi, BASF, Rolls-Royce, Ciba-Geigy etc. Joint ventures are becoming increasingly important, not only to spread costs but also to make use of existing distribution and marketing networks. Witness ICI's collaboration with Japan's Ube, to market the latter's Upilex film.

Major application areas for polyimides include aerospace, automotive, general engineering and microelectronics/ electrical. In aerospace, polyimides are used in structural composites and as high temperature adhesives. They also find

use in load bearing structures in the automotive industry.
General engineering applications include high temperature
bearings and seals. Polyimides find widespread use in
microelectronics and electrical applications, including as
dielectrics for integrated circuits (ICs) and in electrical
insulation. The use of chemistry to solve some of the
practical problems associated with polyimides will be
illustrated by two important examples, one from aerospace
and the other from microelectronics.

Aerospace - PMR-15

The chemistry of PMR-15 was outlined in section 2. The
material was first developed by NASA in 1972.[32] It is most
commonly used as a 'prepreg', with the monomers impregnated
on carbon or glass fibre. This prepreg can then be fabricated
into a composite. PMR-15 is currently the leading high
temperature resin for use in aerospace composites. GE of the
United States, for example, fabricates the F404 engine duct
for the F18 fighter aircraft from PMR-15 prepreg. The market
for PMR-15 prepreg in the late 1990's has been estimated as
$50 - 100 million per annum. Current suppliers of PMR-15
prepreg include ICI, BP, American Cyanamid, Ferro and Hexcel.
Companies either manufacturing from or demonstrating
manufacturing technology using PMR-15 include GE, Rohr and
Rolls-Royce.

The pre-eminent position of PMR-15 is a consequence of
its excellent high temperature properties and the advantages
offered by its high stiffness/weight and strength/weight
ratios compared to metals. Nonetheless, it suffers from a
number of drawbacks which to date have limited its more
widespread use. These include microcracking due to thermal
cycling, irreproducibility and the toxicity of the constituent
MDA monomer.[5] A variety of chemical approaches have been
taken in an attempt to solve one or more of these problems.

For example, BP, partly in conjunction with Rolls-Royce,
carried out studies of the PMR-15 crosslinking mechanism to
gain a better understanding of the effect of cure temperature
on composite properties. This included studies on PMR-15
itself [26] as well as model compound studies on the polymerisation
of N-phenylnadimide. [33] This work was conducted to help
explain the occurrence of microcracking, the formation of
small cracks in the matrix between the fibres in PMR-15
composites when they are repeatedly cycled between low and
high temperatures. Similar studies of N-phenylnadimide
polymerisation have subsequently been reported by Ciba-
Geigy.[34]

The problem of PMR-15 irreproducibility has been addressed in several ways. Considerable effort has been expended by several groups to develop HPLC methods for monitoring the quality of both monomers and prepreg. Work in the area has been conducted by BP, [27] NASA,[28] GE and Rohr. Development of such techniques is a prerequisite for defining strict quality specifications. A detailed study of PMR-15 imidisation chemistry has been carried out in an attempt to identify processing causes for the irreproducibility.[21] In a similar vein, Dexter Corporation in conjunction with BASF has tried to circumvent this problem by preparing prepreg from oligomer which has been pre-imidised in a controlled way. This approach also solves the problem of MDA toxicity. Another solution to the toxicity problem is to replace the MDA with an alternative diamine of low toxicity. This route has been taken by BP, Dornier, MTU and ERA technology in a project part funded by the European Communities' BRITE initiative.[35] The resultant resin system, B1, may replace PMR-15 in some applications.

In conclusion, it should be noted that successful application of PMR-15 in the aerospace industry depends on close control of the chemistry. Both suppliers and end-users have been working actively to optimise the chemistry of PMR-15, with varying degrees of success. The ultimate realisation of the predicted market for PMR-15 may depend on a detailed understanding of the chemistry being obtained.

Polyimides in Microelectronics

The growth of the world market in electronic chemicals in the 1980's is illustrated in Figure 11.[36] Sales of electronics components have been estimated as $400 billion in 1990, an increase of 25% per annum through the 1980's. Polymers have many advantages as electronics materials compared to conventionally used inorganic chemicals. Polyimides are finding increasing use in a range of applications in the microelectronics industry. These include dielectric layers for integrated circuits, integrated circuit packaging and as die attach materials (adhesives).

The use of polyimides as dielectric layers for ICs is expanding due to the increasing complexity of the ICs. This is particularly true of Very Large Scale Integration (VLSI) devices, where there may be more than 10^5 components per chip. The dielectric layers need to be able to even out (planarise) the topographical features of the underlying conduction layers to improve line definition. This is more easily achieved using polyimides than conventional dielectrics such as silicon dioxide. The planarisation ability of the polyimides is perhaps their most important characteristic, but other advantages include low dielectric constant and good thermal and chemical stability. A number of companies are active in this field, including Hitachi, Du Pont, Ciba-Geigy and Bell.

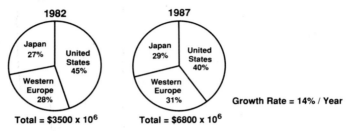

1982

Japan 27%

United States 45%

Western Europe 28%

Total = $3500 x 10^6

1987

Japan 29%

United States 40%

Western Europe 31%

Growth Rate = 14% / Year

Total = $6800 x 10^6

<u>Figure 11</u>   World market for electronic chemicals

Despite the advantages of polyimides, a number of problems remain to be solved. These include thermal stresses arising between different layers in the IC, high processing temperature and processing irreproducibilty (cf. PMR-15) and long term durability. Much work has been carried out in recent years to develop new polyimides with improved processing characteristics, mechanical properties and electrical properties. A few examples will serve to illustrate some of the chemistry employed.

Thermal stresses arise due to mismatch of the thermal expansion coefficients (TECs) of adjacent organic (polymer) and inorganic layers. Hitachi's solution to this has been to synthesise polyimides with low TECs.[37] The polymers are prepared by reacting inflexible dianhydrides such as PMDA (pyromellitic dianhydride) or BPDA (biphenyl dianhydride) with rigid p-linked diamines. This leads to essentially linear chain conformations which in turn lead to low values of TEC.

Photosensitive polyimides have been developed as a means of avoiding the use of additional lithographic techniques, thereby reducing the overall number of processing steps and the processing time. One method involves esterification of the polyamic acid with photosensitive (meth)acrylic groups which form crosslinks on irradiation. Subsequent thermal treatment will generate the polyimide by polyamic acid cyclisation, which is then followed by photo-induced depolymerisation and volatilisation of the crosslinks. A more elegant alternative involves the use of intrinsically photosensitive polyimides, such as those prepared from cyclobutane dianhydride.[38] Irradiation results in cleavage of the cyclobutane ring leading to smaller polymer/oligomer fragments.

GE have reported the synthesis of polyimide-siloxane block copolymers, where the presence of the siloxane permits good adhesion to silicon-based substrates without the need for an adhesion promoter.[39] These polymers are also readily patternable. Synthesis can be carried out by reaction of a diamine with a dianhydride and either an amino- or anhydride-functionalised siloxane.

The use of polyimides in microelectronics will continue to grow as further improvements are made to the chemistry and properties of the systems.

Other Recent Trends

There is no doubt that further applications of polyimides will open up as a better understanding of their chemistry and properties is obtained. Two examples are mentioned below of recent work which could lead to commercial applications in the future.

Research groups at Du Pont, NASA and Ube are trying to develop high temperature gas separation membranes based on polyimides.[40,41] These membranes have potential for the separation of important gas mixtures such as methane/carbon dioxide and oxygen/nitrogen (air). Workers at Brooklyn Polytechnic are attempting to synthesise 'recoverable' polymers, such as polyimides and epoxies.[42] Success in this area would perhaps allow reprocessing and would lessen the detrimental environmental impact of such intractable materials. The current approach uses polymers prepared from monomers containing disulphide linkages. The polymers can be reduced to thiol-terminated fragments which can then in principle be oxidised to regenerate the polymer. A disadvantage of such linkages in polyimides is their likely deleterious effect on thermo-oxidative stability.

6   CONCLUSIONS

Polyimides are one of the most important classes of high temperature polymers available today. Despite their frequently outstanding properties, a number of problems remain to be solved before the polymers achieve their full potential. Important targets include improvements in quality, reproducibility and cost reduction.

An improved understanding of the mechanisms involved in

polyimide synthesis will lead to better control of the manufacturing process and the structure of the resultant polymer. Better characterisation techniques are required to facilitate this. Success in achieving a better understanding of the chemistry should allow tailoring of the polymer structure to obtain the desired properties. This in turn will lead to more widespread use of the polyimides industrially. The range of companies involved in various aspects of polyimide chemistry clearly demonstrates their current and future commercial importance.

REFERENCES

1.   T.M. Bogert and R.R. Renshaw, J. Am. Chem. Soc., 1908, 30, 1140.
2.   European Plastics News, March 1990, 3.
3.   Y. Oishi, M. Kakimoto and Y. Imai, Proc. Third Int. Conf. Polyimides, Ellenville, NY, USA, 1988, 14.
4.   T. Takekoshi, Polymer J., 1987, 19, 191.
5.   D. Wilson, Br. Polym. J., 1988, 20, 405.
6.   H Stenzenberger, Br. Polym. J., 1988, 20 383.
7.   F.E. Arnold and L.S Tan, Proc. 31st Int. SAMPE Symp., 1986, 31 968.
8.   A.L. Landis, ACS Polymer Preprints, 1974, 15(2), 537.
9.   A.L. Landis and A.B. Naselow, Proc. Nat. SAMPE Tech. Conf., 1982, 14, 236.
10. M.I. Bessonov, M.M. Koton, V.V. Kudryavtsev and L.A. Laius, Polyimides - Thermally Stable Polymers, Consultants Bureau, New York, 1987.
11. A.Y. Ardashnikov, I.Y. Kardash and A.N. Pravednikov, Polym. Sci. USSR, 1971, 13, 2092
12. P.P. Nechayev, Y.S. Vygodskii, G.Y. Zaikov and S.V. Vinogradova, Polym. Sci. USSR, 1976, 18, 1903.
13. Y.S. Vygodskii, T.N. Spirina, P.P. Nechayev, L.I. Chudina, G.Y. Zaikov, V.V. Korshak and S.V. Vinogradova, Polym. Sci. USSR, 1977, 19, 1738.
14. R.L. Kaas, J. Polym. Sci., Polym. Chem. Ed., 1981, 19, 2255.
15. C.C. Walker, J. Polym. Sci., Polym. Chem. Ed., 1988, 26, 1649.
16. P.R. Young and A.C. Chang, SAMPE J., March/April 1986, 70.
17. J.A. Kreuz, A.L. Endrey, F.P. Gay and C.E. Sroog, J. Polym.Sci., A-1, 1966, 4, 2607.
18. L.A. Laius, M.I. Bessonov, Y.V. Kallistova, N.A. Adrova and F.S. Florinskii, Polym. Sci. USSR, 1967, A9, 2470.
19. R.J. Angelo, R.C. Golike, W.E. Tatum and J.A. Kreuz, Proc. Second Int. Conf. Polyimides, Ellenville, NY, USA, 1985, 631.

20. S.V. Vinogradova, Z.V. Gersashehenko, Y.S. Vygodskii, F.B. Sherman and V.V. Korshak, Dokl. Akad. Nauk SSSR, 1972, 203, 285.
21. J.N. Hay, J.D. Boyle, P.G. James, J.R. Walton, K.J. Bare, M. Konarski and D. Wilson, High Perf. Polym., 1989, 1, 145.
22. C.A. Pryde, J. Polym. Sci., Polym. Chem. Ed., 1989, 27, 711.
23. R.W. Lauver, Ibid, 1979, 17, 2529.
24. R.W. Snyder, C.W. Sheen and P.C. Painter, Appl. Spectrosc., 1988, 42, 503.
25. D.E. Kranbuehl, S.E. Delos and P.K. Jue, Polymer., 1986, 27, 11.
26. D. Wilson, J.K. Wells, J.N. Hay, D. Lind, G.A. Owens and F. Johnson, SAMPE J., May/June 1987, 35.
27. R.E.A. Escott, Proc. 19th Int. SAMPE Tech. Conf., 1987, 19, 398.
28. G.D. Roberts and R.W. Lauver, J. Appl. Polym. Sci., 1987, 33, 2893.
29. L.P. Buchwalter, J. Adhesion, 1987, 1, 341.
30. C.J. Lee, Proc. Third Int. Conf. Polyimides, Ellenville, NY, USA, 1988, 93.
31. J.D. Summers, C.A. Arnold and J.E. McGrath, Mod. Plast. Int., March 1989, 19, 47.
32. T.T. Serafini, P. Delvigs and G.R. Lightsey, J. Appl. Polym. Sci., 1972, 16, 905.
33. J.N. Hay, J.D. Boyle, S.F. Parker and D. Wilson, Polymer., 1989, 30, 1032.
34. C.T. Vijayakumar, K. Lederer and A. Kramer, J. Polym. Sci., Polym. Chem. Ed., 1989, 27, 2723.
35. N.D. Hoyle, N.J. Stewart, J.N. Hay, D. Wilson, M. Baschant, J. Greenwood, G.D. Small, H. Merz and S. Sikorski, Proc. 11th Int. European SAMPE Conf., 1990, 11, 519.
36. D.S. Soane and Z. Martynenko, 'Polymers in Microelectronics', Elsevier, 1989, Chapter 1, p. 2.
37. S. Numata, S. Oohara, K. Fujisaki, J. Imaizumi and N. Kinjo, J. Appl. Polym. Sci., 1986, 31, 101.
38. J.A. Moore and A.N. Dasheff, Proc. Third Int. Conf. Polyimides, Ellenville, NY, USA, 1988, 163.
39. D.A. Bolon, J.E. Hallgren, V.J. Eddy, P.J. Codella and K.A. Regh, Ibid, 1988, 168.
40. S.A. Stern, Y. Mi, H. Yamamoto and A.K. St. Clair, J. Polym. Sci., Polym. Phys. Ed., 1989, 27, 1887.
41. K. Okamoto, K. Tanaka, A. Nakamura, Y. Kusuki and H. Kita, J. Polym. Sci., Polym. Phys. Ed., 1989, 27, 1221.
42. V.R. Sastri and G.C. Tesoro, J. Polym. Sci., Polym. Lett., 1989, 27, 153.

Subject Index